Y PEOPLE BELIEVE WEIRD THINGS

Why People Believe
Weird Things

Pseudoscience, Superstition,
and other confusions of our time

•

Michael Shermer

Foreword by
Stephen Jay Gould

SOUVENIR PRESS

First published in the USA by Henry Holt and Company LLC

First published in Great Britain in 2007 by Souvenir Press Ltd
43 Great Russell Street, London WC1B 3PD

Reprinted 2008, 2009, 2011

'Science Defended, Science Defined' originally appeared in the journal Science, Technology and Human Values, 16, no.4 (Autumn 1991), 517 - 539.

All artwork and illustrations, except as noted in the text, are by Pat Linse, are copyrighted by Pat Linse, and are reprinted with permission.

ISBN 9780285638037

Printed and bound in the UK by
CPI Antony Rowe, Chippenham, Wiltshire

To the memory of Carl Sagan, 1934–1996,
colleague and inspiration,
whose lecture on "The Burden of Skepticism" ten years ago
gave me a beacon when I was intellectually and professionally
adrift, and ultimately inspired the birth of the Skeptics Society,
Skeptic magazine, and this book, as well as my commitment
to skepticism and the liberating possibilities of science

It seems to me what is called for is an exquisite balance between two conflicting needs: the most skeptical scrutiny of all hypotheses that are served up to us and at the same time a great openness to new ideas. If you are only skeptical, then no new ideas make it through to you. You never learn anything new. You become a crotchety old person convinced that nonsense is ruling the world. (There is, of course, much data to support you.)

On the other hand, if you are open to the point of gullibility and have not an ounce of skeptical sense in you, then you cannot distinguish useful ideas from the worthless ones. If all ideas have equal validity then you are lost, because then, it seems to me, no ideas have any validity at all.

—Carl Sagan, "The Burden of Skepticism,"
Pasadena lecture, 1987

CONTENTS

FOREWORD

The Positive Power of Skepticism

Stephen Jay Gould

Skepticism or debunking often receives the bad rap reserved for activities—like garbage disposal—that absolutely must be done for a safe and sane life, but seem either unglamorous or unworthy of overt celebration. Yet the activity has a noble tradition, from the Greek coinage of "skeptic" (a word meaning "thoughtful") to Carl Sagan's last book, *The Demon-Haunted World*. (Since I also wrote a book in this genre—*The Mismeasure of Man*—I must confess my own belief in this enterprise.)

The need—both intellectual and moral—for skepticism arises from Pascal's famous metaphorical observation that humans are "thinking reeds," that is, both gloriously unique and uniquely vulnerable. Consciousness, vouchsafed only to our species in the history of life on earth, is the most god-awfully potent evolutionary invention ever developed. Although accidental and unpredictable, it has given *Homo sapiens* unprecedented power both over the history of our own species and the life of the entire contemporary biosphere.

But we are thinking reeds, not rational creatures. Our patterns of thought and action lead to destruction and brutality as often as to kindness and enlightenment. I do not wish to speculate about the sources of our dark side: Are they evolutionary legacies of "nature red in tooth and claw," or just nonadaptive quirks in the operation of a brain designed to perform quite different functions from the ones that now regulate our collective lives? In any case, we are capable both of the most unspeakable horrors and the most heartrending acts of courage and nobility—both done in the name of some ideal like religion, the absolute, national pride, and the like. No one has ever exposed this human dilemma, caught between the two poles of our nature, better than Alexander Pope in the mid-eighteenth century:

> *Placed on this isthmus of a middle state,*
> *A being darkly wise and rudely great . . .*
> *He hangs between; in doubt to act or rest;*
> *In doubt to deem himself a god, or beast;*
> *In doubt his mind or body to prefer;*
> *Born but to die, and reasoning but to err.*

Only two possible escapes can save us from the organized mayhem of our dark potentialities—the side that has given us crusades, witch hunts, enslavements, and holocausts. Moral decency provides one necessary ingredient, but not nearly enough. The second foundation must come from the rational side of our mentality. For, unless we rigorously use human reason both to discover and acknowledge nature's factuality, and to follow the logical implications for efficacious human action that such knowledge entails, we will lose out to the frightening forces of irrationality, romanticism, uncompromising "true" belief, and the apparent resulting inevitability of mob action. Reason is not only a large part of our essence; reason is also our potential salvation from the vicious and precipitous mass action that rule by emotionalism always seems to entail. Skepticism is the agent of reason against organized irrationalism—and is therefore one of the keys to human social and civic decency.

Michael Shermer, as head of one of America's leading skeptic organizations, and as a powerful activist and essayist in the service of this operational form of reason, is an important figure in American public life. This book on his methods and experiences and his analysis of the attractions of irrational belief provides an important perspective on the needs and successes of skepticism.

The old cliché that eternal vigilance is the price of liberty must be the watchword of this movement, for if the apparently benign cult maintains the same structure of potentially potent irrationality as the overtly militant witch hunt, then we must be watchful and critical of all movement based on suppression of thought. I was most impressed, on this theme, by Shermer's analysis of the least likely candidate for potent harm—Ayn Rand's "Objectivist" movement, which would seem, at first glance, to be part of the solution rather than the problem. But Shermer shows that this sect, despite its brave words about logic and rational belief, acts as a true cult on two key criteria—first, the social phenomenon of demanding unquestioned loyalty to a leader (the cult of personalities), and second, the intellectual failure of a central irrationalism used as a criterion of potential membership (the false belief that morality can have a unique and objective state—to be determined and dictated, of course, by the cult leaders).

Shermer's book moves from this powerful case in minimalism, through the more "conceptual" (however empty of logic and empirical content) irrationalisms of creationism and Holocaust denial, to the scarier forms of activity represented in ages past by crusades and witch hunts and, today, by hysteria about Satanic cults and the sexual abuse of children (a real and tragic problem, of course) on a scale simply inconceivable and therefore resting on an unwitting conspiracy of false accusations, however deeply felt.

We really hold only one major weapon against such irrationality—reason itself. But the cards are stacked against us in contemporary America, where even a well-intentioned appearance on *Oprah* or *Donahue* (both of which Shermer has attempted with troubling results, as described herein) only permits a hyped-up sound bite rather than a proper analysis. So we have to try harder. We can, we have, we will. We have also won great victories, big and small—from Supreme Court decisions against creationism to local debunkings of phony psychics and faith healers.

Our best weapons come from the arsenals of basic scientific procedures—for nothing can beat the basic experimental technique of the double-blind procedure and the fundamental observational methods of statistical analysis. Almost every modern irrationalism can be defeated by these most elementary of scientific tools, when well applied. For example, in a case close to my heart (for I am the father of an autistic young man), the poignant but truly unreasonable hope for communication by non-speaking autists via the use of "facilitators" (people who claim that they can guide the fingers of non-speaking autists over a computer keyboard to type out messages) met with insufficient skepticism (it always looked like the old Ouija board trick to me!) when most facilitators were typing out messages that parents wanted to hear ("Dad I love you; I'm sorry I've never been able to say so"). But when several facilitators, swept up in the witch hunting craze of childhood sexual abuse as the source of all problems, decided (probably unconsciously) that autism must have a similar cause, and then started to type out messages of accusation with their phony "facilitation," then a "harmless" sop to hope turned into a nightmare, as several loving parents were falsely and judicially charged. The issue was resolved by classic double-blind experiments—information known only to the autist and not to the facilitator never showed up in messages, while information known only to the facilitator and not to the autist usually did appear in the supposed messages—but not before the lives of loving parents (who had suffered enough already from the basic circumstance) had been tragically twisted, perhaps permanently (for one never quite overcomes such a heinous charge, even when it has been absolutely proven untrue—a fact well appreciated by all cynical witch hunters).

Skepticism's bad rap arises from the impression that, however necessary the activity, it can only be regarded as a negative removal of false claims. Not so—as this book shows so well. Proper debunking is done in the interest of an alternate model of explanation, not as a nihilistic exercise. The alternate model is rationality itself, tied to moral decency—the most powerful joint instrument for good that our planet has ever known.

Introduction to the Paperback Edition

Magical Mystery Tour

The Whys and Wherefores of Weird Things

The bane of hypocrisy is not its visibility to others, it is its invisibility to the practitioner. In his Sermon on the Mount, Jesus pointed out both the problem and the solution:

> Thou hypocrite, first cast out the beam out of thine own eye; and then shalt thou see clearly to cast out the mote out of thy brother's eye. (Matthew 7:5)

While winding down a national publicity tour in the summer of 1997 for the hardcover edition of this book, I witnessed just such an example. I was scheduled to appear on a radio program hosted by Ayn Rand's hand-picked intellectual heir, Leonard Peikoff, the Objectivist philosopher who, like a medieval monk, has carried on Rand's flame of Truth through books, articles, and now his own radio show. We were told that Peikoff was interested in having me on because I had written a book praising the value of reason, the highest virtue in Objectivist philosophy. I assumed I was actually booked because I had written a chapter (8) critical of Ayn Rand, and that Peikoff did not intend to allow this critique to go unchallenged. Frankly, I was a bit nervous about the appearance because, although I know Rand's philosophy fairly well (I have read all her major works and most of her minor ones) Peikoff is a bright, acerbic man who knows Rand's works chapter and verse and can quote them from memory. I have seen him reduce debate opponents to intellectual mush through wit and steel-cold logic. But I wrote what I wrote so I figured I would buck up and take it like a man.

Imagine my surprise, then, when my publicist informed me that the interview had been canceled because they took exception to my criticism of Rand's personality, movement, and followers, they objected to my classification of them as a cult, and they would not acknowledge a book that "contains libelous statements about Ms. Rand." Obviously, someone from the show had finally gotten around to reading the book. They said they would be happy to debate me on the metaphysics of absolute morality (they believe there is such a thing and that Rand discovered it), but not in a forum that would give recognition to my libelous book. The real irony of all this is that my chapter on Rand focuses on showing how one of the telltale signs of a cult is its inability or unwillingness to consider criticisms of the leader or the leader's beliefs. So, while denying they are a cult, Peikoff and his Ayn Rand Institute did precisely what a cult would do by squelching criticism.

Amazed that anyone could be this blind to such obvious hypocrisy, I called the producer myself and pointed out to him the two important caveats I included in that chapter: "One, criticism of the founder or followers of a philosophy does not, by itself, constitute a negation of any part of the philosophy. Two, criticism of part of a philosophy does not gainsay the whole." I explained to him that on many levels I have great respect for Rand. She is the embodiment of rugged individualism and unsullied rationalism. I embrace many of her economic philosophies. In a pluralistic age in search of nontraditional heroes, she stands out as one of the few women in a field dominated by men. I told him that I even have a picture of her on my wall. This got his attention for a moment so I asked him for a specific example of libel, since this is a mighty strong word that implies purposeful defamation. "*Everything* in the chapter is a libel of Ms. Rand," he concluded. "Give me just *one* example," I insisted. Did she not cuckold her husband? Did she not excommunicate followers who breached her absolute morality, even over such trivial matters as choice of music? He replied that he would have to reread the chapter. He never called back. (It is only fair to note that a very reasonable group of scholars at The Institute for Objectivist Studies, headed by David Kelly, are very open to criticism of Rand and do not hold her in worshipful esteem as "the greatest human being who ever lived," in the words of an earlier intellectual heir, Nathaniel Branden.)

Ayn Rand seems to generate strong emotions in anyone who encounters her work, both for and against. In addition to libel, I was accused of presenting nothing more than an *ad hominem* attack on Rand. I meant to do neither. I wanted merely to write a chapter on cults. So much has already been written on cults in general, and on specific cults such as the Church of Scientology or the Branch Davidians, that I did not wish to repeat the work of others. At one time I considered myself an Objectivist and an enthusiastic follower of Ayn Rand. To put it bluntly she was something of a hero, or at least the characters in her novels were, especially those in *Atlas Shrugged*. Thus, it was somewhat painful for me to examine my hero through the lens of skepticism, and to apply a cultic analysis to a group I would have never considered as such. However, like my other forays into Christianity, New Age claims, and other belief systems (recounted in these pages), as time offered distance and perspective I recognized in Objectivism the type of certainty and Truth claims typically found in cults and religions, including and especially the veneration, inerrancy, and omniscience of the leader, and the belief one has absolute truth, particularly with regard to moral questions. These are the characteristics of a cult as defined by most cult experts, not me; I simply examined the Objectivist movement to see how well it fit these criteria. After reading this chapter you be the judge.

"Judgment" is the appropriate word here. I purposefully chose to open this Introduction with an excerpt on hypocrisy from the Sermon on the Mount, because that chapter in Matthew (7) begins as such: "Judge not, that ye be not judged." Nathaniel Branden begins his memoirs of his years with Rand, appropriately titled *Judgment Day*, with this same quote as well as an analysis from Ayn Rand:

> The precept: "Judge not, that ye be not judged" is an abdication of moral responsibility: it is a moral blank check one gives to others in exchange for a moral blank check one expects for oneself. There is no escape from the fact that men have to make choices, there is no escape from moral values; so long as moral values are at stake, no moral neutrality is possible. To abstain from condemning a torturer, is to become an accessory to the torture and murder of his victims. The moral principle to adopt is: "*Judge, and be prepared to be judged.*"

Actually, what Jesus says in full is:

> Judge not, that ye be not judged.

> For with what judgment ye judge, ye shall be judged: and with what measure ye mete, it shall be measured to you again.
> And why beholdest thou the mote that is in thy brother's eye, but considerest not the beam that is in thine own eye?
> Or how wilt thou say to thy brother, Let me pull out the mote out of thine eye; and, behold, a beam is in thine own eye?
> Thou hypocrite, first cast out the beam out of thine own eye; and then shalt thou see clearly to cast out the mote out of they brother's eye. (Matthew 7:1-5)

Rand has completely misread Jesus. The principle he extols is not moral neutrality or a moral blank check, but a warning against self-righteous severity and a "rush to judgment." There is a long tradition of this line of thinking found in the Talmudic collection of commentary on Jewish custom and law called the *Mishnah*: "Do not judge your fellow until you are in his position" (Aboth 2:5); "When you judge any man weight the scales in his favor" (Aboth 1:6). (See *The Interpreter's Bible*, Vol. 7, pp. 324–326, for a lengthy discussion of this issue.) Jesus wants us to be cautious not to cross the line between legitimate and hypocritical moral judgment. The "mote" and "beam" metaphor is purposeful hyperbole. The man who lacks virtue feels morally smug in judging the virtue of his neighbor. The "hypocrite" is the critic who disguises his own failings by focusing attention on the failings of others. Jesus is, perhaps, offering insight into human psychology where, for example, the adulterer is obsessed with judging other peoples' sexual offenses, the homophobe secretly wonders about his own sexuality, or, perhaps, the accuser of libel is himself guilty of the charge.

As insightful as this experience was for me, my exchange with the Objectivists was just one avenue of what I consider to be a form of data collection to discover more about why people believe weird things. Writing first the book, later doing hundreds of radio, newspaper, and television interviews, and reading the hundreds of reviews and letters in response to it has given me the opportunity to get a fair sampling of what interests people and what sets them off. It has been a magical mystery tour.

Why People Believe Weird Things was reviewed in most major publications with mostly minor criticisms, and some readers were kind enough to point out a handful of spelling, grammatical, and other minute errors that managed to slip past the otherwise outstanding editors at my publisher (and so corrected in this edition). But a few reviewers had more substantive critical comments that are worth noting because they help us refine our thinking about the many controversies in this book. So in the spirit of healthy acceptance of criticism, it is worth examining a few of these critiques.

Perhaps the most worthwhile criticism in terms of self-review came from the *Toronto Globe and Mail* (June 28, 1997). The reviewer brought up an important problem for all skeptics and scientists to ponder. After first observing that "rational reflection does not end with the tenets of the scientific method, themselves subject to various forms of weird belief now and then," he concludes: "Skepticism of the aggressively debunking sort sometimes has a tendency to become a cult of its own, a kind of fascistic scientism, even when it is undertaken for the best of rational motives." Excusing the exaggerated rhetoric (I have never encountered a fellow skeptic who would qualify as a cultist or a fascist), he does have a point that there are limitations to science (which I do not deny) and that occasionally skepticism has its witchhunts. This is why I emphasize in this book, and in virtually every public lecture I give, that *skepticism is not a position; skepticism is an approach to claims*, in the same way that *science is not a subject but a method.*

In a very intelligent and thoughtful review, *Reason* magazine (November, 1997) took me to task for the statement that it is our job "to investigate and refute bogus claims." That is wrong: we should not go into an investigation with the preconceived idea that we are going to refute a given claim, but rather "investigate claims to discover if they are bogus" (as the text has now been corrected). After examining the evidence, one may be skeptical of the claim, or skeptical of the skeptics. The creationists are skeptical of the theory of evolution. Holocaust "revisionists" are skeptical of the traditional historiography of the Holocaust. I am skeptical of these skeptics. In other cases, such as recovered memories or alien abductions, I am skeptical of the claims themselves. It is the evidence that matters, and as limited as it may be, the scientific method is the best tool we have for determining which claims are true and which are false (or at least offering probabilities of the likelihood of a claim being true or false).

The reviewer in *The New York Times* (August 4, 1997) was himself skeptical of the Gallup Poll data I present in Chapter 2 about percentages of Americans who believe in astrology, ESP, ghosts, etc., and wondered "how this alarming poll was conducted and whether it measured real conviction or a casual flirtation with notions of the invisible." Actually, I too have wondered about this and other such polls, and I am concerned with the phrasing of some questions, as well as with the potential shortcomings of such surveys to measure the level of commitment someone has to a particular claim. But self-report data can be reliable when it is corroborated with other independent polls, and these figures of belief have been consistent over many decades by many pollsters. Our own informal polls conducted through *Skeptic* magazine also confirm these statistics as being alarmingly high. Depending on the claims, anywhere from one out of four to three out of four Americans believes in the paranormal. Although our society is a lot less superstitious than, say, that of medieval Europe, we obviously have a long, long way to go before publications like *Skeptic* become obsolete.

Of all the reviews, I got the biggest laugh out of Ev Cochrane's opening paragraph in the November, 1997 edition of *Aeon*, a "Journal of Myth, Science, and Ancient History." It is amusing not only because of his analogy but also because if there were a journal one might consider the antithesis of *Skeptic*, it is *Aeon*. Nevertheless, Cochrane concluded: "For me to praise Michael Shermer's new book is a bit like O.J. Simpson applauding the closing statement of Marcia Clark, inasmuch as the author would probably include the Saturn-thesis, to which I subscribe, amongst the pseudosciences he revels in exposing. Yet praise it I must, for this is a damned entertaining and provocative book." Praise from Brutus indeed, yet Cochrane, along with other reviewers and numerous correspondents (some good friends), have taken me to task for my chapter on *The Bell Curve* (15).

Some accused me of indulging in *ad hominem* assaults in my analysis of Wycliffe Draper, founder of the Pioneer Fund, an agency that, since 1937, has funded research into the heritability and racial differences in IQ. In this chapter I show the historical connection between racial theories of IQ (that blacks' lower IQs are largely inherited and thus immutable) and racial theories of history (the Holocaust is Jewish propaganda) through the Pioneer Fund that also has a direct connection to Willis Carto, one of the founders of the modern Holocaust denial movement. However, I am by training a psychologist and a historian of science, so I am interested in extrascientific issues like who does the funding and therefore what biases might be created in one's research. In other words, I am not only interested in examining data, I am interested in exploring the motives and biases that go into data collection and interpretation. So, the question is, how can one explore this interesting and (I think) important aspect of science without being accused of the *ad hominem* attack?

In the end, however, this chapter is about race, not IQ, nor Charles Murray and Richard Herrnstein's controversial book *The Bell Curve*. The subject is similar to what is known as the "demarcation problem" in discriminating between science and pseudoscience, physics and metaphysics: Where do we draw the line in the gray areas? Similarly, where does one race begin and another leave off? Any formal definition must be arbitrary in the sense that there is no "correct" answer. I am willing to concede that races might be thought of as "fuzzy sets," where my colleagues can (and do) say "come on Shermer, you can't tell the difference between a white, black, Asian, and Native American?" Okay, often, in some general way, I can, as long as the individual in question falls squarely in the middle, between the fuzzy boundaries. But it seems to me that the fuzzy boundaries of the numerous sets (and no one agrees on how many there are) are becoming so broad and overlapping that this distinction is mostly dictated by cultural factors and not biological ones. What race is Tiger Woods? Today we may view him as an unusual blending of ethnic backgrounds, but a thousand years from now all humans may look like this, and historians will look back upon this brief period of racial segregation as a tiny blip on the screen of the human career spanning hundreds of thousands of years.

If the "Out of Africa" theory holds true, then it appears a single race migrated out of Africa (probably "black") that then branched out into geographically isolated populations and races with unique features to each, and finally merged back into a single race with the onset of global exploration and colonization beginning in the late fifteenth century. From the sixteenth through the twentieth centuries the racial sets became fuzzier through interracial marriages and other forms of sexual interaction, and some time over the next millennium the fuzzy boundaries will be so blurred that we will have to abandon race altogether as a means of discrimination (in both uses of the word). Unfortunately, the human mind is so good at finding patterns that other criteria for dividing people will no doubt find their way into our lexicon.

One of the more interesting developments since *Why People Believe Weird Things* was first published is the rise of what might be called the "New Creationism" (to be distinguished from the old creationism that dates back centuries that I discuss in the book). New Creationism comes in two parts:

1. *Intelligent Design Creationism*: arguments made by those on the conservative religious right, where they believe that the "irreducible complexity" of life indicates it was created by an intelligent designer, i.e., God.

2. *Cognitive Behavioral Creationism*: arguments made by those on the liberal, multicultural left, where they believe that the theory of evolution cannot or should not be applied to human thought and behavior.

Imagine that: the marriage of the conservative right and liberal left. How did this come about?

In Chapter 11, I outline the three major strategies of the creationists in the twentieth century, including banning the teaching of evolution, the demand that Genesis get equal time as Darwin, and the demand that "creation-science" and "evolution-science" also get equal time, the former being an attempt to skirt the First Amendment by labeling their religious doctrines as "science," as if the name alone will make it so. All three of these strategies were defeated in court cases, starting with the famed Scopes "Monkey Trial" in 1925, and ending with the Louisiana trial that went all the way to the United States Supreme Court and was defeated in 1987 by a vote of 7 to 2. This ended what I have called the "top down" strategies of the creationists to legislate their beliefs into culture through public schools. This New Creationism, regardless of how long it lasts before it mutates into another form, is supportive of my claim that the creationists are not going to go away and that scientists cannot afford to ignore them.

1. *Intelligent Design Creationism.* With these defeats the creationists have turned to "bottom up" strategies of mass mailings of creationist literature to schools, debates at schools and colleges, and enlisting the aid of people like University of California, Berkeley law professor Phillip Johnson, biochemist Michael Behe, and even the conservative commentator William F. Buckley, who hosted a PBS Firing Line debate in December, 1997, where it was resolved: "Evolutionists should acknowledge creation." The "newness" of this creationism is really in the language, where creationists now talk about "intelligent design," i.e. where life had to have been created by an intelligent designer because it shows "irreducible complexity." A favorite example is the human eye, a very complex organ where, so the argument goes, all the parts must be working at the same time or vision is not possible. The eye, we are told, is irreducibly complex: take out any one part and the whole collapses. How could natural selection have created the human eye when none of the individual parts themselves have any adaptive significance?

First of all, it is not true that the human eye is irreducibly complex such that the removal of any part results in blindness. Any form of light detection is better than none, and lots of people are visually impaired with a variety of different diseases and injuries to the eyes, yet they are able to function reasonably well and lead a full life. (This argument falls into the "either-or fallacy" discussed in Chapter 3 on how thinking goes wrong.) But the deeper answer to the argument is that natural selection did not create the human eye out of a warehouse of used parts laying around with nothing to do, any more than Boeing created the 747 without the ten million halting steps and jerks and starts from the Wright Brothers to the present. Natural selection simply does not work that way. The human eye is the result of a long and complex pathway that goes back hundreds of millions of years to a *simple eyespot* where a handful of light sensitive cells provide information to the organism about an important source of the light—the sun; to a *recessed eyespot* where a small surface indentation filled with light sensitive cells provides additional data in the form of direction; to a *deep recession eyespot* where additional cells at greater depth provide more accurate information about the environment; to a *pinhole camera eye* that is actually able to focus an image on the back of a deeply recessed layer of light-sensitive cells; to a *pinhole lens eye* that is actually able to focus the image; to a *complex eye* found in such modern mammals as humans. In addition, the eye has evolved independently a dozen different times through its own unique pathways, so this alone tells us that no creator had a single, master plan.

The "Intelligent Design" argument also suffers from another serious flaw: the world is simply not always so intelligently designed! We can even use the human eye as an example. The configuration of the retina is in three layers, with the light-sensitive rods and cones at the bottom, facing away from the light, and underneath a layer of bipolar, horizontal, and amacrine cells, themselves underneath a layer of ganglion cells that help carry the signal from the eye to the brain. And this entire structure sits beneath a layer of blood vessels. For optimal vision why would an intelligent designer have built an eye backwards and upside down? Because an intelligent designer did not build the eye from scratch. Natural selection built the eye from simple to complex using whatever materials were available, and in the particular configuration of the ancestral organism.

2. *Cognitive Behavioral Creationism.* The aberrant marriage between the conservative right and liberal left comes in this odd new form of creationism that accepts evolutionary theory for everything below the human head. The idea that our thoughts and behaviors might be influenced by our evolutionary past is politically and ideologically unacceptable to many on the left who fear (admittedly with some justification) the misuse of the theory in the past in a form known as Social Darwinism. The eugenics programs that led to everything from sterilizations in America to mass exterminations in Nazi Germany have, understandably, put off many thoughtful people from exploring how natural selection, in addition to selecting for eyes, also selected for brains and behavior. These evolutionary critics argue that the theory is nothing more than a socially-constructed ideology meant to suppress the poor and marginalized and justify the status quo of those in power. Social Darwinism is the ultimate confirmation of Hume's naturalistic "is-ought fallacy": whatever is ought to be. If nature has granted certain races or a certain sex with "superior" genes, then so should society be structured.

But in their understandable zeal, these critics go too far. One can find in the literature such ideological terms as "oppressive," "sexist," "imperialist," "capitalist," "control," and "order" being attached to physical concepts as DNA, genetics, biochemistry, and evolution. The nadir of this secular form of creationism came at a 1997 interdisciplinary conference in which a psychologist was defending science against a beating by science critics by praising the advances in modern genetics, beginning with the 1953 discovery of DNA. He was asked rhetorically: "You believe in DNA?"

Certainly this is about as ridiculous as it gets, yet I can understand the concerns of the left, given the checkered history of abuse of evolutionary theory in general, and eugenics in particular. I am equally horrified at how some people have used Darwin to control, subjugate, or even destroy others. One of the underlying motives for William Jennings Bryan to take up the anti-evolution cause in the Scopes trial was the application of Social Darwinism by the German militia during the First World War to justify their militarism. The public recognition of the misuses of science is a valuable enterprise which I endorse and participate in (see Chapters 15 and 16). But here again the creationists are succumbing to the "either-or fallacy" where, because of occasional errors, biases, and even gross misuses of science, the entire enterprise must be abandoned. Babies and bathwater comes to mind.

It may prove useful to wrap up this introduction with an example of what I think is proper and cautious application of evolutionary theory to human behavior. Specifically, I wish to inquire why people believe weird things from an evolutionary perspective.

Humans are pattern-seeking animals. We search for meaning in a complex, quirky, and contingent world. But we are also storytelling animals, and for thousands of years our myths and religions have sustained us with stories of meaningful patterns—of gods and God, of supernatural beings and mystical forces, of the relationship between humans with other humans and their creators, and of our place in the cosmos. One of the reasons why humans continue thinking magically is that the modern, scientific way of thinking is a couple of hundred years old, whereas humanity has existed for a couple of hundred thousand years. What were we doing all those long gone millennia? How did our brains evolve to cope with the problems in that radically different world?

This is a problem tackled by evolutionary psychologists—scientists who study brain and behavior from an evolutionary perspective. They make the very reasonable argument that the brain (and along with it the mind and behavior) evolved over a period of two million years from the small fist-sized brain of the *Australopithecine* to the melon-sized brain of modern *Homo sapiens*. Since civilization arose only about 13,000 years ago with the domestication of plants and animals, 99.99% of human evolution took place in our ancestral environment (called the EEA—environment of evolutionary adaptation). The conditions of *that* environment are what shaped our brains, not what happened over the past thirteen millennia. Evolution does not work that fast. Leda Cosmides and John Tooby, Co-Directors of the Center for Evolutionary Psychology at the University of California, Santa Barbara, have summarized the field this way in a 1994 descriptive brochure:

> Evolutionary psychology is based on the recognition that the human brain consists of a large collection of functionally specialized computational devices that evolved to solve the adaptive problems regularly encountered by our hunter-gatherer ancestors. Because humans share a universal evolved architecture, all ordinary individuals reliably develop a distinctively human set of preferences, motives, shared conceptual frameworks, emotion programs, content-specific reasoning procedures, and specialized interpretation systems—programs that operate beneath the surface of expressed cultural variability, and whose designs constitute a precise definition of human nature.

In his new book, *How the Mind Works* (W. W. Norton, 1997), Steven Pinker describes these specialized computational devices as "mental modules." The "module" is a metaphor, and is not necessarily located in a single spot in the brain, and should not be confused with the nineteenth century notion of phrenologists who allocated specific bumps on the head for specific brain functions. A module, says Pinker, "may be broken into regions that are interconnected by fibers that make the regions act as a unit." A bundle of neurons here connected to another bundle of neurons there, "sprawling messily over the bulges and crevasses of the brain" might form a module (pp. 27–31). Their interconnectedness is the key to the module's function, not its location.

While most mental modules are thought of as quite specific, however, evolutionary psychologists argue about mental modules being "domain-specific" vs. "domain-general." Tooby, Cosmides, and Pinker, for example, reject the idea of a domain-general processor, whereas many psychologists accept the notion of a global intelligence, called "g." Archaeologist Steven Mithen, in his book *The Prehistory of the Mind* (Thames and Hudson, 1996) goes so far as to say that it is a domain-general processor that makes us modern humans: "The critical step in the evolution of the modern mind was the switch from a mind designed like a Swiss army knife to one with cognitive fluidity, from a specialized to a generalized type of mentality. This enabled people to design complex tools, to create art and believe in religious ideologies. Moreover, the potential for other types of thought which are critical to the modern world can be laid at the door of cognitive fluidity" (p. 163).

Instead of the metaphor of a module, then, I would like to suggest that we evolved a more general *Belief Engine*, which is Janus-faced—under certain conditions it leads to magical thinking—a *Magic Belief Engine*; under different circumstances it leads to scientific thinking. We might think of the Belief Engine as the central processor that sits beneath more specific modules. Allow me to explain.

We evolved to be skilled, pattern-seeking, causal-finding creatures. Those who were best at finding patterns (standing upwind of game animals is bad for the hunt, cow manure is good for the crops) left behind the most offspring. We are their descendants. The problem in seeking and finding patterns is knowing which ones are meaningful and which ones are not. Unfortunately our brains are not always good at determining the difference. The reason is that discovering a meaningless pattern (painting animals on a cave wall before a hunt) usually does no harm and may even do some good in reducing anxiety in uncertain situations. So we are left with the legacy of two types of thinking errors: *Type 1 Error: believing a falsehood* and *Type 2 Error: rejecting a truth*. Since these errors will not necessarily get us killed, they persist. The Belief Engine has evolved as a mechanism for helping us to survive because in addition to committing Type 1 and Type 2 Errors, we also commit what we might call a *Type 1 Hit: not believing a falsehood* and a *Type 2 Hit: believing a truth*.

It seems reasonable to argue that the brain consists of both specific and general modules, and the Belief Engine is a domain-general processor. It is, in fact, one of the most general of all modules because at its core it is the basis of all learning. After all, we have to believe something about our environment, and these beliefs are learned through experience. But the *process of forming beliefs* is genetically hardwired. To account for the fact that the Belief Engine is capable of both Type 1 and 2 Errors along with Type 1 and 2 Hits, we have to consider two conditions under which it evolved:

1. *Natural Selection*: The Belief Engine is a *useful mechanism* for survival, not just for learning about dangerous and potentially lethal environments (where Type 1 and 2 Hits help us survive), but in reducing anxiety about that environment through magical thinking—there is psychological evidence that magical thinking reduces anxiety in uncertain environments, medical evidence that prayer, meditation, and worship may lead to greater physical and mental health, and anthropological evidence that magicians, shamans, and the kings who use them have more power and win more copulations, thus spreading their genes for magical thinking.

2. *Spandrel*: The magical thinking part of the Belief Engine is also a spandrel—Stephen Jay Gould's and Richard Lewontin's metaphor for a necessary by-product of an evolved mechanism. In their influential 1979 paper, "The Spandrels of San Marco and the Panglossian Paradigm: A Critique of the Adaptationist Programme" (*Proceedings of the Royal Society*, V. B205: 581-598), Gould and Lewontin explain that in architecture a spandrel is "the tapering triangular spaces formed by the intersection of two rounded arches at right angle." This leftover space in medieval churches is filled with elaborate, beautiful designs so purposeful looking "that we are tempted to view it as the starting point of any analysis, as the cause in some sense of the surrounding architecture. But this would invert the proper path of analysis." To ask "what is the purpose of the spandrel" is to ask the wrong question. It would be like asking "why do males have nipples?" The correct question is "why do *females* have nipples?" The answer is that females need them to nurture their babies, and males and females are built on the same architectural frame. It was simply easier for nature to construct males with worthless nipples rather than reconfigure the underlying genetic architecture.

In this sense the magical thinking component of the Belief Engine is a spandrel. We think magically because we have to think causally. We make Type 1 and 2 Errors because we need to make Type 1 and 2 Hits. We have magical thinking and superstitions because we need critical thinking and pattern-finding. The two cannot be separated. Magical thinking is a necessary by product of the evolved mechanism of causal thinking. In my next book, *Why People Believe in God*, can be found an expanded version of this theory in which I present abundant historical and anthropological evidence, but here I will allow the "weird things" written about in this book to serve as examples of such ancestral magical thinking in fully modern humans. Believers in UFOs, alien abductions, ESP, and psychic phenomena have committed a Type 1 Error in thinking: they are believing a falsehood. Creationists and Holocaust deniers have made a Type 2 Error in thinking: they are rejecting a truth. It is not that these folks are ignorant or uninformed; they are intelligent but misinformed. Their thinking has gone wrong. Type 1 and 2 Errors are squelching Type 1 and 2 Hits. Fortunately there is an abundance of evidence that the Belief Engine is malleable. Critical thinking can be taught. Skepticism is learnable. Type 1 and 2 Errors are tractable. I know. I became a skeptic after being a sucker for a lot of these beliefs (recounted in detail in this book). I am a born-again skeptic, as it were.

Having offered this deeper answer to the "why" question, allow me to close with the final exchange in an interview I had with Georgea Kovanis, in the *Detroit Free Press* (May 2, 1997), who understood the bigger skeptical picture when she printed my two-word answer to her final question: "Why should we believe anything you say?" My response: "You shouldn't."

Cogita tute—think for yourself.

A Note on the Revised and Expanded Edition

For years skeptics have been asked by detractors and the media: "What's the harm in believing in UFOs, ESP, astrology, and pseudoscience in general? Aren't you skeptics just taking the fun out of people's lives?" A striking answer by way of example was provided by the Heaven's Gate UFO cult on March 27, 1997, when the mass suicide story broke and a media feeding frenzy lasting two full days flooded the Skeptics Society office. One week later the first edition of *Why People Believe Weird Things* was released, so the publicity tour for the book was heavily slanted toward explaining how such intelligent and educated people as the members of this group could come to believe in something so strongly that they would give up their lives.

The question has renewed relevance, in light of the recent wave of suicidal terrorism on our shores and around the world, and of the sometimes incendiary responses to those attacks. Understanding the psychology of belief systems is the primary focus of this book, and the new chapter that appears at the end of this revised and expanded edition, "Why *Smart* People Believe Weird Things," addresses this question head on, bringing to light the latest research on belief systems, particularly considering how it is that educated and intelligent people also believe that which is apparently irrational. My answer is deceptively simple: *Smart people believe weird things because they are skilled at defending beliefs they arrived at for non-smart reasons.*

Humans are pattern-seeking, storytelling animals, in search of deep meaning behind the seemingly random events of day-to-day life. I hope that this book in some small way helps you navigate a path through the often confusing array of claims and beliefs presented to us as meaningful stories and patterns.

—Altadena, California
December 2001

WHY PEOPLE BELIEVE WEIRD THINGS

PROLOGUE

Next on Oprah

On Monday, October 2, 1995, for the first time in its ten-year history, the *Oprah Winfrey Show* offered a psychic as the featured guest. She was Rosemary Altea (a nom de plume), who claims to communicate with the dead. Her book about this extraordinary assertion—*The Eagle and the Rose: A Remarkable True Story*—had been on the *New York Times* and the *Wall Street Journal* best-seller lists for several weeks. ("The eagle" is a Native American Indian—Altea's spirit guide—and Altea is "the rose.") Oprah began with the disclaimer that she was doing this show only because several trusted friends had described Altea as the class act of the psychic world. Next, the producers rolled several minutes of video, taped the previous day, that showed Altea working a small audience in a Chicago flat, asking countless questions, making numerous generalizations, and providing occasional specifics about their dearly departed. Altea then began working the audience in the studio. "Did someone here lose a loved one in a drowning accident?" "I see a man standing behind you." "Was there a boat involved?" And so on.

Unlike most psychics I have seen, Altea was bombing. The audience was not feeding her the cues she needed to "divine" her information. Finally, well into the program, she struck pay dirt. Calling out to a middle-aged woman partially hidden behind a studio camera, Altea said the woman had lost her mother to cancer. The woman screamed and started crying. Furthermore, Altea noted, the young man next to the woman was her son, who was troubled by school and career decisions. He acknowledged the

1

observation and recounted his tale of woe. The audience was stunned. Oprah was silenced. Altea pumped out more details and predictions. After the taping, one woman stood up and announced that she had come to the studio to debunk Altea but was now a believer.

Enter the skeptic. Three days before the taping of the show, one of Oprah's producers called me. Shocked that the publisher of *Skeptic* magazine had never heard of Rosemary Altea, the producer was preparing to call someone else to do the show when I told her, sight unseen, exactly how Altea operated. The producer mailed me an airline ticket. In my allotted few minutes, I explained that what the audience had just witnessed could be seen at the Magic Castle in Hollywood on any night that a mentalist who knows how to work a crowd is appearing. By "work," I mean the time-proven technique of cold-reading, where the mentalist asks general questions until he or she finds someone who gives generous doses of feedback. Continued questioning eventually finds targets. "Was it lung cancer? Because I'm getting a pain here in the chest." Subject says, "It was a heart attack." "Heart attack? Yes, that explains the chest pains." Or, "I'm sensing a drowning. Was there a boat involved? I'm seeing a boat of some kind on a body of water, maybe a lake or river." And so on. In an audience of two hundred fifty people, every major cause of death will be represented.

The principles of cold-reading are simple: start general (car accidents, drownings, heart attacks, cancer), keep it positive ("He wants you to know he loves you very much," "She says to tell you that she is no longer suffering," "His pain is gone now"), and know that your audience will remember the hits and forget the misses ("How did she know it was cancer?" "How did he get her name?"). But how did Rosemary Altea, without asking, know that the woman's mother had died of cancer and that her son was having doubts about his career? For Oprah, two hundred fifty studio eyewitnesses, and millions of television viewers, Altea appeared to have a direct line to the spirit world.

The explanation is very much of this world, however. Mentalists call this a hot reading where you actually obtain information on your subject ahead of time. Earlier that day, I had shared a limousine from the hotel to the studio with several guests on the show, two of whom were this woman and her son. During the drive, they mentioned that they had met with Altea before and had been invited by Oprah's producers to share their experience with the television audience. Since almost no one knew this little fact, Altea could use her prior knowledge of the woman and her son to snatch victory from the jaws of defeat. Naturally I pointed out this fact but, incredibly, the woman denied having previously met with Altea and the exchange was simply edited out of the show.

I doubt that Altea deliberately deceives her audiences by consciously using cold-reading techniques. Rather, I believe she innocently developed a belief in her own "psychic powers" and innocently learned cold-reading by trial and error. She says it all began in November 1981, when "I woke early one morning to find him standing by the bed, looking down at me. Although I was still half asleep, I knew he was no apparition, no specter in the night" (1995, p. 56). From there, as her book reveals, it was a long process of becoming open to the possibility of a spirit world through what psychologists call *hypnopompic hallucinations*—visions of ghosts, aliens, or loved ones that occur as one emerges from deep sleep—and mystical interpretations of unusual experiences.

But whether we are talking about rats pressing a bar to get food or humans playing a Las Vegas slot machine, it only takes an occasional hit to keep them coming back for more. Altea's belief and behavior were shaped by operant conditioning on a variable-ratio schedule of reinforcement—lots of misses but just enough hits to shape and maintain the behavior. Positive feedback in the form of happy customers paying up to $200 per session was a mechanism sufficient to reinforce her own belief in her powers and to encourage her to hone her mentalist skills.

The same explanation probably holds for the master of cold-reading in the psychic world—James Van Praagh—who wowed audiences for months on NBC's New Age talk show *The Other Side*, until he was debunked on *Unsolved Mysteries*. Here's how. I was asked to sit in a room with nine other people. Van Praagh was asked to do a reading on each of us, all of whom had lost a loved one. I worked closely with the producers to ensure that Van Praagh would have no prior knowledge of any of us. (In addition to subscribing to demographic marketing journals so that they can make statistically educated guesses about subjects based on age, gender, race, and residence, mentalists have been known to go as far as running a name through a detective agency.) His readings would have to be "cold" indeed. The session lasted eleven hours and included several snack breaks, an extended lunch break, and numerous pauses in the filming while technicians reloaded the cameras. Van Praagh opened with a half-hour of New Age music and astrological mumbo jumbo to "prepare" us for our journey to the other side. His mannerisms were somewhat effeminate, and he came off as quite empathic, as if he could "feel our pain."

With most of us, Van Praagh figured out the cause of death through a technique I had not seen before. He would rub either his chest or his head and say "I'm getting a pain here," watching the subject's face for feedback. After the third time, it suddenly struck me why: most people die from heart, lung, or brain failure, regardless of the specific cause (such as, heart attack, stroke, lung cancer, drowning, falling, or automobile accident). With several subjects, he got nothing and said so. "I'm not getting anything. I'm sorry. If it's not there, it's not there." For most of us, however, he got many details as well as the specific cause of death—but not without lots and lots of misses. For the first two hours, I kept track of the number of "no's" and negative head shakes. There were well over a hundred misses for only a dozen or so hits. Given time and enough questions, anyone with a little training could become sensitive enough to do exactly what Van Praagh does.

I also noticed that during the film-changing breaks, Van Praagh would make small talk with the people in the room. "Who are you here for?" he asked one woman. She told him it was her mother. Several readings later, Van Praagh turned to the woman and said, "I see a woman standing behind you. Is that your mother?" At all times he kept it positive. There was redemption for all—our loved ones forgive us for any wrongdoing; they still love us; they suffer no more; they want us to be happy. What else would he say? "Your father wants you to know that he will never forgive you for wrecking his car"? One young woman's husband had been run over by a car. Van Praagh told her, "He wants you to know you will be married again." It turned out that she was engaged to be married, and, of course, she credited Van Praagh with a hit. But, as I explained on camera, Van Praagh said nothing of the sort. He gave his usual positive generalization with no specifics. He did not tell her she was presently engaged to be married. He just said that someday she would marry again. So what? His alternative was to tell the young lady that she would be a lonely widow the rest of her life, which is both statistically unlikely and depressing.

The most dramatic moment of the day came when Van Praagh got the name of a couple's son who had been killed in a drive-by shooting. "I'm seeing the letter K," he proclaimed. "Is it Kevin or Ken?" The mother responded tearfully in a cracking voice, "Yes, Kevin." We were all astonished. Then I noticed around the mother's neck a large, heavy ring with the letter "K" inscribed in diamonds on a black background. Van Praagh denied having seen the ring when I pointed it out on camera. In eleven hours of taping and small talk during breaks, surely he saw the ring. I did, and he's the professional.

The reactions of the audience members I found even more intriguing than the mentalist techniques of Altea and Van Praagh. Anyone can learn cold-reading techniques in half an hour. They work because subjects *want* them to work. Every person at the *Unsolved Mysteries* taping except me wanted Van Praagh to be successful. They came there to speak with their loved ones. In the post-session interviews, all nine subjects gave Van Praagh a positive evaluation, even the few for whom he obviously missed. One woman's daughter had been raped and murdered many years ago, and the police still have no clues to the perpetrator or even to how the crime was committed. The mother had been making the rounds on talk shows, desperately seeking help in finding her daughter's killer. Van Praagh went to her heart like salt into a wound. He reconstructed the murder scene, describing a man on top of the young woman raping her and stabbing her with a knife, and left this grieving mother in tears. (Van Praagh was credited by all with getting this cause of death correct, but earlier, in the morning session, while he was fishing around by rubbing his chest and head, the mother slashed her fingers across her throat, indicating that her daughter's throat had been cut. Everyone but me had forgotten this clue by the time Van Praagh used it.)

After the *Unsolved Mysteries* taping, it became clear that everyone but me was impressed with Van Praagh. The others challenged me to explain all his amazing hits. When I finally told them who I am, what I was doing there, and how cold-reading works, most were uninterested but several walked away. One woman glared at me and told me it was "inappropriate" to destroy these people's hopes during their time of grief.

Herein lies the key to understanding this phenomenon. Life is contingent and filled with uncertainties, the most frightening of which is the manner, time, and place of our own demise. For a parent, an even worse fear is the death of one's child, which makes those who have suffered such a loss especially vulnerable to what "psychics" offer. Under the pressure of reality, we become credulous. We seek reassuring certainties from fortune-tellers and palm-readers, astrologers and psychics. Our critical faculties break down under the onslaught of promises and hopes offered to assuage life's great anxieties. Wouldn't it be marvelous if we did not really die? Wouldn't it be wonderful if we could speak with our lost loved ones again? Of course it would. Skeptics are no different from believers when it comes to such desires. This is an ancient human drive. In a world where one's life was as uncertain as the next meal, our ancestors all over the globe developed beliefs in an afterlife and spirit world. So, when we are vulnerable and afraid, the provider of hope has only to make the promise of an afterlife and offer the flimsiest of proofs. Human credulity will do the rest, as poet Alexander Pope observed in his 1733 *Essay on Man* (Epistle i, 1. 95):

> *Hope springs eternal in the human breast;*
> *Man never Is, but always To be blest.*
> *The soul, uneasy, and confin'd from home,*
> *Rests and expatiates in a life to come.*

This hope is what drives all of us—skeptics and believers alike—to be compelled by unsolved mysteries, to seek spiritual meaning in a physical universe, desire immortality, and wish that our hopes for eternity may be fulfilled. It is what pushes many people to spiritualists, New Age gurus, and television psychics, who offer a Faustian bargain: eternity in exchange for the willing suspension of disbelief (and usually a contribution to the provider's coffers).

But hope springs eternal for scientists and skeptics as well. We are fascinated by mysteries and awed by the universe and the ability of humans to achieve so much in so little time. We seek immortality through our cumulative efforts and lasting achievements; we too wish that our hopes for eternity might be fulfilled.

This book is about people who share similar beliefs and hopes yet pursue them by very dissimilar methods. It is about the distinction between science and pseudoscience, history and pseudohistory, and the difference it makes. Although each chapter can be read independently, cumulatively they show the allure of psychic power and extrasensory perception, UFOs and alien abductions, ghosts and haunted houses. But more than this, the book deals with controversies not necessarily on the margins of society which may have pernicious social consequences: creation-science and biblical literalism, Holocaust denial and freedom of speech, race and IQ, political extremism and the radical right, modern witch crazes prompted by moral panics and mass hysterias, including the recovered memory movement, Satanic ritual abuse, and facilitated communication. Here the difference in thinking makes all the difference.

But more than this—much more—the book is a celebration of the scientific spirit and of the joy inherent in exploring the world's great mysteries even when final answers are not forthcoming. The intellectual journey matters, not the destination. We live in the age of science. It is the reason pseudosciences flourish—pseudoscientists know that their ideas must at least *appear* scientific because science is the touchstone of truth in our culture. Most of us harbor a type of faith in science, a confidence that somehow science will solve our major problems—AIDS, overpopulation, cancer, pollution, heart disease, and so on. Some even entertain scientistic visions of a future without aging, where we will ingest nanotechnological computers that will repair cells and organs, eradicate life-threatening diseases, and maintain us at our chosen age.

So hope springs eternal not just for spiritualists, religionists, New Agers, and psychics, but for materialists, atheists, scientists, and, yes, even skeptics. The difference is in where we find hope. The first group uses science and rationality when convenient, and dumps them when they are not. For this group, any thinking will do, as long as it fulfills that deeply rooted human need for certainty. Why?

Humans evolved the ability to seek and find connections between things and events in the environment (snakes with rattles should be avoided), and those who made the best connections left behind the most offspring. We are their descendants. The problem is that causal thinking is not infallible. We make connections whether they are there or not. These misidentifications come in two varieties: false negatives get you killed (snakes with rattles are okay); false positives merely waste time and energy (a rain dance will end a drought). We are left with a legacy of false positives—hypnopompic hallucinations become ghosts or aliens; knocking noises in an empty house indicate spirits and poltergeists; shadows and lights in a tree become the Virgin Mary; random mountain shadows on Mars are seen as a face constructed by aliens. The belief influences the perception. "Missing" fossils in geological strata become evidence of divine creation. The lack of a written order by Hitler to exterminate the Jews means that perhaps there was no such order . . . or no such extermination. Coincidental configurations of subatomic particles and astronomical structures indicate an intelligent designer of the universe. Vague feelings and memories evoked through hypnosis and guided-imagery in therapy evolve into crystal-clear memories of childhood sexual abuse, even when no corroborating evidence exists.

Scientists have their false positives—but the methods of science were specifically designed to weed them out. Had the cold fusion findings, to take a recent spectacular example of a false positive, not been made so public before corroboration from other scientists, they would have been nothing out of the ordinary. This is precisely how science progresses—countless identified false negatives and false positives. The public, however, does not usually hear about them because negative findings are not usually published. That silicon breast implants might cause serious health problems was big news; that there has been no corroborative and replicable scientific evidence that they do has gone almost unnoticed.

What, then, you may ask, does it mean to be a skeptic? Some people believe that skepticism is rejection of new ideas or, worse, they confuse *skeptic* with *cynic* and think that skeptics are a bunch of grumpy curmudgeons unwilling to accept any claim that challenges the status quo. This is wrong. Skepticism is a provisional approach to claims. *Skepticism is a method, not a position.* Ideally, skeptics do not go into an investigation closed to the possibility that a phenomenon might be real or that a claim might be true. For example, when I investigated the claims of the Holocaust deniers, I ended up being skeptical of these skeptics (see chapters 13 and 14). In the case of recovered memories, I came down on the side of the skeptics (see chapter 7). One may be skeptical of a belief or of those who challenge it.

The analyses in this book explain in three tiers why people believe weird things: (1) because hope springs eternal; (2) because thinking can go wrong in general ways; (3) because thinking can go wrong in particular ways. I mix specific examples of "weird beliefs" with general principles about what we can learn from examining such beliefs. To this end, I have taken Stephen Jay Gould's style as a model for a healthy blend of the particular and the universal, the details and the big picture; and as inspiration James Randi's mission to understand some of the more perplexing mysteries of our age and ages past.

In the five years since we founded the Skeptics Society and *Skeptic* magazine, my partner, friend, and wife, Kim Ziel Shermer, has provided countless hours of feedback during meals, while driving in the car and riding bikes, and on our daily jaunt up the mountain with the dogs and our daughter, Devin. My other *Skeptic* partner, Pat Linse, has proved to be far more than just a brilliant art director. She is one of a rare species, an artistic and scientific polymath, whose prolific reading (she doesn't own a television) enables her not only to converse on virtually any subject but to make original and constructive contributions to the skeptic movement.

I also wish to acknowledge those who have been most helpful in producing *Skeptic* magazine and putting on our lecture series at Caltech, without which this book would not exist. Jaime Botero has been there with me since I taught the evening course in introductory psychology at Glendale College a decade ago. Diane Knudtson has worked nearly every Skeptics Society lecture at Caltech for nothing more than a meal and food for thought. Brad Davies has produced videos of every lecture and provided valuable feedback on the speakers' many and diverse ideas. Jerry Friedman constructed our database, organized the Skeptics Society survey, and provided valuable information on the animal rights movement. Terry Kirker continues to contribute to the promotion of science and skepticism in her own unique way.

Most of the chapters began as essays originally published in *Skeptic* magazine, which I edit. Skeptical readers may then reasonably ask, Who edits the editor? Who is skeptical of the skeptic? Every essay in this volume has been read and edited by my publisher's editors, Elizabeth Knoll, Mary Louise Byrd, and Michelle Bonnice; by my partners, Kim and Pat; by one or more of *Skeptic* magazine's contributing editors; and, where appropriate, by a member of *Skeptic* magazine's editorial board or by an expert in the field. For this, I heartily thank David Alexander, Clay Drees, Gene Friedman, Alex Grobman, Diane Halpern, Steve Harris, Gerald Larue, Jim Lippard, Betty McCollister, Tom McDonough, Paul McDowell, Tom McIver, Sara Meric, John Mosley, Richard Olson, D'art Phares, Donald Prothero, Rick Shaffer, Elie Shneour, Brian Siano, Jay Snelson, Carol Tavris, Kurt Wochholtz, and especially Richard Hardison, Bernard Leikind, Frank Miele, and Frank Sulloway, for not allowing friendship to get in the way of brutal honesty when editing my essays. At W. H. Freeman I wish to thank Simone Cooper who brilliantly organized my national book tour and made it a joy rather than a chore; Peter McGuigan for bringing the book to audio so people can hear it as well as read it; John Michel for his critical feedback on this and the transition to my next book, *Why People Believe in God*. A special thanks to Sloane Lederer who maintained the progress of the publishing and promotion of this book throughout numerous personnel changes at the publisher, as well as for understanding the deeper importance of what we skeptics are trying to accomplish through writing books such as this. Thanks to my agents Katinka Matson and John Brockman, and their foreign rights director Linda Wollenberger, for helping to bring about the book in this and other languages. Finally, Bruce Mazet has made it possible for the Skeptics Society, *Skeptic* magazine, and Millennium Press to battle ignorance and misunderstanding; he has pushed us well beyond what I ever dreamed we were capable of accomplishing.

In his 1958 masterpiece, *The Philosophy of Physical Science*, physicist and astronomer Sir Arthur Stanley Eddington asked about observations made by scientists, "*Quis custodiet ipsos custodes?—Who will observe the observers?*" "The epistemologist," answered Eddington. "He watches them to see what they really observe, which is often quite different from what they say they observe. He examines their procedure and the essential limitations of the equipment they bring to their task, and by so doing becomes aware before-hand of limitations to which the results they obtain will have to conform" (1958, p. 21). Today the observers' observers are the skeptics. But who will observe the skeptics? You. So have at it and have fun.

PART 1

SCIENCE

AND

SKEPTICISM

Science is founded on the conviction that experience, effort, and reason are valid; magic on the belief that hope cannot fail nor desire deceive.

—Branislaw Malinowski, *Magic, Science, and Religion*, 1948

1

I Am Therefore I Think

A Skeptic's Manifesto

O n the opening page of his splendid little book *To Know a Fly*, biologist Vincent Dethier makes this humorous observation about how children grow up to be scientists: "Although small children have taboos against stepping on ants because such actions are said to bring on rain, there has never seemed to be a taboo against pulling off the legs or wings of flies. Most children eventually outgrow this behavior. Those who do not either come to a bad end or become biologists" (1962, p. 2). In their early years, children are knowledge junkies, questioning everything in their purview, though exhibiting little skepticism. Most never learn to distinguish between skepticism and credulity. It took me a long time.

In 1979, unable to land a full-time teaching job, I found work as a writer for a cycling magazine. The first day on the job, I was sent to a press conference held in honor of a man named John Marino who had just ridden his bicycle across America in a record 13 days, 1 hour, 20 minutes. When I asked him how he did it, John told me about special vegetarian diets, megavitamin therapy, fasting, colonics, mud baths, iridology, cytotoxic blood testing, Rolfing, acupressure and acupuncture, chiropractic and massage therapy, negative ions, pyramid power, and a host of weird things with which I was unfamiliar. Being a fairly inquisitive fellow, when I took up cycling as a serious sport I thought I would try these things to see for myself whether they worked. I once fasted for a week on nothing but a strange mixture of water, cayenne pepper, garlic, and lemon. At the end of the week, John and I rode from Irvine to Big Bear Lake and back, some seventy miles each way. About halfway up the mountain I collapsed, violently ill from the concoction. John and I once rode out to a health spa near Lake Elsinore for a mud bath that was supposed to suck the toxins out of my body. My skin was dyed red for a week. I set up a negative ion generator in my bedroom to charge the air to give me more energy. It turned the walls black with dust. I got my iris read by an iridologist, who told me that the little green flecks in my eyes meant something was wrong with my kidneys. To this day my kidneys are functioning fine.

I really got into cycling. I bought a racing bike the day after I met John and entered my first race that weekend. I did my first century ride (100 miles) a month later, and my first double century later that year. I kept trying weird things because I figured I had nothing to lose and, who knows, maybe they would increase performance. I tried colonics because supposedly bad things clog the plumbing and thus decrease digestive efficiency, but all I got was an hour with a hose in a very uncomfortable place. I installed a pyramid in my apartment because it was supposed to focus energy. All I got were strange looks from guests. I starting getting massages, which were thoroughly enjoyable and quite relaxing. Then my massage therapist decided that "deep tissue" massage was best to get lactic acid out of the muscles. That wasn't so relaxing. One guy massaged me with his feet. That was even less relaxing. I tried Rolfing, which is *really* deep tissue massage. That was so painful that I never went back.

In 1982 John and I and two other men competed in the first Race Across America, the 3,000-mile, nonstop, transcontinental bike race from Los Angeles to New York. In preparation, we went for cytotoxic blood testing because it was supposed to detect food allergies that cause blood platelets to clump together and block capillaries, thus decreasing blood flow. By now we were a little skeptical of the truth of these various claims, so we sent in one man's blood under several names. Each sample came back with different food allergies, which told us that there was a problem with their testing, not with our blood. During the race, I slept with an "Electro-Acuscope," which was to measure my brain waves and put me into an alpha state for better sleeping. It was also supposed to rejuvenate my muscles and heal any injuries. The company swore that it helped Joe Montana win the Super Bowl. Near as I can figure, it was totally ineffective.

The Electro-Acuscope was the idea of my chiropractor. I began visiting a chiropractor not because I needed one but because I had read that energy flows through the spinal cord and can get blocked at various places. I discovered that the more I got adjusted, the more I needed to get adjusted because my neck and back kept going "out." This went on for a couple of years until I finally quit going altogether, and I've never needed a chiropractor since.

All told, I raced as a professional ultra-marathon cyclist for ten years, all the while trying anything and everything (except drugs and steroids) that might improve my performance. As the Race Across America got bigger—it was featured for many years on ABC's *Wide World of Sports*—I had many offers to try all sorts of things, which I usually did. From this ten-year experiment with a subject pool of one, I drew two conclusions: nothing increased performance, alleviated pain, or enhanced well-being other than long hours in the saddle, dedication to a consistent training schedule, and a balanced diet; and it pays to be skeptical. But what does it mean to be skeptical?

What Is a Skeptic?

I became a skeptic on Saturday, August 6, 1983, on the long, climbing road to Loveland Pass, Colorado. It was Day 3 of the second Race Across America, and the nutritionist on my support crew believed that if I followed his megavitamin therapy program, I would win the race. He was in a Ph.D. program and was trained as a nutritionist, so I figured he knew what he was doing. Every six hours I would force down a huge handful of assorted vitamins and minerals. Their taste and smell nearly made me sick, and they went right through me, producing what I thought had to be the most expensive and colorful urine in America. After three days of this, I decided that megavitamin therapy, along with colonics, iridology, Rolfing, and all these other alternative, New Age therapies were a bunch of hooey. On that climb up Loveland Pass, I dutifully put the vitamins in my mouth and then spit them out up the road when my nutritionist wasn't looking. Being skeptical seemed a lot safer than being credulous.

After the race I discovered that the nutritionist's Ph.D. was to be awarded by a nonaccredited nutrition school and, worse, *I* was the subject of his doctoral dissertation! Since that time I have noticed about extraordinary claims and New Age beliefs that they tend to attract people on the fringes of academia—people without formal scientific training, credentialed (if at all) by nonaccredited schools, lacking research data to support their claims, and excessively boastful about what their particular elixir can accomplish. This does not automatically disprove all claims made by individuals exhibiting these characteristics, but it would be wise to be especially skeptical when encountering them.

Being skeptical is nothing new, of course. Skepticism dates back 2,500 years to ancient Greece and Plato's Academy. But Socrates' quip that "All I know is that I know nothing" doesn't get us far. Modern skepticism has developed into a science-based movement, beginning with Martin Gardner's 1952 classic, *Fads and Fallacies in the Name of Science*. Gardner's numerous essays and books over the next four decades, such as *Science: Good, Bad, and Bogus* (1981), *The New Age: Notes of a Fringe Watcher* (1991a), and *On the Wild Side* (1992), established a pattern of incredulity about a wide variety of bizarre beliefs. Skepticism joined pop culture through magician James "the Amazing" Randi's countless psychic challenges and media appearances in the 1970s and 1980s (including thirty-six appearances on the *Tonight Show*). Philosopher Paul Kurtz helped create dozens of skeptics groups throughout the United States and abroad, and publications such as *Skeptic* magazine have national and international circulation. Today, a burgeoning group of people calling themselves skeptics—scientists, engineers, physicians, lawyers, professors, teachers, and the intellectually curious from all walks of life—conduct investigations, hold monthly meetings and annual conferences, and provide the media and the general public with natural explanations for apparently supernatural phenomena.

Modern skepticism is embodied in the scientific method, which involves gathering data to test natural explanations for natural phenomena. A claim becomes factual when it is confirmed to such an extent that it would be reasonable to offer temporary agreement. But all facts in science are provisional and subject to challenge, and therefore skepticism is a *method* leading to provisional conclusions. Some things, such as water dowsing, extrasensory perception, and creationism, have been tested and have failed the tests often enough that we can provisionally conclude that they are false. Other things, such as hypnosis, lie detectors, and vitamin C, have been tested but the results are inconclusive, so we must continue formulating and testing hypotheses until we can reach a provisional conclusion. The key to skepticism is to navigate the treacherous straits between "know nothing" skepticism and "anything goes" credulity by continuously and vigorously applying the methods of science.

The flaw in pure skepticism is that when taken to an extreme, the position itself cannot stand. If you are skeptical about everything, you must be skeptical of your own skepticism. Like the decaying subatomic particle, pure skepticism spins off the viewing screen of our intellectual cloud chamber.

There is also a popular notion that skeptics are closed-minded. Some even call us cynics. In principle, skeptics are not closed-minded or cynical. What I mean by a skeptic is *one who questions the validity of a particular claim by calling for evidence to prove or disprove it.* In other words, skeptics are from Missouri—the "show me" state. When we hear a fantastic claim, we say, "That's nice, prove it."

Here is an example. For many years I had heard stories about the "Hundredth Monkey phenomenon" and was fascinated with the possibility that there might be some sort of collective consciousness that we could tap into to decrease crime, eliminate wars, and generally unite as a single species. In the 1992 presidential election, in fact, one candidate—Dr. John Hagelin from the Natural Law Party—claimed that if elected he would implement a plan that would solve the problems of our inner cities: meditation. Hagelin and others (especially proponents of Transcendental Meditation, or TM) believe that thought can somehow be transferred between people, especially people in a meditative state; if enough people meditate at the same time, some sort of critical mass will be reached, thereby inducing significant planetary change. The Hundredth Monkey phenomenon is commonly cited as empirical proof of this astonishing theory. In the 1950s, so the story goes, Japanese scientists gave monkeys on Koshima Island potatoes. One day one of the monkeys learned to wash the potatoes and then taught the skill to others. When about one hundred monkeys had learned the skill—the so-called critical mass—suddenly all the monkeys knew it, even those on other islands hundreds of miles away. Books about the phenomenon have spread this theory widely in New Age circles. Lyall Watson's *Lifetide* (1979) and Ken Keyes's *The Hundredth Monkey* (1982), for example, have been through multiple printings and sold millions of copies; Elda Hartley even made a film called *The Hundredth Monkey*.

As an exercise in skepticism, start by asking whether events really happened as reported. They did not. In 1952, primatologists began providing Japanese macaques with sweet potatoes to keep the monkeys from raiding local farms. One monkey did learn to wash dirt off the sweet potatoes in a stream or the ocean, and other monkeys did learn to imitate the behavior. Now let's examine Watson's book more carefully. He admits that "one has to gather the rest of the story from personal anecdotes and bits of folklore among primate researchers, because most of them are still not quite sure what happened. So I am forced to improvise the details." Watson then speculates that "an unspecified number of monkeys on Koshima were washing sweet potatoes in the sea"—hardly the level of precision one expects. He then makes this statement: "Let us say, for argument's sake, that the number was ninety-nine and that at 11:00 A.M. on a Tuesday, one further convert was added to the fold in the usual way. But the addition of the hundredth monkey apparently carried the number across some sort of threshold, pushing it through a kind of critical mass." At this point, says Watson, the habit "seems to have jumped natural barriers and to have appeared spontaneously on other islands" (1979, pp. 2–8).

Let's stop right there. Scientists do not "improvise" details or make wild guesses from "anecdotes" and "bits of folklore." In fact, some scientists *did* record *exactly* what happened (for example, Baldwin et al. 1980; Imanishi 1983; Kawai 1962). The research began with a troop of twenty monkeys in 1952, and every monkey on the island was carefully observed. By 1962, the troop had increased to fifty-nine monkeys and exactly thirty-six of the fifty-nine monkeys were washing their sweet potatoes. The "sudden" acquisition of the behavior actually took ten years, and the "hundred monkeys" were actually only thirty-six in 1962. Furthermore, we can speculate endlessly about what the monkeys knew, but the fact remains that not all of the monkeys in the troop were exhibiting the washing behavior. The thirty-six monkeys were not a critical mass even at home. And while there are some reports of similar behavior on other islands, the observations were made between 1953 and 1967. It was not sudden, nor was it necessarily connected to Koshima. The monkeys on other islands could have discovered this simple skill themselves, for example, or inhabitants on other islands might have taught them. In any case, not only is there no evidence to support this extraordinary claim, there is not even a real phenomenon to explain.

Science and Skepticism

Skepticism is a vital part of science, which I define as *a set of methods designed to describe and interpret observed or inferred phenomena, past or present, and aimed at building a testable body of knowledge open to rejection or confirmation.* In other words, science is a specific way of analyzing information with the goal of testing claims. Defining the *scientific method* is not so simple, as philosopher of science and Nobel laureate Sir Peter Medawar observed: "Ask a scientist what he conceives the scientific method to be and he will adopt an expression that is at once solemn and shifty-eyed: solemn, because he feels he ought to declare an opinion; shifty-eyed, because he is wondering how to conceal the fact that he has no opinion to declare" (1969, p. 11).

A sizable literature exists on the scientific method, but there is little consensus among authors. This does not mean that scientists do not know what they are doing. Doing and explaining may be two different things. However, scientists agree that the following elements are involved in thinking scientifically:

> *Induction:* Forming a hypothesis by drawing general conclusions from existing data.
>
> *Deduction:* Making specific predictions based on the hypotheses.
>
> *Observation:* Gathering data, driven by hypotheses that tell us what to look for in nature.
>
> *Verification:* Testing the predictions against further observations to confirm or falsify the initial hypotheses.

Science, of course, is not this rigid; and no scientist consciously goes through "steps." The process is a constant interaction of making observations, drawing conclusions, making predictions, and checking them against evidence. And data-gathering observations are not made in a vacuum. The hypotheses shape what sorts of observations you will make of nature, and these hypotheses are themselves shaped by your education, culture, and particular biases as an observer.

This process constitutes the core of what philosophers of science call the *hypothetico-deductive* method, which, according to the *Dictionary of the History of Science*, involves "(a) putting forward a hypothesis, (b) conjoining it with a statement of 'initial conditions,' (c) deducing from the two a prediction, and (d) finding whether or not the prediction is fulfilled" (Bynum, Browne, and Porter 1981, p. 196). It is not possible to say which came first, the observation or the hypothesis, since the two are inseparably interactive. But additional observations are what flesh out the hypothetico-deductive process, and they serve as the final arbiter on the validity of predictions. As Sir Arthur Stanley Eddington noted, "For the truth of the conclusions of science, observation is the supreme court of appeal" (1958, p. 9). Through the scientific method, we may form the following generalizations:

> *Hypothesis:* A testable statement accounting for a set of observations.
>
> *Theory:* A well-supported and well-tested hypothesis or set of hypotheses.
>
> *Fact:* A conclusion confirmed to such an extent that it would be reasonable to offer provisional agreement.

A theory may be contrasted with a *construct:* a nontestable statement to account for a set of observations. The living organisms on Earth may be accounted for by the statement "God made them" or the statement "They evolved." The first statement is a construct, the second a theory. Most biologists would even call evolution a fact.

Through the scientific method, we aim for *objectivity:* basing conclusions on external validation. And we avoid *mysticism:* basing conclusions on personal insights that elude external validation.

There is nothing wrong with personal insight as a starting point. Many great scientists have attributed their important ideas to insight, intuition, and other mental leaps hard to pin down. Alfred Russel Wallace said that the idea of natural selection "suddenly flashed upon" him during an attack of malaria. But intuitive ideas and mystical insights do not become objective until they are externally validated. As psychologist Richard Hardison explained,

> Mystical "truths," by their nature, must be solely personal, and they can have no possible external validation. Each has equal claim to truth. Tealeaf reading and astrology and Buddhism; each is equally sound or unsound if we judge by the absence of related evidence. This is not intended to disparage any one of the faiths; merely to note the impossibility of verifying their correctness. The mystic is in a paradoxical position. When he seeks external support for his views he must turn to external arguments, and he denies mysticism in the process. External validation is, by definition, impossible for the mystic. (1988, pp. 259–260)

Science leads us toward *rationalism:* basing conclusions on logic and evidence. For example, how do we know the Earth is round? It is a logical conclusion drawn from observations such as

- The shadow of the Earth on the moon is round.
- The mast of a ship is the last thing seen as it sails into the distance.
- The horizon is curved.
- Photographs from space.

And science helps us avoid *dogmatism:* basing conclusions on authority rather than logic and evidence. For example, how do we know the Earth is round?

- Our parents told us.
- Our teachers told us.
- Our minister told us.
- Our textbook told us.

Dogmatic conclusions are not necessarily invalid, but they do beg other questions: How did the authorities come by their conclusions? Were they guided by science or some other means?

The Essential Tension Between Skepticism and Credulity

It is important to recognize the fallibility of science and the scientific method. But within this fallibility lies its greatest strength: self-correction. Whether a mistake is made honestly or dishonestly, whether a fraud is unknowingly or knowingly perpetrated, in time it will be flushed out of the system by lack of external verification. The cold fusion fiasco is a classic example of the system's swift exposure of error.

Because of the importance of this self-correcting feature, among scientists there is at best what Caltech physicist and Nobel laureate Richard Feynman called "a principle of scientific thought that corresponds to a kind of utter honesty—a kind of leaning over backwards." Said Feynman, "If you're doing an experiment, you should report everything that you think might make it invalid—not only what you think is right about it: other causes that could possibly explain your results" (1988, p. 247).

Despite these built-in mechanisms, science remains subject to problems and fallacies ranging from inadequate mathematical notation to wishful thinking. But, as philosopher of science Thomas Kuhn (1977) noted, the "essential tension" in science is between total commitment to the status quo and blind pursuit of new ideas. The paradigm shifts and revolutions in science depend upon proper balancing of these opposing impulses. When enough of the scientific community (particularly those in positions of power) are willing to abandon orthodoxy in favor of the (formerly) radical new theory, then and only then can a paradigm shift occur (see chapter 2).

Charles Darwin is a good example of a scientist who negotiated the essential tension between skepticism and credulity. Historian of science Frank Sulloway identifies three characteristics in Darwin's thinking that helped Darwin find his balance: (1) he respected others' opinions but was willing to challenge authorities (he intimately understood the theory of special creation, yet he overturned it with his own theory of natural selection); (2) he paid close attention to negative evidence (Darwin included a chapter called "Difficulties on Theory" in the *Origin of Species*—as a result his opponents could rarely present him with a challenge that he had not already addressed); (3) he generously used the work of others (Darwin's collected correspondence numbers over 14,000 letters, most of which include lengthy discussions and question-and-answer sequences about scientific problems). Darwin was constantly questioning, always learning, confident enough to formulate original ideas yet modest enough to recognize his own fallibility. "Usually, it is the scientific community as a whole that displays this essential tension between tradition and change," Sulloway observed, "since most people have a preference for one or the other way of thinking. What is relatively rare in the history of science is to find these contradictory qualities combined in such a successful manner in one individual" (1991, p. 32).

The essential tension in dealing with "weird things" is between being so skeptical that revolutionary ideas pass you by and being so open-minded that flimflam artists take you in. Balance can be found by answering a few basic questions: What is the quality of the evidence for the claim? What are the background and credentials of the person making the claim? Does the thing work as claimed? As I discovered during my personal odyssey in the world of alternative health and fitness therapies and gadgets, often the evidence is weak, the background and credentials of the claimants are questionable, and the therapy or gadget almost never does what it is supposed to.

This last point may well be the crucial one. I regularly receive calls about astrology. Callers usually want to know about the theory behind astrology. They are wondering whether the alignment of planetary bodies can significantly influence human destiny. The answer is no, but the more important point is that one need not understand gravity and the laws governing the motion of the planets to evaluate astrology. All one needs to do is ask, Does it work? That is, do astrologers accurately and specifically predict human destiny from the alignment of the planets? No, they do not. Not one astrologer predicted the crash of TWA flight #800; not one astrologer predicted the Northridge earthquake. Thus, the theory behind astrology is irrelevant, because astrology simply does not do what astrologers claim it can do. It vanishes hand-in-hand with the hundredth monkey.

The Tool of the Mind

Vincent Dethier, in his discussion of the rewards of science, runs through a pantheon of the obvious ones—money, security, honor—as well as the transcendent: "a passport to the world, a feeling of belonging to one race, a feeling that transcends political boundaries and ideologies, religions, and languages." But he brushes all these aside for one "more lofty and more subtle"—the natural curiosity of humans:

> One of the characteristics that sets man apart from all the other animals (and animal he indubitably is) is a need for knowledge for its own sake. Many animals are curious, but in them curiosity is a facet of adaptation. Man has a hunger to know. And to many a man, being endowed with the capacity to know, he has a duty to know. All knowledge, however small, however irrelevant to progress and well-being, is a part of the whole. It is of this the scientist partakes. To know the fly is to share a bit in the sublimity of Knowledge. That is the challenge and the joy of science. (1962, pp. 118–119)

At its most basic level, curiosity about how things work is what science is all about. As Feynman observed, "I've been caught, so to speak—like someone who was given something wonderful when he was a child, and he's always looking for it again. I'm always looking, like a child, for the wonders I know I'm going to find—maybe not every time, but every once in a while" (1988, p. 16). The most important question in education, then, is this: What tools are children given to help them explore, enjoy, and understand the world? Of the various tools taught in school, science and thinking skeptically about all claims should be near the top.

Children are born with the ability to perceive cause-effect relations. Our brains are natural machines for piecing together events that may be related and for solving problems that require our attention. We can envision an ancient hominid from Africa chipping and grinding and shaping a rock into a sharp tool for carving up a large mammalian carcass. Or perhaps we can imagine the first individual who discovered that knocking flint would create a spark that would light a fire. The wheel, the lever, the bow and arrow, the plow—inventions intended to allow us to shape our environment rather than be shaped by it—started us down a path that led to our modern scientific and technological world.

On the most basic level, we must think to remain alive. To think is the most essential human characteristic. Over three centuries ago, the French mathematician and philosopher René Descartes, after one of the most thorough and skeptical purges in intellectual history, concluded that he knew one thing for certain: "*Cogito ergo sum—I think therefore I am.*" But to be human is to think. To reverse Descartes, "*Sum ergo cogito—I am therefore I think.*"

2

The Most Precious Thing We Have

The Difference Between
Science and Pseudoscience

The part of the world known as the Industrial West could, in its entirety, be seen as a monument to the Scientific Revolution, begun over 400 years ago and succinctly captured in a single phrase by one of its initiators, Francis Bacon: "Knowledge itself is power." We live in an age of science and technology. Thirty years ago, historian of science Derek J. De Solla Price observed that "using any reasonable definition of a scientist, we can say that 80 to 90 percent of all the scientists that have ever lived are alive now. Alternatively, any young scientist, starting now and looking back at the end of his career upon a normal life span, will find that 80 to 90 percent of all scientific work achieved by the end of the period will have taken place before his very eyes, and that only 10 to 20 percent will antedate his experience" (1963, pp. 1–2).

There are now, for example, more than six million articles published in well over 100,000 scientific journals each year. The Dewey Decimal Classification now lists more than a thousand different classifications under the heading "Pure Science," and within each of these classifications are dozens of specialty journals. Figure 1 depicts the growth in the number of scientific journals, from the founding of the Royal Society in 1662 when there were two, to the present.

Virtually every field of learning shows such an exponential growth curve. As the number of individuals working in a field grows, so too does

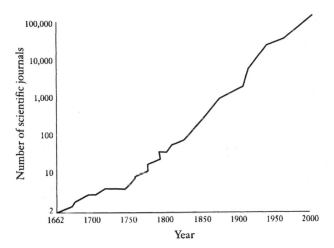

FIGURE 1:
Number of scientific journals, 1662–present. [From De Solla Price 1963.]

the amount of knowledge, which creates more jobs, attracts more people, and so on. The membership growth curves for the American Mathematical Society (founded in 1888) and the Mathematical Association of America (founded in 1915), which are shown in figure 2, dramatically demonstrate this phenomenon. In 1965, observing the accelerating rate at which individuals were entering the sciences, the junior minister of science and education of Great Britain concluded, "For more than 200 years scientists everywhere were a significant minority of the population. In Britain today they outnumber the clergy and the officers of the armed forces. If the rate of progress which has been maintained ever since the time of Sir Isaac Newton were to continue for another 200 years, every man, woman and child on Earth would be a scientist, and so would every horse, cow, dog, and mule" (in Hardison 1988, p. 14).

Transportation speed has also shown geometric progression, with most of the change being made in the last 1 percent of human history. French historian Fernand Braudel tells us, for example, that "Napoleon moved no faster than Julius Caesar" (1981, p. 429). But in the twentieth century the speed of transportation has increased astronomically (figuratively and literally), as the following list shows:

1784	Stagecoach.	10 mph
1825	Steam locomotive	13 mph
1870	Bicycle	17 mph
1880	Steam-powered train	100 mph
1906	Steam-powered automobile	127 mph
1919	Early aircraft	164 mph
1938	Airplane	400 mph
1945	Combat airplane	606 mph
1947	Bell X-1 rocket-plane	769 mph
1960	Rocket	4,000 mph
1985	Space shuttle	18,000 mph
2000	TAU deep-space probe	225,000 mph

One final example of technological change based on scientific research will serve to drive the point home. Timing devices in various forms—dials, watches, and clocks—have improved exponentially in accuracy, as illustrated in figure 3.

If we are living in the Age of Science, then why do so many pseudoscientific and nonscientific beliefs abound? Religions, myths, superstitions, mysticisms, cults, New Age ideas, and nonsense of all sorts have penetrated every nook and cranny of both popular and high culture. A 1990 Gallup poll of 1,236 adult Americans showed percentages of belief in the paranormal that are alarming (Gallup and Newport 1991, pp. 137–146).

Astrology	52%
Extrasensory perception	46%
Witches	19%
Aliens have landed on Earth	22%
The lost continent of Atlantis	33%
Dinosaurs and humans lived simultaneously	41%
Noah's flood	65%
Communication with the dead	42%
Ghosts	35%
Actually had a psychic experience	67%

Other popular ideas of our time that have little to no scientific support include dowsing, the Bermuda Triangle, poltergeists, biorhythms, creationism, levitation, psychokinesis, astrology, ghosts, psychic detectives, UFOs, remote viewing, Kirlian auras, emotions in plants, life after death, monsters, graphology, crypto-zoology, clairvoyance, mediums, pyramid power, faith healing, Big Foot, psychic prospecting, haunted houses, perpetual motion machines, antigravity locations, and, amusingly, astrological birth control. Belief in these phenomena is not limited to a quirky handful on the lunatic fringe. It is more pervasive than most of us like to think, and this is curious considering how far science has come since the Middle Ages. Shouldn't we know by now that ghosts cannot exist unless the laws of science are faulty or incomplete?

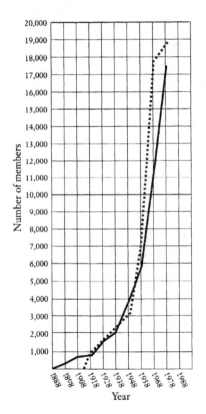

FIGURE 2:
Growth in membership of (*solid line*) the American Mathematical Society and its predecessor, the New York Mathematical Society, founded 1888; and (*dashed line*) the Mathematical Association of America, founded 1915. [Courtesy Mathematical Association of America.]

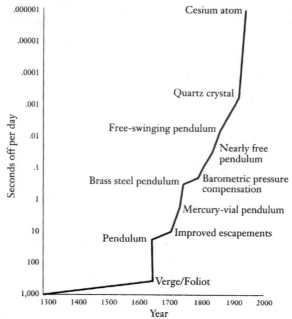

FIGURE 3:
Accuracy of timing devices, 1300–present.

Pirsig's Paradox

There is a priceless dialogue between father and son in Robert Pirsig's classic 1974 intellectual adventure story, *Zen and the Art of Motorcycle Maintenance*, that takes place during a cross-country motorcycle tour that included many late-night discussions. The father tells his son that he does not believe in ghosts because "they are unscientific. They contain no matter and have no energy and therefore according to the laws of science, do not exist except in people's minds. Of course, the laws of science contain no matter and have no energy either and therefore do not exist except in people's minds. It's best to refuse to believe in either ghosts or the laws of science." The son, now confused, wonders if his father has wandered off into nihilism (1974, pp. 38–39):

"So you don't believe in ghosts or science?"
"No, I do believe in ghosts."
"What?"

"The laws of physics and logic, the number system, the principle of algebraic substitution. These are ghosts. We just believe in them so thoroughly they seem real. For example, it seems completely natural to presume that gravitation and the law of gravity existed before Isaac Newton. It would sound nutty to think that until the seventeenth century there was no gravity."

"Of course."

"So, before the beginning of the Earth, before people, etc., the law of gravity existed. Sitting there, having no mass of its own, no energy, and not existing in anyone's mind."

"Right."

"Then what has a thing to do to be nonexistent? It has just passed every test of nonexistence there is. You cannot think of a single attribute of nonexistence that the law of gravity didn't have, or a single scientific attribute of existence it did have. I predict that if you think about it long enough, you will go round and round until you realize that the law of gravity did not exist before Isaac Newton. So the law of gravity exists nowhere except in people's heads. It is a ghost!"

This is what I call *Pirsig's Paradox*. One of the knottier problems for historians and philosophers of science over the past three decades has been resolving the tension between the view of science as a progressive, culturally independent, objective quest for Truth and the view of science as a non-progressive, socially constructed, subjective creation of knowledge. Philosophers of science label these two approaches *internalist* and *externalist*, respectively. The *internalist* focuses on the internal workings of science independent of its larger cultural context: the development of ideas, hypotheses, theories, and laws, and the internal logic within and between them. The Belgian-American George Sarton, one of the founders of the history of science field, launched the internalist view. Sarton's discussion of the internalist approach may be summarized as follows:

1. The study of the history of science is only justified by its relevance to present and future science. Therefore, historians must understand present science in order to see how past science has shaped its development.

2. Science is "systematized positive knowledge," and "the acquisition and systematization of positive knowledge are the only human activities which are truly cumulative and progressive" (Sarton 1936, p. 5). Therefore, the historian should consider each historical step in terms of progressive or regressive effects.

3. Although science is embedded in culture, it is not influenced by culture to any significant degree. Thus, the historian need not worry about external context and should concentrate on the internal workings of science.

4. Science, because it is positive, cumulative, and progressive, is the most important contribution to the history of humanity. Therefore, it is the most important thing a historian can study. Doing so will help prevent wars and build bridges between peoples and cultures.

By contrast, the *externalist* concentrates on placing science within the larger cultural context of religion, politics, economics, and ideologies and considers the effect these have on the development of scientific ideas, hypotheses, theories, and laws. Philosopher of science Thomas Kuhn began the externalist tradition in 1962, with the publication of his *The Structure of Scientific Revolutions*. In this book, he introduced the concepts of scientific paradigms and paradigm shifts. Reflecting upon the internalist tradition, Kuhn concluded, "Historians of science owe the late George Sarton an immense debt for his role in establishing their profession, but the image of their specialty which he propagated continues to do much harm even though it has long since been rejected" (1977, p. 148).

Science historian Richard Olson, who switched from physics to the history of science, strikes a balance between these positions. Olson opens his 1991 book, *Science Deified and Science Defied*, with a quotation from psychologist B. F. Skinner that succinctly states the internalist position: "No theory changes what it's a theory about." Olson goes on to reject such strict internalism: "There is a serious question about whether such a statement can be interpreted in a way that could be true even if the objects of the theory were inanimate; but there is no question that it is false when it is applied to humans and other living organisms." A more balanced position, says Olson, is seeing science as both product and producer of culture: "In many ways science has merely justified the successive substitutions of more modern myths for obsolete ones as the basis for our understanding of the world. Scientific theory itself arises only out of and under the influence of its social and intellectual milieu; that is, it is a product as well as a determinant of culture" (p. 3). Such a balance is required because strict internalism is impossible but if all knowledge is socially constructed and a product of culture, the externalist position is subject to itself and must then collapse. The belief that all knowledge is culturally determined and therefore lacks certainty is largely the product of an uncertain cultural milieu.

Extreme externalism (sometimes called *strong relativism*) cannot be right. Yet those of us trained by Olson's generation of historians (Olson was one of my thesis advisers) know all too well that social phenomena and cultural traditions *do* influence theories, which, in turn, determine how facts are interpreted; the facts then reinforce theories, and round and round we go until, for some reason, a paradigm shifts. Yet if culture *determines* science—if ghosts and the laws of nature exist nowhere but in people's minds—then is science no better than pseudoscience? Is there no difference between ghosts and the laws of science?

We can get out of this circle of questions by recognizing this about science: despite being influenced by culture, science can be considered cumulative and progressive when these terms are used in a precise and nonjudgmental way. Scientific progress is *the cumulative growth of a system of knowledge over time, in which useful features are retained and nonuseful features are abandoned, based on the rejection or confirmation of testable knowledge.* By this definition, science (and technology by extension) are the only cultural traditions that are progressive, not in any moralistic or hierarchical way but in an actual and definable manner. Whether it is deified or defied, science is progressive in this cumulative sense. This is what sets science apart from all other traditions, especially pseudoscience.

Resolution of the internalist-externalist problem—Pirsig's Paradox—follows from semantic precision and study of historical examples. One example will serve to illustrate the fascinating connections between science and politics. Most political theoreticians regard Thomas Hobbes' *Leviathan* (1651) as one of the most important political tracts of the modern age. Most do not realize, however, how much Hobbes' politics built upon the scientific ideas of his time. Hobbes, in fact, fancied himself as the Galileo Galilei and William Harvey of the science of society. The dedicatory letter to his *De Corpore Politico* (1644) has to be one of the most immodest statements in the history of science: "Galileus . . . was the first that opened to us the gate of natural philosophy universal, which is the knowledge of the nature of motion. . . . The science of man's body, the most profitable part of natural science, was first discovered with admirable sagacity by our countryman, Doctor Harvey. Natural philosophy is therefore but young; but civil philosophy is yet much younger, as being no older . . . than my own *de Cive*" (1839–1845, vol. 1, pp. vii–ix).

Hobbes' introduction to scientific thinking came at the age of forty, when he happened upon a copy of Euclid's *Elements* at a friend's home and turned to a theorem he could not understand until he examined the preceding definitions and postulates. In one of those flashes of insight so important in the annals of science, Hobbes began to apply geometrical logic to social theory. Just as Euclid built a science of geometry, Hobbes would build a science of society, beginning with the first principle that the universe is composed of material matter in motion. His second principle was that all life depends on "vital motion," just as, in Hobbes' words, "the motion of the blood, perpetually circulating (as hath been shown from many infallible signs and marks by Dr. Harvey, the first observer to it) in the veins and arteries" (1839–1845, vol. 4, p. 407). Through the senses, the brain detects the mechanical motion of objects in the environment. Since all simple ideas come from these basic sense movements, complex ideas must come from combinations of simple ideas. Thus, all thought is a type of motion in the brain called memories. As the motion fades, the memory fades.

Humans are also in motion, driven by passions—appetites (pleasure) and aversions (pain)—to maintain the vital motion of life itself. To gain pleasure and avoid pain, one needs power. In the state of nature everyone is free to exert power over others in order to gain greater pleasure. This Hobbes calls the *right of nature*. Unequal passions among individuals living in nature lead to a state of "war of all against all." In the most famous passage in political theory, Hobbes imagines life without government and the state: "In such condition there is no place for industry because the fruit thereof is uncertain . . . no account of time, no arts, no letters, no society, and which is worst of all, continual fear and danger of violent death and the life of man, solitary, poor, nasty, brutish, and short" ([1651] 1968, p. 76). Fortunately, Hobbes argues, humans have reason and can alter the right of nature in favor of the *law of nature*, out of which comes the social contract. The contract calls for individuals to surrender *all* rights (except self-defense) to the sovereign who, like the biblical Leviathan, is responsible only to God. Compared to a war of all against all, a sovereign presiding over the state is far superior and forms the basis for a rational society in which peace and prosperity are available on a mass scale.

I have oversimplified the steps in Hobbes' complex theory, but the point is that his reasoning was Euclidean and his system mechanical. He began with metaphysical first principles and ended with an entire social structure. Moreover, because many political theorists consider Hobbes the most influential thinker of the modern age, the connection Hobbes made between politics and science is not dead yet. Science and culture are interactive, not separate and independent, despite attempts by scientists to keep them separate. One of the founders of modern science, Isaac Newton, in the third edition (1726) of his great work, the *Principia*, claimed, "Hitherto I have not been able to discover the cause of properties of gravity from phenomena, and I feign no hypothesis; and hypotheses, whether metaphysical or physical, whether of occult qualities or mechanical, have no place in experimental philosophy" ([1729] 1962, vol. 2, p. 547). Yet Olson has demonstrated just how often Newton did feign hypotheses, "such as the conjecture that light is globular and resembles tennis balls, which is clearly presented in the first optics paper" (1991, p. 98). Moreover, says Olson, even with regard to the law of gravity—Newton's greatest achievement—he feigned hypotheses: "It is undeniable that he did speculate about the cause of gravity—not only privately, but also in print. It has even been argued very convincingly that, so far as the study of experimental natural philosophy in the eighteenth century is concerned, Newton's conjectures and hypotheses . . . were more important than the antihypothetical tradition of the *Principia*" (1991, p. 99). What could be more occult and metaphysical, in fact, than the "action at a distance" gravity produces. What is gravity? It is the tendency for objects to be attracted to one another. Why are objects attracted to one another? Because of gravity. In addition to being tautological, this explanation sounds rather ghostly, which brings us to the resolution of Pirsig's Paradox.

Do ghosts exist? Do scientific laws exist? Is there no difference between ghosts and scientific laws? Of course there is, and most scientists believe in scientific laws but not ghosts. Why? Because a scientific law is *a description of a regularly repeating action that is open to rejection or confirmation*. A scientific law describes some action in nature that can be tested. The description is in the mind. The repeating action is in nature. The test confirms or rejects it as a law. The law of gravity, for example, describes the repeating attraction between objects, and it has been tested over and over against external reality, and thus it has been confirmed. Ghosts have never been successfully tested against external reality (I do not count blurry photographs with smudges on them that can be explained and replicated by lens distortions or light aberrations). The law of gravity can be considered factual, meaning that it has been confirmed to such an extent that it would be reasonable to offer temporary agreement. Ghosts can be considered nonfactual because they have never been confirmed to *any* extent. Finally, although the law of gravity did not exist before Newton, gravity did. Ghosts never exist apart from their description by believers. The difference between ghosts and scientific laws is significant and real. Pirsig's Paradox is resolved: all description is in the mind, but scientific laws describe repeating natural phenomena while pseudoscientific claims are idiosyncratic.

Pseudoscience and Pseudohistory

Okay, so ghosts are bunk, along with most claims that fall under the heading of pseudoscience, by which I mean *claims presented so that they appear scientific even though they lack supporting evidence and plausibility*. The search for extraterrestrial life is not pseudoscience because it is plausible, even though the evidence for it thus far is nonexistent (the SETI—Search for

Extraterrestrial Intelligence—program looks for extraterrestrial radio signals). Alien abduction claims, however, are pseudoscience. Not only is physical evidence lacking but it is highly implausible that aliens are beaming thousands of people into spaceships hovering above the Earth without anyone detecting the spacecrafts or reporting the people missing.

But what about historical events? How do we know they happened since they do not repeat, either in nature or in the laboratory? As we shall see in chapters 13 and 14, there is a significant difference between history and pseudohistory. Most people would argue that history is not a science. Yet they would agree that Holocaust deniers and extreme Afrocentrists are doing something different from what historians are doing. What is that difference? In chapter 1, I emphasized that external validation through observation and testing is one of the key characteristics of science. We are told by believers in alien abductions that there is no way to test their claims because the experience was, in a way, a historical event, and we were not there to observe for ourselves. Further, the abduction experience itself is often a memory reconstructed through "regression hypnosis," which makes external validation even more difficult.

Yet historical events can be tested. External validation is possible. For example, classicist Mary Lefkowitz has written a thoughtful reply to Afrocentric claims that Western civilization, philosophy, science, art, literature, and so on came out of Africa, not Greece and Rome. Her book, *Not Out of Africa*, raised storms across America, and she was accused of being everything from racist to politically incorrect. Lefkowitz wrote her book after attending a lecture given in February 1993 at Wellesley College (where she teaches) by Dr. Yosef A. A. ben-Jochannan, a noted extreme Afrocentrist. Among the outrageous statements made in the lecture was the claim that Aristotle stole the ideas that became the foundation of Western philosophy from the library of Alexandria, where Black Africans had deposited their philosophical works. During the question-and-answer session, Lefkowitz asked ben-Jochannan how this could be since the library was built after Aristotle was dead. The response was enlightening:

> Dr. ben-Jochannan was unable to answer the question, and said that he resented the tone of the inquiry. Several students came up to me after the lecture and accused me of racism, suggesting that I had been brainwashed by white historians. . . .
> . . . As if that were not disturbing enough in itself, there was also the strange silence on the part of many of my faculty colleagues. Several of them were well aware that what Dr. ben-Jochannan was saying was factually wrong. One of them said later that she found the lecture so "hopeless" that she decided to say nothing. . . . When I went to the then dean of the college to explain that there was no factual evidence behind some Afrocentric claims about ancient

history, she replied that each of us had a different but equally valid view of
history. . . .
. . . When I stated at a faculty meeting that Aristotle could not have stolen his
philosophy from the library of Alexandria in Egypt, because that library had
not been built until after his death, another colleague responded, "I don't care
who stole what from whom." (1996, pp. 2, 3, 4)

Therein lies the problem. Each of us may have a different view of history,
but they are not all equally valid. Some are historical, and some are
pseudohistorical, namely, *without supporting evidence and plausibility and presented primarily for political or ideological purposes.*

A variety of sources independently attest to the life span of Aristotle
(384–322 B.C.E.) and to the earliest date for the library of Alexandria (after
323 B.C.E.). It is a fact that Aristotle died before the library of Alexandria
was built. One would have to posit a massive and widespread campaign of
denial and fabrication to change this fact, which is exactly what extreme
Afrocentrists do. True, humans are capable of almost anything and historical inferences have been wrong. Nonetheless, as Lefkowitz points out,
"There is no reason why claims of conspiracy should be credited, if no real
evidence can be produced to support it" (p. 8). Which brings us to another
important point: pseudohistorians and historians do not treat their audiences equally and they use data differently. If Dr. ben-Jochannan wanted to
argue that Aristotle was influenced by or acquainted with certain ideas circulating between Greece and Africa, he could examine the evidence for
and against such a theory. Indeed, Lefkowitz does just that. But Dr. ben-
Jochannan is not as interested in historical facts as he is in historical flavor-
ing, not as interested in teaching the nuances of historiography as he is in
instilling an Afrocentrist agenda. He takes a valid point about the influence
of ideology on knowledge, stirs in the ignorance or apathy of an audience
about historical events, adds a few historical facts and series of eccentric
inferences about the past, and makes pseudohistory.

The historical sciences are rooted in the rich array of data from the
past that, while nonreplicable, are nevertheless valid as sources of information for piecing together specific events and confirming general hypotheses. The inability to actually observe past events or set up controlled
experiments is no obstacle to a sound science of paleontology or geology,
so why should it be for a sound science of human history? The key is the
ability to test one's hypothesis. Based on data from the past the historian
tentatively constructs a hypothesis, then checks it against "new" data
uncovered from the historical source.

Here is an example of this. I once had the opportunity to dig up a
dinosaur with Jack Horner, curator of paleontology at the Museum of the

Rockies in Bozeman, Montana. In *Digging Dinosaurs*, Horner reflected on the historical process in describing the two phases of the famous dig in which he exposed the first dinosaur eggs found in North America. The initial stage was "getting the fossils out of the ground; the second was to look at the fossils, study them, make hypotheses based on what we saw and try to prove or disprove them" (Horner and Gorman 1988, p. 168). The first phase of unsheathing the bones from the surrounding stone is backbreaking work. As you move from jack hammers and pickaxes to dental tools and small brushes, however, the historical interpretation accelerates as a function of the rate of bone unearthed, as does one's enthusiasm to keep digging. "Paleontology is not an experimental science; it's an historical science," Horner explained. "This means that paleontologists are seldom able to test their hypotheses by laboratory experiments, but they can still test them" (p. 168). How?

In 1981 Horner discovered a site in Montana that contained approximately thirty million fossil fragments of *Maiasaur* bones, from which he concluded "at a conservative estimate, we had discovered the tomb of ten thousand dinosaurs" (p. 128). Horner and his team did not dig up thirty million fossil fragments. Rather, they extrapolated from selected exposed areas how many bones there were in the 1.25 by 0.25 mile bed. The hypothesizing began with a question: "What could such a deposit represent?" (p. 129). There was no evidence that predators had chewed the bones, yet many were broken in half, lengthwise. Further, the bones were all arranged from east to west—the long dimension of the bone deposit. Small bones had been separated from bigger bones, and there were no bones of baby *Maiasaurs*, just those of *Maiasaurs* between nine and twenty-three feet long. The find revealed more questions than answers. What would cause the bones to splinter lengthwise? Why would the small bones be separated from the big bones? Was this one giant herd, all killed at the same time, or was it a dying ground over many years?

An early hypothesis that a mudflow buried the herd alive was rejected as "it didn't make sense that even the most powerful flow of mud could break bones lengthwise . . . nor did it make sense that a herd of living animals buried in mud would end up with all their skeletons disarticulated." Applying the hypothetico-deductive method, Horner formulated a second hypothesis: "It seemed that there had to be a twofold event, the dinosaurs dying in one incident and the bones being swept away in another." Since there was a layer of volcanic ash a foot and a half above the bone bed, volcanic activity was implicated in the death of the herd. Deduction: because the fossil bones split only lengthwise, the damage to the bones came long after the event that caused death, which might have been a volcanic eruption, especially since volcanoes "were a dime a dozen in the Rockies back in the late Cretaceous." Conclusion: "A herd of *Maiasaura* were killed by the gases, smoke and ash of a volcanic eruption. And if a huge eruption killed them all at once, then it might have also killed everything else around," including scavengers or predators. Then perhaps there was a flood, maybe from a breached lake, that carried the rotting bodies downstream, separated the big bones from the small bones (which are lighter), and gave them a uniform orientation. "Finally the ash, being light, would have risen to the top in this slurry, as it settled, just as the bones sank to the bottom." What about the baby *Maiasaurs*? "Perhaps the babies of that year were still in the egg or in nests when the volcano erupted, or perhaps nesting had not even begun." But what about babies from the previous season who would now be juveniles? Horner admits "that nobody knows for sure that these dinosaurs would have produced young each year" (pp. 129–133).

Even in the first stage of a dig while fossils are being released from their rocky shroud, the hypothetico-deductive method is constantly applied. When I arrived at Horner's camp, I expected to find the busy director of a fully sponsored dig barking out orders to his staff. I was surprised to come upon a patient historical scientist sitting cross-legged before a cervical vertebra from a 140-million-year-old *Apatosaurus* and wondering just what to make of it. Soon a reporter from a local paper arrived (apparently a common occurrence as no one took notice) and inquired of Horner what this discovery meant for the history of dinosaurs. Did it change any of his theories? Where was the head? Was there more than one body at this site? And so on. Horner's answers were consistent with those of the cautious scientist: "I don't know yet." "Beats me." "We need more evidence." "We'll have to wait and see."

This was historical science at its best. For example, after two long days of exposing nothing but solid rock and my own ineptness at seeing bone within stone, one of the preparators pointed out that the rock I was about to toss was a piece of bone that appeared to be part of a rib. *If* it was a rib, *then* the bone should retain its rib-like shape as more of the overburden was chipped away. This it did for about a foot, until it suddenly flared to the right. Was it a rib or something else? Jack moved in to check. "It could be part of the pelvis," he suggested. *If* it was part of the pelvis, *then* it should also flare out to the left when more was uncovered. Sure enough, Jack's prediction was verified by further empirical evidence. And so it went day after day. The whole dig depends on such hypothetico-deductive reasoning. In a sense, historical science becomes experimental when predictions based on initial evidence are verified or rejected by later evidence. The digging up of history, whether bones or letters, is the experimental procedure of the historical scientist interested in putting a hypothesis to the test.

I should note that there are differences between paleontological evidence and human historical evidence. The former is mostly first-order evidence—strictly physical, natural, and interpreted by extrapolating how natural laws apply now and in the past. The latter typically is second-order evidence—documents written by highly selective humans who add, delete, and alter the evidence. Historians have learned to treat historical evidence differently from archeological or paleontological evidence, to acknowledge that the gaps in historical evidence often have something to do with the fact that humans write about what interests them and what they think is important at the time. Nature does not delete the record of the socially marginalized. Still, as historian of science Frank Sulloway has shown in his controversial 1996 book, *Born to Rebel*, historical hypotheses can be tested (see chapter 16 for discussion of Sulloway's model). For the past hundred years, for example, historians have hypothesized that social class and social class conflict have been the driving forces behind revolutions, both political and scientific. Sulloway has tested this Marxian hypothesis by coding thousands of individuals in dozens of revolutions for their social class and then doing statistical analyses to see whether there really are significant differences in social class on opposing sides in revolutions. It turns out there is not. Marx was wrong, but it took a historian trained in the sciences to discover this fact by running a simple historical experiment.

How Science Changes

Science is different from pseudoscience, and history is different from pseudohistory, not only in evidence and plausibility but in how they change. Science and history are cumulative and progressive in that they continue to improve and refine knowledge of our world and our past based on new observations and interpretations. Pseudohistory and pseudoscience, if they change at all, change primarily for personal, political, or ideological reasons. But *how* do science and history change?

One of the most useful theories of how science changes is Thomas Kuhn's (1962) concept of "paradigm shift." The paradigm defines the "normal science" of an age—as accepted by the majority of the practicing scientists in a field—and a shift (or revolution) may occur when enough renegade and heretical scientists gain enough evidence and enough power to overthrow the existing paradigm. "Power" is made visible in the social and political aspects of science: research and professorial positions at major universities, influence within funding agencies, control of journals and conferences, prestigious books, and so forth. I define a paradigm as *a model shared by most but not all members of a scientific community, designed to describe and interpret observed or inferred phenomena, past or present, and aimed at building a testable body of knowledge open to rejection or confirmation.* In other words, a paradigm captures the scientific thinking of the majority but most of the time it coexists with competing paradigms—as is necessary if new paradigms are to displace old paradigms.

Philosopher of science Michael Ruse, in *The Darwinian Paradigm* (1989), identified at least four usages of the word.

1. *Sociological*, focusing on "a group of people who come together, feeling themselves as having a shared outlook (whether they do really, or not), and to an extent separating themselves off from other scientists" (pp. 124–125). Freudian psychoanalysts within psychology are a good example of science guided by a sociological paradigm.

2. *Psychological*, where individuals within the paradigm literally see the world differently from those outside the paradigm. We have all seen the reversible figures in perceptual experiments, such as the old woman/young woman shifting figure where the perception of one precludes the perception of the other. In this particular perceptual experiment, presenting subjects with a strong "young woman" image followed by the ambiguous figure always produces the perception of the young woman, while presenting a strong "old woman" image followed by the ambiguous figure produces the perception of the old woman 95 percent of the time (Leeper 1935).

   Similarly, some researchers view aggression in humans primarily as biologically innate and essential, while others view it primarily as culturally induced and dispensable. Those who focus their research on proving one or the other of these views would be doing science guided by a psychological paradigm: both views have support, but the choice of which to believe more is influenced by psychological factors.

3. *Epistemological*, where "one's ways of doing science are bound up with the paradigm" because the research techniques, problems, and solutions are determined by the hypotheses, theories, and

models. A theory of phrenology that leads to the development of phrenological equipment for measuring bumps on the skull would be an example of science guided by an epistemological paradigm.

4. *Ontological*, where in the deepest sense "what there is depends crucially on what paradigm you hold. For Priestley, there literally was no such thing as oxygen. . . . In the case of Lavoisier, he not only believed in oxygen: oxygen existed" (pp. 125–126). Similarly, for Georges Buffon and Charles Lyell, varieties in a population were merely degenerates from the originally created kind; nature eliminated them to preserve the essence of the species. For Charles Darwin and Alfred Russel Wallace, varieties were the key to evolutionary change. Each view depends on a different ontological paradigm: Buffon and Lyell could not see varieties as evolutionary engines because evolution did not exist for them; Darwin and Wallace did not view varieties as degenerative because degeneration is irrelevant to evolution.

My definition of a paradigm holds for the *sociological, psychological,* and *epistemological* uses. To make it wholly *ontological*, however, would mean that any paradigm is as good as any other paradigm because there is no outside source for corroboration. Tealeaf reading and economic forecasting, sheep's livers and meteorological maps, astrology and astronomy, all equally determine reality under an ontological paradigm. This is not even wrong. It is ridiculous. As difficult as it is for economists and meteorologists to predict the future, they are still better at it than tealeaf readers and sheep's liver diviners. Astrologers cannot explain the interior workings of a star, predict the outcome of colliding galaxies, or chart the course of a spacecraft to Jupiter. Astronomers can, for the simple reason that they operate within a scientific paradigm that is constantly refined against the harsh realities of nature itself.

Science is progressive because its paradigms depend upon the cumulative knowledge gained through experimentation, corroboration, and falsification. Pseudoscience, nonscience, superstition, myth, religion, and art are not progressive because they do not have goals or mechanisms that allow the accumulation of knowledge that builds on the past. Their paradigms either do not shift or coexist with other paradigms. Progress, in the cumulative sense, is not their purpose. This is not a criticism, just an observation. Artists do not improve upon the styles of their predecessors; they invent new styles. Priests, rabbis, and ministers do not attempt to improve upon the sayings of their masters; they repeat, interpret, and teach them. Pseudoscientists do not correct the errors of their predecessors; they perpetuate them.

By cumulative change I mean, then, that when a paradigm shifts, scientists do not abandon the entire science. Rather, what remains useful in the paradigm is retained as new features are added and new interpretations given. Albert Einstein emphasized this point in reflecting upon his own contributions to physics and cosmology: "Creating a new theory is not like destroying an old barn and erecting a skyscraper in its place. It is rather like climbing a mountain, gaining new and wider views, discovering unexpected connections between our starting point and its rich environment. But the point from which we started out still exists and can be seen, although it appears smaller and forms a tiny part of our broad view gained by the mastery of the obstacles on our adventurous way up" (in Weaver 1987, p. 133). Even though Darwin replaced the theory of special creation with that of evolution by natural selection, much of what came before was retained in the new theory—Linnean classification, descriptive geology, comparative anatomy, and so forth. What changed was how these various fields were linked to one another through history—the theory of evolution. There was cumulative growth *and* paradigmatic change. This is scientific progress, defined as *the cumulative growth of a system of knowledge over time, in which useful features are retained and nonuseful features are abandoned, based on the rejection or confirmation of testable knowledge.*

The Triumph of Science

Though I have defined science as progressive, I admit it is not possible to know whether the knowledge uncovered by the scientific method is absolutely certain because we have no place outside—no Archimedean point—from which to view Reality. There is no question but that science is heavily influenced by the culture in which it is embedded, and that scientists may all share a common bias that leads them to think a certain way about nature. But this does not take anything away from the progressive nature of science, in the cumulative sense.

In this regard, philosopher Sydney Hook makes an interesting comparison between the arts and sciences: "Raphael's Sistine Madonna without Raphael, Beethoven's sonatas and symphonies without Beethoven, are inconceivable. In science, on the other hand, it is quite probable that most of the achievements of any given scientist would have been attained by other individuals working in the field" (1943, p. 35). The reason for this is that science, with progress as one of its primary goals, seeks understanding through objective methods (even though it rarely attains it). The arts seek provocation of emotion and reflection through subjective means. The more subjective the endeavor, the more individual it becomes, and therefore difficult if not impossible for someone else to produce. The more objective the pursuit, the more likely it is that someone else will duplicate the achievement. Science actually depends upon duplication for verification. Darwin's theory of natural selection would have occurred to another scientist—and, in fact, did occur to Alfred Russel Wallace simultaneously—because the scientific process is empirically verifiable.

In the Industrial West, the emphasis on scientific and technological progress has affected Western cultures deeply—so much so that we now define a culture as progressive if it encourages the development of science and technology. In science, useful features are retained and nonuseful features are abandoned through the confirmation or rejection of testable knowledge by the community of scientists. The scientific method, in this way, is constructed to be progressive. In technology, useful features are retained and nonuseful features are abandoned based on the rejection or acceptance of the technologies by the consuming public. Technologies, then, are also constructed to be progressive. Cultural traditions (art, myth, religion) may exhibit some of the features found in science and technology, such as being accepted or rejected within their own community or by the public, but none have had as their primary goal cumulative growth through an indebtedness to the past. But in the Industrial West, culture has taken on a new guise: *it has as a primary goal the accumulation of cultural traditions and artifacts, and it uses, ignores, and returns to cultural traditions and artifacts as needed to aid the progress of science and technology.* We cannot, in any absolute sense, equate happiness with progress, or progress with happiness, but an individual who finds happiness in a variety of knowledge and artifacts, cherishes novelty and change, and esteems the living standards set by the Industrial West will view a culture driven by scientific and technological progress as progressive.

Lately the word *progress* has taken on a pejorative meaning, implying superiority over those who "have not progressed as far," namely, they have not adopted the values or the standard of living defined by the Industrial West, because they are either not able or not willing to encourage the development of science and technology. I do not mean *progress* to have this pejorative sense. Whether or not a culture pursues science and technology does not make one culture better than another or one way of life more moral than another or one people happier than another. Science and technology have plenty of limitations, and they are double-edge swords. Science has made the modern world, but it may also unmake it. Our advances in the physical sciences have given us plastics and plastic explosives, cars and tanks, supersonic transports and B-1 bombers; they have also put men on the moon and missiles in silos. We travel faster and further, but so do our destructive agents. Medical advances allow us to live twice as long as our ancestors did a mere 150 years ago, and now we have a potentially devastating overpopulation problem without a corresponding overproduction solution. Discoveries in anthropology and cosmology have given us insight into the origins of species and the workings of the universe. But for many people, these insights and their corresponding ideologies are an insult to personal and religious beliefs and a provocative threat to the comfortable status quo. Our scientific and technological progress has, for the first time in history, given us many ways of causing the extinction of our own species. This is neither good nor bad. It is simply the outcome of a cumulative system of knowledge. But flawed as it may be, science is at present the best method we have for doing what we want it to do. As Einstein observed, "One thing I have learned in a long life: that all our science, measured against reality, is primitive and childlike—and yet it is the most precious thing we have."

3

How Thinking Goes Wrong

Twenty-five Fallacies That Lead Us to Believe Weird Things

I n 1994 NBC began airing a New Age program called *The Other Side* that explored claims of the paranormal, various mysteries and miracles, and assorted "weird" things. I appeared numerous times as the token skeptic—the "other side" of *The Other Side*, if you will. On most talk shows, a "balanced" program is a half-dozen to a dozen believers and one lone skeptic as the voice of reason or opposition. *The Other Side* was no different, even though the executive producer, many of the program producers, and even the host were skeptical of most of the beliefs they were covering. I did one program on werewolves for which they flew in a fellow from England. He actually looked a little like what you see in werewolf movies—big bushy sideburns and rather pointy ears—but when I talked to him, I found that he did not actually remember becoming a werewolf. He recalled the experience under hypnosis. In my opinion, his was a case of false memory, either planted by the hypnotist or fantasized by the man.

Another program was on astrology. The producers brought in a serious, professional astrologer from India who explained how it worked using charts and maps with all the jargon. But, because he was so serious, they ended up featuring a Hollywood astrologer who made all sorts of predictions about the lives of movie starts. He also did some readings for members of the audience. One young lady was told that she was having problems staying in long-term relationships with men. During the break, she told me that she was fourteen years old and was there with her high-school class to see how television programs were produced.

In my opinion, most believers in miracles, monsters, and mysteries are not hoaxers, flimflam artists, or lunatics. Most are normal people whose normal thinking has gone wrong in some way. In chapters 4, 5, and 6, I will discuss in detail psychic power, altered states of consciousness, and alien abductions, but I would like to round out part 1 of this book by looking at twenty-five fallacies of thinking that can lead anyone to believe weird things. I have grouped them in four categories, listing specific fallacies and problems in each. But as an affirmation that thinking can go right, I begin with what I call Hume's Maxim and close with what I call Spinoza's Dictum.

Hume's Maxim

Skeptics owe a lot to the Scottish philosopher David Hume (1711–1776), whose *An Enquiry Concerning Human Understanding* is a classic in skeptical analysis. The work was first published anonymously in London in 1739 as *A Treatise of Human Nature*. In Hume's words, it "fell dead-born from the press, without reaching such distinction as even to excite a murmur among the zealots." Hume blamed his own writing style and reworked the manuscript into *An Abstract of a Treatise of Human Nature*, published in 1740, and then into *Philosophical Essays Concerning the Human Understanding*, published in 1748. The work still garnered no recognition, so in 1758 he brought out the final version, under the title *An Enquiry Concerning Human Understanding*, which today we regard as his greatest philosophical work.

Hume distinguished between "antecedent skepticism," such as René Descartes' method of doubting everything that has no "antecedent" infallible criterion for belief; and "consequent skepticism," the method Hume employed, which recognizes the "consequences" of our fallible senses but corrects them through reason: "A wise man proportions his belief to the evidence." Better words could not be found for a skeptical motto.

Even more important is Hume's foolproof, when-all-else-fails analysis of miraculous claims. For when one is confronted by a true believer whose apparently supernatural or paranormal claim has no immediately apparent natural explanation, Hume provides an argument that he thought so important that he placed his own words in quotes and called them a maxim:

> The plain consequence is (and it is a general maxim worthy of our attention), "That no testimony is sufficient to establish a miracle, unless the testimony be of such a kind, that its falsehood would be more miraculous than the fact which it endeavors to establish."

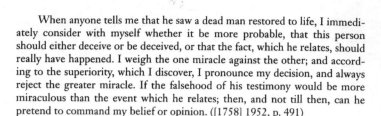

> When anyone tells me that he saw a dead man restored to life, I immedi-
> ately consider with myself whether it be more probable, that this person
> should either deceive or be deceived, or that the fact, which he relates, should
> really have happened. I weigh the one miracle against the other; and accord-
> ing to the superiority, which I discover, I pronounce my decision, and always
> reject the greater miracle. If the falsehood of his testimony would be more
> miraculous than the event which he relates; then, and not till then, can he
> pretend to command my belief or opinion. ([1758] 1952, p. 491)

Problems in Scientific Thinking

1. Theory Influences Observations

About the human quest to understand the physical world, physicist and
Nobel laureate Werner Heisenberg concluded, "What we observe is not
nature itself but nature exposed to our method of questioning." In quantum mechanics, this notion has been formalized as the "Copenhagen
interpretation" of quantum action: "a probability function does not pre-
scribe a certain event but describes a continuum of possible events until
a measurement interferes with the isolation of the system and a single
event is actualized" (in Weaver 1987, p. 412). The Copenhagen interpretation eliminates the one-to-one correlation between theory and reality.
The theory in part *constructs* the reality. Reality exists independent of
the observer, of course, but our perceptions of reality are influenced by the
theories framing our examination of it. Thus, philosophers call science
theory laden.

That theory shapes perceptions of reality is true not only for quantum physics but also for all observations of the world. When Columbus
arrived in the New World, he had a theory that he was in Asia and proceeded to perceive the New World as such. Cinnamon was a valuable
Asian spice, and the first New World shrub that smelled like cinnamon
was declared to *be* it. When he encountered the aromatic gumbo-limbo
tree of the West Indies, Columbus concluded it was an Asian species similar to the mastic tree of the Mediterranean. A New World nut was
matched with Marco Polo's description of a coconut. Columbus's surgeon even declared, based on some Caribbean roots his men uncovered,
that he had found Chinese rhubarb. A theory of Asia produced observations of Asia, even though Columbus was half a world away. Such is the
power of theory.

2. The Observer Changes the Observed

Physicist John Archibald Wheeler noted, "Even to observe so minuscule an object as an electron, [a physicist] must shatter the glass. He must reach in. He must install his chosen measuring equipment. . . . Moreover, the measurement changes the state of the electron. The universe will never afterward be the same" (in Weaver 1987, p. 427). In other words, the act of studying an event can change it. Social scientists often encounter this phenomenon. Anthropologists know that when they study a tribe, the behavior of the members may be altered by the fact they are being observed by an outsider. Subjects in a psychology experiment may alter their behavior if they know what experimental hypotheses are being tested. This is why psychologists use blind and double-blind controls. Lack of such controls is often found in tests of paranormal powers and is one of the classic ways that thinking goes wrong in the pseudosciences. Science tries to minimize and acknowledge the effects of the observation on the behavior of the observed; pseudoscience does not.

3. Equipment Constructs Results

The equipment used in an experiment often determines the results. The size of our telescopes, for example, has shaped and reshaped our theories about the size of the universe. In the twentieth century, Edwin Hubble's 60- and 100-inch telescopes on Mt. Wilson in southern California for the first time provided enough seeing power for astronomers to distinguish individual stars in other galaxies, thus proving that those fuzzy objects called nebulas that we thought were in our own galaxy were actually separate galaxies. In the nineteenth century, craniometry defined intelligence as brain size and instruments were designed that measured it as such; today intelligence is defined by facility with certain developmental tasks and is measured by another instrument, the IQ test. Sir Arthur Stanley Eddington illustrated the problem with this clever analogy:

> Let us suppose that an ichthyologist is exploring the life of the ocean. He casts a net into the water and brings up a fishy assortment. Surveying his catch, he proceeds in the usual manner of a scientist to systematize what it reveals. He arrives at two generalizations:
>
> (1) No sea-creature is less than two inches long.
>
> (2) All sea-creatures have gills.
>
> In applying this analogy, the catch stands for the body of knowledge which constitutes physical science, and the net for the sensory and intellectual

equipment which we use in obtaining it. The casting of the net corresponds to observations.

An onlooker may object that the first generalization is wrong. "There are plenty of sea-creatures under two inches long, only your net is not adapted to catch them." The ichthyologist dismisses this objection contemptuously. "Anything uncatchable by my net is *ipso facto* outside the scope of ichthyological knowledge, and is not part of the kingdom of fishes which has been defined as the theme of ichthyological knowledge. In short, what my net can't catch isn't fish." (1958, p. 16)

Likewise, what my telescope can't see isn't there, and what my test can't measure isn't intelligence. Obviously, galaxies and intelligence exist, but how we measure and understand them is highly influenced by our equipment.

Problems in Pseudoscientific Thinking

4. Anecdotes Do Not Make a Science

Anecdotes—stories recounted in support of a claim—do not make a science. Without corroborative evidence from other sources, or physical proof of some sort, ten anecdotes are no better than one, and a hundred anecdotes are no better than ten. Anecdotes are told by fallible human storytellers. Farmer Bob in Puckerbrush, Kansas, may be an honest, church-going, family man not obviously subject to delusions, but we need physical evidence of an alien spacecraft or alien bodies, not just a story about landings and abductions at 3:00 A.M. on a deserted country road. Likewise with many medical claims. Stories about how your Aunt Mary's cancer was cured by watching Marx brothers movies or taking a liver extract from castrated chickens are meaningless. The cancer might have gone into remission on its own, which some cancers do; or it might have been misdiagnosed; or, or, or. . . . What we need are controlled experiments, not anecdotes. We need 100 subjects with cancer, all properly diagnosed and matched. Then we need 25 of the subjects to watch Marx brothers movies, 25 to watch Alfred Hitchcock movies, 25 to watch the news, and 25 to watch nothing. Then we need to deduct the average rate of remission for this type of cancer and then analyze the data for statistically significant differences between the groups. If there are statistically significant differences, we better get confirmation from other scientists who have conducted their own experiments separate from ours before we hold a press conference to announce the cure for cancer.

5. Scientific Language Does Not Make a Science

Dressing up a belief system in the trappings of science by using scientistic language and jargon, as in "creation-science," means nothing without evidence, experimental testing, and corroboration. Because science has such a powerful mystique in our society, those who wish to gain respectability but do not have evidence try to do an end run around the missing evidence by looking and sounding "scientific." Here is a classic example from a New Age column in the *Santa Monica News*: "This planet has been slumbering for eons and with the inception of higher energy frequencies is about to awaken in terms of consciousness and spirituality. Masters of limitation and masters of divination use the same creative force to manifest their realities, however, one moves in a downward spiral and the latter moves in an upward spiral, each increasing the resonant vibration inherent in them." How's that again? I have no idea what this means, but it has the language components of a physics experiment: "higher energy frequencies," "downward and upward spirals," and "resonant vibration." Yet these phrases mean nothing because they have no precise and operational definitions. How do you measure a planet's higher energy frequencies or the resonant vibration of masters of divination? For that matter, what *is* a master of divination?

6. Bold Statements Do Not Make Claims True

Something is probably pseudoscientific if enormous claims are made for its power and veracity but supportive evidence is scarce as hen's teeth. L. Ron Hubbard, for example, opens his *Dianetics: The Modern Science of Mental Health*, with this statement: "The creation of Dianetics is a milestone for man comparable to his discovery of fire and superior to his invention of the wheel and arch" (in Gardner 1952, p. 263). Sexual energy guru Wilhelm Reich called his theory of Orgonomy "a revolution in biology and psychology comparable to the Copernican Revolution" (in Gardner 1952, p. 259). I have a thick file of papers and letters from obscure authors filled with such outlandish claims (I call it the "Theories of Everything" file). Scientists sometimes make this mistake, too, as we saw at 1:00 P.M., on March 23, 1989, when Stanley Pons and Martin Fleischmann held a press conference to announce to the world that they had made cold nuclear fusion work. Gary Taubes's excellent book about the cold fusion debacle, appropriately named *Bad Science* (1993), thoroughly examines the implications of this incident. Maybe fifty years of physics will be proved wrong by one experiment, but don't throw out your furnace until that experiment has been replicated. The moral is that the more extraordinary the claim, the more extraordinarily well-tested the evidence must be.

7. Heresy Does Not Equal Correctness

They laughed at Copernicus. They laughed at the Wright brothers. Yes, well, they laughed at the Marx brothers. Being laughed at does not mean you are right. Wilhelm Reich compared himself to Peer Gynt, the unconventional genius out of step with society, and misunderstood and ridiculed as a heretic until proven right: "Whatever you have done to me or will do to me in the future, whether you glorify me as a genius or put me in a mental institution, whether you adore me as your savior or hang me as a spy, sooner or later necessity will force you to comprehend that I have discovered the laws of the living" (in Gardner 1952, p. 259). Reprinted in the January/February 1996 issue of the *Journal of Historical Review*, the organ of Holocaust denial, is a famous quote from the nineteenth-century German philosopher Arthur Schopenhauer, which is quoted often by those on the margins: "All truth passes through three stages. First, it is ridiculed. Second, it is violently opposed. Third, it is accepted as self-evident." But "all truth" does not pass through these stages. Lots of true ideas are accepted without ridicule or opposition, violent or otherwise. Einstein's theory of relativity was largely ignored until 1919, when experimental evidence proved him right. He was not ridiculed, and no one violently opposed his ideas. The Schopenhauer quote is just a rationalization, a fancy way for those who are ridiculed or violently opposed to say, "See, I must be right." Not so.

History is replete with tales of the lone scientist working in spite of his peers and flying in the face of the doctrines of his or her own field of study. Most of them turned out to be wrong and we do not remember their names. For every Galileo shown the instruments of torture for advocating a scientific truth, there are a thousand (or ten thousand) unknowns whose "truths" never pass muster with other scientists. The scientific community cannot be expected to test every fantastic claim that comes along, especially when so many are logically inconsistent. If you want to do science, you have to learn to play the game of science. This involves getting to know the scientists in your field, exchanging data and ideas with colleagues informally, and formally presenting results in conference papers, peer-reviewed journals, books, and the like.

8. Burden of Proof

Who has to prove what to whom? The person making the extraordinary claim has the burden of proving to the experts and to the community at large that his or her belief has more validity than the one almost everyone else accepts. You have to lobby for your opinion to be heard. Then you have to marshal experts on your side so you can convince the majority to support your claim over the one that they have always supported. Finally, when you are in the majority, the burden of proof switches to the outsider who wants to challenge you with his or her unusual claim. Evolutionists had the burden of proof for half a century after Darwin, but now the burden of proof is on creationists. It is up to creationists to show why the theory of evolution is wrong and why creationism is right, and it is not up to evolutionists to defend evolution. The burden of proof is on the Holocaust deniers to prove the Holocaust did not happen, not on Holocaust historians to prove that it did. The rationale for this is that mountains of evidence prove that both evolution and the Holocaust are facts. In other words, it is not enough to have evidence. You must convince others of the validity of your evidence. And when you are an outsider this is the price you pay, regardless of whether you are right or wrong.

9. Rumors Do Not Equal Reality

Rumors begin with "I read somewhere that . . . " or "I heard from someone that. . . . " Before long the rumor becomes reality, as "I know that . . ." passes from person to person. Rumors *may* be true, of course, but usually they are not. They do make for great tales, however. There is the "true story" of the escaped maniac with a prosthetic hook who haunts the lover's lanes of America. There is the legend of "The Vanishing Hitchhiker," in which a driver picks up a hitchhiker who vanishes from his car along with his jacket; locals then tell the driver that his hitchhiking woman had died that same day the year before, and eventually he discovers his jacket on her grave. Such stories spread fast and never die.

Caltech historian of science Dan Kevles once told a story he suspected was apocryphal at a dinner party. Two students did not get back from a ski trip in time to take their final exam because the activities of the previous day had extended well into the night. They told their professor that they had gotten a flat tire, so he gave them a makeup final the next day. Placing the students in separate rooms, he asked them just two questions: (1) "For 5 points, what is the chemical formula for water?" (2) "For 95 points, which tire?" Two of the dinner guests had heard a vaguely similar story. The next day I repeated the story to my students and before I got to the punch line, three of them simultaneously blurted out, "Which tire?" Urban legends and persistent rumors are ubiquitous. Here are a few:

- The secret ingredient in Dr. Pepper is prune juice.
- A woman accidentally killed her poodle by drying it in a microwave oven.
- Paul McCartney died and was replaced by a look-alike.
- Giant alligators live in the sewers of New York City.
- The moon landing was faked and filmed in a Hollywood studio.
- George Washington had wooden teeth.
- The number of stars inside the "P" on *Playboy* magazine's cover indicates how many times publisher Hugh Hefner had sex with the centerfold.
- A flying saucer crashed in New Mexico and the bodies of the extra-terrestrials are being kept by the Air Force in a secret warehouse.

How many have you heard . . . and believed? None have ever been confirmed.

10. Unexplained Is Not Inexplicable

Many people are overconfident enough to think that if *they* cannot explain something, it must be inexplicable and therefore a true mystery of the paranormal. An amateur archeologist declares that because he cannot figure out how the pyramids were built, they must have been constructed by space aliens. Even those who are more reasonable at least think that if the *experts* cannot explain something, it must be inexplicable. Feats such as the bending of spoons, firewalking, or mental telepathy are often thought to be of a paranormal or mystical nature because most people cannot explain them. When they are explained, most people respond, "Yes, of course" or "That's obvious once you see it." Firewalking is a case in point. People speculate endlessly about supernatural powers over pain and heat, or mysterious brain chemicals that block the pain and prevent burning. The simple explanation is that the capacity of light and fluffy coals to contain heat is very low, and the conductivity of heat from the light and fluffy coals to your feet is very poor. As long as you don't stand around on the coals, you will not get burned. (Think of a cake in a 450°F oven. The air, the cake, and the pan are all at 450°F, but only the metal pan will burn your hand. Air has very low heat capacity and also low conductivity, so you can put your hand in the oven long enough to touch the cake and pan. The heat capacity of the cake is a lot higher than air, but since it has low conductivity you can briefly touch it without getting burned. The metal pan has a heat capacity similar to the cake, but high conductivity too. If you touch it, you will get burned.) This is why magicians do not tell their secrets. Most of their tricks are, in principle, relatively simple (although many are extremely difficult to execute) and knowing the secret takes the magic out of the trick.

There are many genuine unsolved mysteries in the universe and it is okay to say, "We do not yet know but someday perhaps we will." The problem is that most of us find it more comforting to have certainty, even if it is premature, than to live with unsolved or unexplained mysteries.

11. Failures Are Rationalized

In science, the value of negative findings—failures—cannot be overemphasized. Usually they are not wanted, and often they are not published. But most of the time failures are how we get closer to truth. Honest scientists will readily admit their errors, but all scientists are kept in line by the fact that their fellow scientists will publicize any attempt to fudge. Not pseudo-scientists. They ignore or rationalize failures, especially when exposed. If they are actually caught cheating—not a frequent occurrence—they claim that their powers usually work but not always, so when pressured to perform on television or in a laboratory, they sometimes resort to cheating. If they simply fail to perform, they have ready any number of creative explanations: too many controls in an experiment cause negative results; the powers do not work in the presence of skeptics; the powers do not work in the presence of electrical equipment; the powers come and go, and this is one of those times they went. Finally, they claim that if skeptics cannot explain everything, then there must be something paranormal; they fall back on the *unexplained is not inexplicable* fallacy.

12. After-the-Fact Reasoning

Also known as *"post hoc, ergo propter hoc,"* literally "after this, therefore because of this." At its basest level, it is a form of superstition. The baseball player does not shave and hits two home runs. The gambler wears his lucky shoes because he has won wearing them in the past. More subtly, scientific studies can fall prey to this fallacy. In 1993 a study found that breast-fed children have higher IQ scores. There was much clamor over what ingredient in mother's milk increased intelligence. Mothers who bottle-fed their babies were made to feel guilty. But soon researchers began to wonder whether breast-fed babies are attended to differently. Maybe nursing mothers spend more time with their babies and motherly vigilance was the cause behind the differences in IQ. As Hume taught us, the fact that two events follow each other in sequence does not mean they are connected causally. Correlation does not mean causation.

13. Coincidence

In the paranormal world, coincidences are often seen as deeply significant. "Synchronicity" is invoked, as if some mysterious force were at work behind the scenes. But I see synchronicity as nothing more than a type of contingency—a conjuncture of two or more events without apparent design. When the connection is made in a manner that seems impossible according to our intuition of the laws of probability, we have a tendency to think something mysterious is at work.

But most people have a very poor understanding of the laws of probability. A gambler will win six in a row and then think he is either "on a hot streak" or "due to lose." Two people in a room of thirty people discover that they have the same birthday and conclude that something mysterious is at work. You go to the phone to call your friend Bob. The phone rings and it is Bob. You think, "Wow, what are the chances? This could not have been a mere coincidence. Maybe Bob and I are communicating telepathically." In fact, such coincidences are not coincidences under the rules of probability. The gambler has predicted both possible outcomes, a fairly safe bet! The probability that two people in a room of thirty people will have the same birthday is .71. And you have forgotten how many times Bob did not call under such circumstances, or someone else called, or Bob called but you were not thinking of him, and so on. As the behavioral psychologist B. F. Skinner proved in the laboratory, the human mind seeks relationships between events and often finds them even when they are not present. Slot-machines are based on Skinnerian principles of intermittent reinforcement. The dumb human, like the dumb rat, only needs an occasional payoff to keep pulling the handle. The mind will do the rest.

14. Representativeness

As Aristotle said, "The sum of the coincidences equals certainty." We forget most of the insignificant coincidences and remember the meaningful ones. Our tendency to remember hits and ignore misses is the bread and butter of the psychics, prophets, and soothsayers who make hundreds of predictions each January 1. First they increase the probability of a hit by predicting mostly generalized sure bets like "There will be a major earthquake in southern California" or "I see trouble for the Royal Family." Then, next January, they publish their hits and ignore the misses, and hope no one bothers to keep track.

We must always remember the larger context in which a seemingly unusual event occurs, and we must always analyze unusual events for their representativeness of their class of phenomena. In the case of the "Bermuda Triangle," an area of the Atlantic Ocean where ships and planes "mysteriously" disappear, there is the assumption that something strange or alien is at work. But we must consider how representative such events are in that area. Far more shipping lanes run through the Bermuda Triangle than its surrounding areas, so accidents and mishaps and disappearances are more likely to happen in the area. As it turns out, the accident rate is actually *lower* in the Bermuda Triangle than in surrounding areas. Perhaps this area should be called the "Non-Bermuda Triangle." (See Kusche 1975 for a full explanation of this solved mystery.) Similarly, in investigating haunted houses, we must have a baseline measurement of noises, creaks, and other events before we can say that an occurrence is unusual (and therefore mysterious). I used to hear rapping sounds in the walls of my house. Ghosts? Nope. Bad plumbing. I occasionally hear scratching sounds in my basement. Poltergeists? Nope. Rats. One would be well-advised to first thoroughly understand the probable worldly explanation before turning to other-worldly ones.

Logical Problems in Thinking

15. Emotive Words and False Analogies

Emotive words are used to provoke emotion and sometimes to obscure rationality. They can be positive emotive words—*motherhood, America, integrity, honesty.* Or they can be negative—*rape, cancer, evil, communist.* Likewise, metaphors and analogies can cloud thinking with emotion or steer us onto a side path. A pundit talks about inflation as "the cancer of society" or industry "raping the environment." In his 1992 Democratic nomination speech, Al Gore constructed an elaborate analogy between the story of his sick son and America as a sick country. Just as his son, hovering on the brink of death, was nursed back to health by his father and family, America, hovering on the brink of death after twelve years of Reagan and Bush, was to be nurtured back to health under the new administration. Like anecdotes, analogies and metaphors do not constitute proof. They are merely tools of rhetoric.

16. *Ad Ignorantiam*

This is an appeal to ignorance or lack of knowledge and is related to the *burden of proof* and *unexplained is not inexplicable* fallacies, where someone argues that if you cannot disprove a claim it must be true. For example, if you cannot prove that there isn't any psychic power, then there must be. The absurdity of this argument comes into focus if one argues that if you cannot prove that Santa Claus does not exist, then he must exist. You can argue the opposite in a similar manner. If you cannot prove Santa Claus exists, then he must not exist. In science, belief should come from positive evidence in support of a claim, not lack of evidence for or against a claim.

17. Ad Hominem and Tu Quoque

Literally "to the man" and "you also," these fallacies redirect the focus from thinking about the idea to thinking about the person holding the idea. The goal of an *ad hominem* attack is to discredit the claimant in hopes that it will discredit the claim. Calling someone an atheist, a communist, a child abuser, or a neo-Nazi does not in any way disprove that person's statement. It might be helpful to know whether someone is of a particular religion or holds a particular ideology, in case this has in some way biased the research, but refuting claims must be done directly, not indirectly. If Holocaust deniers, for example, are neo-Nazis or anti-Semites, this would certainly guide their choice of which historical events to emphasize or ignore. But if they are making the claim, for example, that Hitler did not have a master plan for the extermination of European Jewry, the response "Oh, he is saying that because he is a neo-Nazi" does not refute the argument. Whether Hitler had a master plan or not is a question that can be settled historically. Similarly for *tu quoque*. If someone accuses you of cheating on your taxes, the answer "Well, so do you" is no proof one way or the other.

18. Hasty Generalization

In logic, the hasty generalization is a form of improper induction. In life, it is called prejudice. In either case, conclusions are drawn before the facts warrant it. Perhaps because our brains evolved to be constantly on the lookout for connections between events and causes, this fallacy is one of the most common of all. A couple of bad teachers mean a bad school. A few bad cars mean that brand of automobile is unreliable. A handful of members of a group are used to judge the entire group. In science, we must carefully gather as much information as possible before announcing our conclusions.

19. Overreliance on Authorities

We tend to rely heavily on authorities in our culture, especially if the authority is considered to be highly intelligent. The IQ score has acquired nearly mystical proportions in the last half century, but I have noticed that belief in the paranormal is not uncommon among Mensa members (the high-IQ club for those in the top 2 percent of the population); some even argue that their "Psi-Q" is also superior. Magician James Randi is fond of lampooning authorities with Ph.D.s—once they are granted the degree, he says, they find it almost impossible to say two things: "I don't know" and "I was wrong." Authorities, by virtue of their expertise in a field, may have a better chance of being right in that field, but correctness is certainly not guaranteed, and their expertise does not necessarily qualify them to draw conclusions in other areas.

In other words, *who* is making the claim makes a difference. If it is a Nobel laureate, we take note because he or she has been right in a big way before. If it is a discredited scam artist, we give a loud guffaw because he or she has been wrong in a big way before. While expertise is useful for separating the wheat from the chaff, it is dangerous in that we might either (1) accept a wrong idea just because it was supported by someone we respect (false positive) or (2) reject a right idea just because it was supported by someone we disrespect (false negative). How do you avoid such errors? Examine the evidence.

20. Either-Or

Also known as the *fallacy of negation* or the *false dilemma*, this is the tendency to dichotomize the world so that if you discredit one position, the observer is forced to accept the other. This is a favorite tactic of creationists, who claim that life *either* was divinely created *or* evolved. Then they spend the majority of their time discrediting the theory of evolution so that they can argue that since evolution is wrong, creationism must be right. But it is not enough to point out weaknesses in a theory. If your theory is indeed superior, it must explain both the "normal" data explained by the old theory and the "anomalous" data not explained by the old theory. A new theory needs evidence in favor of it, not just against the opposition.

21. Circular Reasoning

Also known as the *fallacy of redundancy*, *begging the question*, or *tautology*, this occurs when the conclusion or claim is merely a restatement of one of the premises. Christian apologetics is filled with tautologies: *Is there a God? Yes. How do you know? Because the Bible says so. How do you know the Bible is correct? Because it was inspired by God.* In other words, God is because God is. Science also has its share of redundancies: *What is gravity? The tendency for objects to be attracted to one another. Why are objects attracted to one another?*

Gravity. In other words, gravity is because gravity is. (In fact, some of Newton's contemporaries rejected his theory of gravity as being an unscientific throwback to medieval occult thinking.) Obviously, a tautological operational definition can still be useful. Yet, difficult as it is, we must try to construct operational definitions that can be tested, falsified, and refuted.

22. *Reductio ad Absurdum* and the Slippery Slope

Reductio ad absurdum is the refutation of an argument by carrying the argument to its logical end and so reducing it to an absurd conclusion. Surely, if an argument's consequences are absurd, it must be false. This is not necessarily so, though sometimes pushing an argument to its limits is a useful exercise in critical thinking; often this is a way to discover whether a claim has validity, especially if an experiment testing the actual reduction can be run. Similarly, the slippery slope fallacy involves constructing a scenario in which one thing leads ultimately to an end so extreme that the first step should never be taken. For example: *Eating Ben & Jerry's ice cream will cause you to put on weight. Putting on weight will make you overweight. Soon you will weigh 350 pounds and die of heart disease. Eating Ben & Jerry's ice cream leads to death. Don't even try it.* Certainly eating a scoop of Ben & Jerry's ice cream *may* contribute to obesity, which could possibly, in very rare cases, cause death. But the consequence does not necessarily follow from the premise.

Psychological Problems in Thinking

23. Effort Inadequacies and the Need for Certainty, Control, and Simplicity

Most of us, most of the time, want certainty, want to control our environment, and want nice, neat, simple explanations. All this may have some evolutionary basis, but in a multifarious society with complex problems, these characteristics can radically oversimplify reality and interfere with critical thinking and problem solving. For example, I believe that paranormal beliefs and pseudoscientific claims flourish in market economies in part because of the uncertainty of the marketplace. According to James Randi, after communism collapsed in Russia there was a significant increase in such beliefs. Not only are the people now freer to try to swindle each other with scams and rackets but many truly believe they have discovered something concrete and significant about the nature of the world. Capitalism is a lot less stable a social structure than communism. Such uncertainties lead the mind to look for explanations for the vagaries and contingencies of the market (and life in general), and the mind often takes a turn toward the supernatural and paranormal.

Scientific and critical thinking does not come naturally. It takes training, experience, and effort, as Alfred Mander explained in his *Logic for the Millions*: "Thinking is skilled work. It is not true that we are naturally endowed with the ability to think clearly and logically—without learning how, or without practicing. People with untrained minds should no more expect to think clearly and logically than people who have never learned and never practiced can expect to find themselves good carpenters, golfers, bridge players, or pianists" (1947, p. vii). We must always work to suppress our need to be absolutely certain and in total control and our tendency to seek the simple and effortless solution to a problem. Now and then the solutions may be simple, but usually they are not.

24. Problem-Solving Inadequacies

All critical and scientific thinking is, in a fashion, problem solving. There are numerous psychological disruptions that cause inadequacies in problem solving. Psychologist Barry Singer has demonstrated that when people are given the task of selecting the right answer to a problem after being told whether particular guesses are right or wrong, they:

A.   Immediately form a hypothesis and look only for examples to confirm it.

B.   Do not seek evidence to disprove the hypothesis.

C.   Are very slow to change the hypothesis even when it is obviously wrong.

D.   If the information is too complex, adopt overly-simple hypotheses or strategies for solutions.

E.   If there is no solution, if the problem is a trick and "right" and "wrong" is given at random, form hypotheses about coincidental relationships they observed. Causality is always found. (Singer and Abell 1981, p. 18)

If this is the case with humans in general, then we all must make the effort to overcome these inadequacies in solving the problems of science and of life.

25. Ideological Immunity, or the Planck Problem

In day-to-day life, as in science, we all resist fundamental paradigm change. Social scientist Jay Stuart Snelson calls this resistance an *ideological immune system*: "educated, intelligent, and successful adults rarely change their most fundamental presuppositions" (1993, p. 54). According to Snelson, the more knowledge individuals have accumulated, and the more well-founded their theories have become (and remember, we all tend to look for and remember confirmatory evidence, not counterevidence), the greater the confidence in their ideologies. The consequence of this, however, is that we build up an "immunity" against new ideas that do not corroborate previous ones. Historians of science call this the *Planck Problem*, after physicist Max Planck, who made this observation on what must happen for innovation to occur in science: "An important scientific innovation rarely makes its way by gradually winning over and converting its opponents: it rarely happens that Saul becomes Paul. What does happen is that its opponents gradually die out and that the growing generation is familiarized with the idea from the beginning" (1936, p. 97).

Psychologist David Perkins conducted an interesting correlational study in which he found a strong positive correlation between intelligence (measured by a standard IQ test) and the ability to give reasons for taking a point of view and defending that position; he also found a strong negative correlation between intelligence and the ability to consider other alternatives. That is, the higher the IQ, the greater the potential for ideological immunity. Ideological immunity is built into the scientific enterprise, where it functions as a filter against potentially overwhelming novelty. As historian of science I. B. Cohen explained, "New and revolutionary systems of science tend to be resisted rather than welcomed with open arms, because every successful scientist has a vested intellectual, social, and even financial interest in maintaining the status quo. If every revolutionary new idea were welcomed with open arms, utter chaos would be the result" (1985, p. 35).

In the end, history rewards those who are "right" (at least provisionally). Change does occur. In astronomy, the Ptolemaic geocentric universe was slowly displaced by Copernicus's heliocentric system. In geology, George Cuvier's catastrophism was gradually wedged out by the more soundly supported uniformitarianism of James Hutton and Charles Lyell. In biology, Darwin's evolution theory superseded creationist belief in the immutability of species. In Earth history, Alfred Wegener's idea of continental drift took nearly a half century to overcome the received dogma of fixed and stable continents. Ideological immunity can be overcome in science and in daily life, but it takes time and corroboration.

Spinoza's Dictum

Skeptics have the very human tendency to relish debunking what we already believe to be nonsense. It is fun to recognize other people's fallacious reasoning, but that's not the whole point. As skeptics and critical thinkers, we must move beyond our emotional responses because by understanding how others have gone wrong and how science is subject to social control and cultural influences, we can improve our understanding of how the world works. It is for this reason that it is so important for us to understand the history of both science and pseudoscience. If we see the larger picture of how these movements evolve and figure out how their thinking went wrong, we won't make the same mistakes. The seventeenth-century Dutch philosopher Baruch Spinoza said it best: "I have made a ceaseless effort not to ridicule, not to bewail, not to scorn human actions, but to understand them."

PART 2

PSEUDOSCIENCE
AND
SUPERSTITION

Rule 1

*We are to admit no more causes of natural things than such
as are both true and sufficient to explain their appearances.*

To this purpose the philosophers say that Nature does nothing in vain, and
more is in vain when less will serve; for Nature is pleased with simplicity,
and affects not the pomp of superfluous causes.

—Isaac Newton, "Rules of Reasoning in Philosophy,"
Principia Mathematica, 1687

4

Deviations

The Normal, the Paranormal, and Edgar Cayce

One of the most overused one-liners in the statistical business is Disraeli's classification (and Mark Twain's clarification) of lies into the three taxa "lies, damn lies, and statistics." Of course, the problem really lies in the misuse of statistics and, more generally, in the misunderstanding of statistics and probabilities that most of us have in dealing with the real world. When it comes to estimating the likelihood of something happening, most of us overestimate or underestimate probabilities in a way that can make normal events seem like paranormal phenomena. I saw a classic example of this at in a visit to Edgar Cayce's Association for Research and Enlightenment (A.R.E.), located in Virginia Beach, Virginia. One day when I was in town, Clay Drees, a professor at nearby Virginia Wesleyan College, and I decided to pay them a visit. We were fortunate to arrive on a relatively busy day during which the A.R.E. staff were conducting an ESP "experiment" in extrasensory perception (ESP). Since they were claiming that one's ESP could be proved scientifically, we considered A.R.E. fair game for skeptics.

According to their own literature, A.R.E. was "founded in 1931 to preserve, research, and make available the readings of Edgar Cayce," one of the most prominent "psychics" of the twentieth century. Like many such organizations, A.R.E. has many of the trappings of science: a building whose size and façade suggest modernity and authority; an extensive research library containing both the psychic readings of Edgar Cayce and a fairly good science and pseudoscience collection (though they do not classify their holdings this way); a bookstore selling a full array of writings on the paranormal, including books on spiritual living, self-discovery, inner help, past lives, health, longevity, healing, native wisdom, and the future. A.R.E. describes itself as "a research organization" that "continues to index and catalogue information, to initiate investigation and experiments, and to promote conferences, seminars, and lectures."

The corpus of accepted beliefs reads like an A-to-Z who's who and what's what of the paranormal. The circulating files index of the library includes the following psychic readings from Cayce: angels and archangels, astrological influences on Earth experiences, economic healing, evaluating psychic talent, intuition, visions and dreams, Karma and the law of grace, magnetic healing, the missing years of Jesus, the oneness of life and death, planetary sojourns and astrology, principles of psychic science, reincarnation, soul retrogression, and vibrations, to name just a few. A "reading" consisted of Cayce reclining in a chair, closing his eyes, going into an "altered state," and dictating hours of material. During his lifetime, Cayce dictated no less than fourteen thousand psychic readings on over ten thousand subjects! A separate medical library has its own circulating files index listing Cayce's psychic readings on every imaginable disease and its cure. One is "Edgar Cayce's famous 'Black Book,'" which will give you a "simple scar removal formula," explain "the best hours of sleep," tell you "the best exercise," clarify what "will help the memory," and, on page 209, solve that most mysterious of medical conundrums, "how to get rid of bad breath."

A.R.E. also has its own press—the A.R.E. Publishing Company—and incorporates the Atlantic University of Transpersonal Studies. The latter offers an "independent studies program" that includes the following courses: "TS 501—Introduction to Transpersonal Studies" (the works of Cayce, Abraham Maslow, Victor Frankl, and Buddhism); "TS 503—The Origin and Development of Human Consciousness" (on ancient magicians and the great mother goddess), "TS 504—Spiritual Philosophies and the Nature of Humanity" (on spiritual creation and evolution), "TS 506—The Inner Life: Dream, Meditation, and Imaging" (dreams as problem-solving tools), "TS 508—Religious Traditions" (Hinduism, Buddhism, Judaism, Islam, and Christianity), and "TS 518—Divination as a Way to Measure All" (astrology, tarot, I Ching, handwriting analysis, palmistry, and psychic readings).

A potpourri of lectures and seminars encourages followers' beliefs and provides opportunities for the uninitiated to get involved. A lecture on "Egypt, Myth, and Legend," by Ahmed Fayed, articulates a not-so-hidden agenda: Cayce's life in ancient Egypt. "Naming the Name: Choosing Jesus the Christ as Your Living Master" demonstrates A.R.E.'s openness to more traditional religions and its lack of discrimination between any and all belief systems. A "Sounding and Overtone Chanting" seminar promises to equip you with "tools for empowerment and transformation." A three-day seminar called "The Healing Power of Past-Life Memories" features, among others, Raymond Moody, who claims that the near-death experience is a bridge to the other side.

Who was Edgar Cayce? According to A.R.E. literature, Cayce was born in 1877 on a farm near Hopkinsville, Kentucky. As a youth, he "displayed powers of perception which extended beyond the five senses. Eventually, he would become the most documented psychic of all times." Purportedly, when he was twenty-one, Cayce's doctors were unable to find a cause or cure for a "gradual paralysis which threatened the loss of his voice." Cayce responded by going into a "hypnotic sleep" and recommended a cure for himself, which he claims worked. The discovery of his ability to diagnose illnesses and recommend solutions while in an altered state led him to do this on a regular basis for others with medical problems. This, in turn, expanded into general psychic readings on thousands of different topics covering every conceivable aspect of the universe, the world, and humanity.

Numerous books have been written on Edgar Cayce, some by uncritical followers (Cerminara 1967; Stearn 1967) and others by skeptics (Baker and Nickell 1992; Gardner 1952; Randi 1982). Skeptic Martin Gardner demonstrates that Cayce was fantasy-prone from his youth, often talking with angels and receiving visions of his dead grandfather. Uneducated beyond the ninth grade, Cayce acquired his broad knowledge through voracious reading, and from this he wove elaborate tales and gave detailed diagnoses while in his trances. His early psychic readings were done in the presence of an osteopath, from whom he borrowed much of his terminology. When his wife got tuberculosis, Cayce offered this diagnosis: "The condition in the body is quite different from what we have had before . . . from the head, pains along through the body from the second, fifth and sixth dorsals, and from the first and second lumbar . . . tie-ups here, and floating lesions, or lateral lesions, in the muscular and nerve fibers." As Gardner explains, "This is talk which makes sense to an osteopath, and to almost no one else" (1952, p. 217).

In Cayce, James Randi sees all the familiar tricks of the psychic trade: "Cayce was fond of expressions like 'I feel that . . .' and 'perhaps'—qualifying words used to avoid positive declarations" (1982, p. 189). Cayce's remedies read like prescriptions from a medieval herbalist: for a leg sore, use oil of

FIGURE 4:
ESP machine at the Association for Research and Enlightenment. [Photograph by Michael Shermer.]

smoke; for a baby with convulsions, a peach-tree poultice; for dropsy, bed-bug juice; for arthritis, peanut oil massage; and for his wife's tuberculosis, ash from the wood of a bamboo tree. Were Cayce's readings and diagnoses correct? Did his remedies work? It is hard to say. Testimony from a few patients does not represent a controlled experiment, and among his more obvious failures are several patients who died between the time of writing to Cayce and Cayce's reading. In one such instance, Cayce did a reading on a small girl in which he recommended a complex nutritional program to cure the disease but admonished, "And this depends upon whether one of the things as intended to be done today is done or isn't done, see?" The girl had died the day before, however (Randi 1982, pp. 189–195).

It was, then, with considerable anticipation that we passed under the words "That we may make manifest the love of God and man" and entered into the halls of Edgar Cayce's legacy. Inside there were no laboratory rooms and no scientific equipment save an ESP machine proudly displayed against a wall in the entrance hall (see figure 4). A large sign next to the machine announced that shortly there would be an ESP experiment in an adjacent room. We saw our opportunity.

The ESP machine featured the standard Zener cards (created by K. E. Zener, they display easily distinguished shapes to be interpreted in Psi experiments), with a button to push for each of the five symbols—plus sign, square, star, circle, and wavy lines. One of the directors of A.R.E. began with a lecture on ESP, Edgar Cayce, and the development of psychic powers. He explained that some people are born with a psychic gift while others need practice, but we all have the power to some degree. When he asked for participants, I volunteered to be a receiver. I was given no instruction on how to receive psychic messages, so I asked. The instructor explained that I should concentrate on the sender's forehead. The thirty-four other people in the room were told to do the same thing. We were all given an ESP Testing Score Sheet (see figure 5), with paired columns for our psychic choices and the correct answers, given after the experiment. We ran two trials of 25 cards each. I got 7 right in the first set, for which I honestly tried to receive the message, and 3 right in the second set, for which I marked the plus sign for every card.

The instructor explained that "5 right is average, chance is between 3 and 7, and anything above 7 is evidence of ESP." I asked, "If 3 to 7 is chance, and anything above 7 is evidence of ESP, what about someone who scores below a 3?" The instructor responded, "That's a sign of negative ESP." (He didn't say what that was.) I then surveyed the group. In the first set, three people got 2 right, while another three got 8 right; in the second set, one even got 9 right. So, while I apparently did not have psychic power, at least four other people did. Or did they?

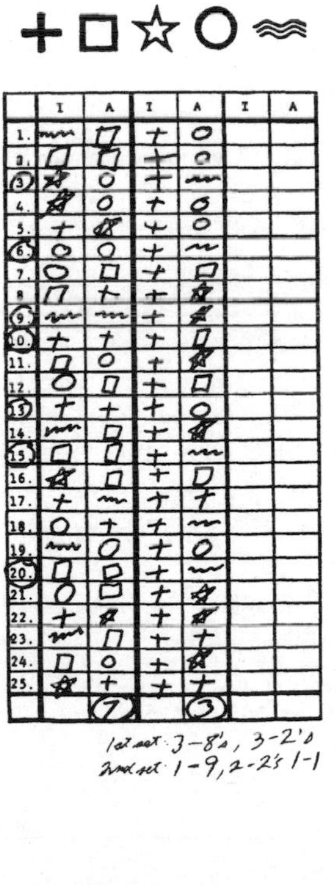

FIGURE 5:
Michael Shermer's ESP Testing Score Sheet.

Before concluding that high scores indicate a high degree of ESP ability, you have to know what kind of scores people would get purely by
chance. The scores expected by chance can be predicted by probability
theory and statistical analysis. Scientists use comparisons between statistically predicted test results and actual test results to determine whether
results are significant, that is, better than what would be expected by
chance. The ESP test results clearly matched the expected pattern for random results.

I explained to the group, "In the first set, three got 2, three got 8, and
everyone else [twenty-nine people] scored between 3 and 7. In the second
set, there was one 9, two 2s, and one 1, *all scored by different people than those
who scored high and low in the first test!* Doesn't that sound like a normal distribution around an average of 5?" The instructor turned and said, with a
smile, "Are you an engineer or one of those statisticians or something?"
The group laughed, and he went back to lecturing about how to improve
your ESP with practice.

When he asked for questions, I waited until no one else had any and
then inquired, "You say you've been working with A.R.E. for several
decades, correct?" He nodded. "And you say that with experience one can
improve ESP, right?" He immediately saw where I was going and said,
"Well . . .," at which point I jumped in and drew the conclusion, "By now
you must be very good at this sort of test. How about we send the signals
to you at the machine. I'll bet you could get at least 15 out of the 25." He
was not amused at my suggestion and explained to the group that he had
not practiced ESP in a long time and, besides, we were out of time for the
experiment. He quickly dismissed the group, upon which a handful of people surrounded me and asked for an explanation of what I meant by "a normal distribution around an average of 5."

On a piece of scrap paper, I drew a crude version of the normal frequency curve, more commonly known as the bell curve (see figure 6). I
explained that the mean, or average number, of correct responses ("hits") is
expected by chance to be 5 (5 out of 25). The amount that the number of
hits will deviate from the standard mean of 5, by chance, is 2. Thus, for a
group this size, we should not put any special significance on the fact that
someone got 8 correct or someone scored only 1 or 2 correct hits. This is
exactly what is expected to happen by chance.

So these test results suggest that nothing other than chance was operating. The deviation from the mean for this experiment was nothing more
than what we would expect. If the audience were expanded into the millions, say on a television show, there would be an even bigger opportunity

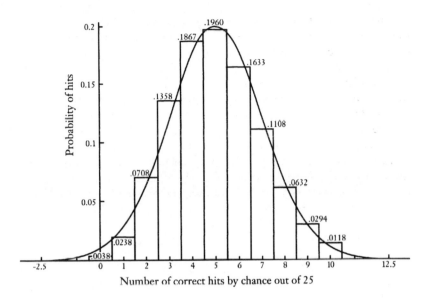

FIGURE 6:
Bell curve for a test of 25 questions with 5 possible answers. If chance is operating, probability predicts that most people (79 percent) will get between 3 and 7 correct, whereas the probability of getting 8 or more correct is 10.9 percent (thus, in a group of 25, several scores in this range will always occur purely by chance), of getting 15 correct is about 1 in 90,000, of getting 20 correct is about 1 in 5 billion, and of getting all 25 correct is about 1 in 300 quadrillion.

for misinterpretation of the high scores. In this scenario, a tiny fraction would be 3 standard deviations above the mean, or get 11 hits, a still smaller percentage would reach 4 standard deviations, or 13 hits, and so on, all as predicted by chance and the randomness of large numbers. Believers in psychic power tend to focus on the results of the most deviant subjects (in the statistical sense) and tout them as the proof of the power. But statistics tells us that given a large enough group, there should be someone who will score fairly high. There may be lies and damned lies, but statistics can reveal the truth when pseudoscience is being flogged to an unsuspecting group.

After the ESP experiment, one woman followed me out of the room and said, "You're one of those skeptics, aren't you?"

"I am indeed," I responded.

"Well, then," she retorted, "how do you explain coincidences like when I go to the phone to call my friend and she calls me? Isn't that an example of psychic communication?"

"No, it is not," I told her. "It is an example of statistical coincidences. Let me ask you this: How many times did you go to the phone to call your friend and she did not call? Or how many times did your friend call you but you did not call her first?"

She said she would have to think about it and get back to me. Later, she found me and said she had figured it out: "I only remember the times that these events happen, and I forget all those others you suggested."

"Bingo!" I exclaimed, thinking I had a convert. "You got it. It is just selective perception."

But I was too optimistic. "No," she concluded, "this just proves that psychic power works sometimes but not others."

As James Randi says, believers in the paranormal are like "unsinkable rubber ducks."

5

Through the Invisible

Near-Death Experiences and the Quest for Immortality

> I sent my Soul through the Invisible,
> some letter of that After-life to spell:
> And by and by my Soul return'd to me,
> And answer'd "I Myself am Heav'n and Hell."
>
> —Omar Khayyám, *The Rubaiyat*

In 1980 I attended a weekend seminar in Klamath Falls, Oregon, on "Voluntary Controls of Internal States," hosted by Jack Schwarz, a man well known to practitioners of alternative medicine and altered states of consciousness. According to literature advertising the seminar, Jack is a survivor of a Nazi concentration camp, where years of isolation, miserable conditions, and physical torture taught him to transcend his body and go to a place where he could not be hurt. Jack's course was intended to teach the principles of mind control through meditation. Mastery of these principles allows one to voluntarily control such bodily functions as pulse rate, blood pressure, pain, fatigue, and bleeding. In a dramatic demonstration, Jack took out a ten-inch-long rusted sail needle and shoved it through his biceps. He didn't wince and after he pulled it out only a tiny drop of blood covered the hole. I was impressed.

The first part of the course was educational. We learned about the color, location, and power of our chakras (energy centers intersecting the physical and psychospiritual realms), the power of the mind to control the body through use of these chakras, the cure of illnesses through visualization, becoming at one with the universe through the interaction of matter and energy, and other remarkable things. The second part of the course was practical. We learned how to meditate, and then we chanted a type of mantra to focus our energies. This went on for quite some time. Jack explained that some people might experience some startling emotions. I didn't, try as I might, but others certainly did. Several women fell off their chairs and began writhing on the floor, breathing heavily and moaning in what appeared to me as an orgasmic state. Even some men really got into it. To help me get in tune with my chakras, one woman took me into a bathroom with a wall mirror, closed the door and shut off the lights, and tried to show me the energy auras surrounding our bodies. I looked as hard as I could but didn't see anything. One night we were driving along a quiet Oregon highway and she started pointing out little light-creatures on the side of the road. I couldn't see these either.

I took a few other seminars from Jack and since this was before I was a "skeptic," I can honestly say I tried to experience what others seemed to— but it always eluded me. In retrospect, I think what was going on had to do with the fact that some people are fantasy-prone, others are open to suggestion and group influence, while still others are good at letting their minds slip into altered states of consciousness. Since I think near-death experiences are a type of altered state of consciousness, let us examine this concept next.

What Is an Altered State of Consciousness?

Most skeptics would agree with me that mystical and spiritual experiences are nothing more than the product of fantasy and suggestion, but many would question my third explanation of altered states of consciousness. James Randi and I have discussed this subject at length. He, along with other skeptics like psychologist Robert Baker (1990, 1996), believes that there is no such thing as an altered state of consciousness because there is nothing you can do in a so-called altered state that you cannot do in an unaltered state (i.e., normal, awake, and conscious). Hypnosis, for example, is often considered a type of altered state, yet hypnotist "The Amazing" Kreskin offers to pay $100,000 to anyone who can get someone to do something under hypnosis that they could not do in an ordinary wakeful state. Baker, Kreskin, Randi, and others think that hypnosis is nothing more than fantasy role-playing. I disagree.

The expression *altered states of consciousness* was coined by parapsychologist Charles Tart in 1969, but mainstream psychologists have been aware for some time of the fact that the mind is more than just conscious awareness. Psychologist Kenneth Bowers argues that experiments prove that "there is something far more pervasive and subtle to hypnotic behavior than voluntary and purposeful compliance with the perceived demands of the situation" and that "the 'faking hypothesis' is an entirely inadequate interpretation of hypnosis" (1976, p. 20). Stanford experimental psychologist Ernest Hilgard discovered through hypnosis a "hidden observer" in the mind aware of what is going on but not on a conscious level, and that there exists a "multiplicity of functional systems that are hierarchically organized but can become dissociated from one another" (1977, p. 17). Hilgard typically instructed his subjects as follows:

> When I place my hand on your shoulder (after you are hypnotized) I shall be able to talk to a hidden part of you that knows things are going on in your body, things that are unknown to the part of you to which I am now talking. The part to which I am now talking will not know what you are telling me or even that you are talking. . . . You will remember that there is a part of you that knows many things that are going on that may be hidden from either your normal consciousness or the hypnotized part of you. (Knox, Morgan, and Hilgard 1974, p. 842)

This dissociation of the hidden observer is a type of altered state.

What exactly do we mean by an altered state or, for that matter, an unaltered state? Here it might be useful to distinguish between *quantitative* differences—those of degree—and *qualitative* differences—those of kind. A pile of six apples and a pile of five apples are quantitatively different. A pile of six apples and a pile of six oranges are qualitatively different. Most differences between states of consciousness are quantitative, not qualitative. In other words, in both states a thing exists, just in different amounts. For example, when sleeping, we think, since we dream; we form memories, since we can remember our dreams; and we are sensitive to our environment, though considerably less so. Some people walk and talk in their sleep, and we can control sleep, planning to get up at a certain time and doing so fairly reliably. In other words, while asleep we just do less of what we do while awake.

Still, sleep is a good example because it is so different that we do not normally mistake it for a waking state. The quantitative difference is so

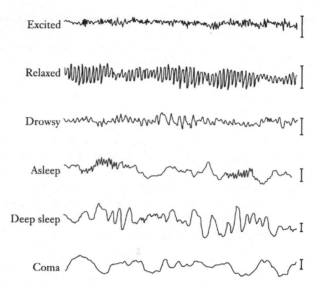

FIGURE 7:
EEG recordings for six different states of consciousness.

great as to be qualitatively different and thus count as an altered state. Though the EEG readings in figure 7 are only quantitatively different, they are so much so that the states they represent may be considered as different in kind. If a coma is not an altered state, I do not know what is. And it cannot be duplicated in a conscious state.

Consciousness has two characteristics: "1. *Monitoring* ourselves and our environment so that perceptions, memories, and thoughts are accurately represented in awareness; 2. *Controlling* ourselves and our environment so that we are able to initiate and terminate behavioral and cognitive activities" (Kihlstrom 1987, p. 1445). Thus, an altered state of consciousness would have to interfere with our accurate monitoring of percepts, memories, and thoughts, as well as disrupt control of our behavior and cognition within the environment. *An altered state of consciousness exists when there is significant interference with our monitoring and control of our environment.* By significant, I mean a dramatic departure from "normal" functioning. Both sleep and hypnosis do this, as do hallucinations, near-death experiences, out-of-body experiences, and other altered states.

Psychologist Barry Beyerstein makes a similar argument in defining altered states of consciousness as the modification of specific neural systems "by disease, repetitive stimulation, mental manipulations, or chemical ingestion" such that "our perception of ourselves and the world can be profoundly altered" (1996, p. 15). Psychologist Andrew Neher (1990) calls them "transcendent states," which he defines as sudden and unexpected alterations of consciousness intense enough to be overwhelming to the person experiencing them. The key here is the *intensity* of the experience and the *profundity* of the alteration of consciousness. Do we do anything in an altered state that we cannot do in an unaltered state of consciousness?

Yes. For example, dreams are significantly different from waking thoughts and daydreams. The fact that we normally never confuse the two is an indication of their qualitative difference. Further, hallucinations are not normally experienced in a stable, awake state unless there is some intervening variable, such as extreme stress, drugs, or sleep deprivation. Near-death experiences and out-of-body experiences are so unusual that they often stand out as life-changing events.

No. The differences are only quantitative. But even here, it could be argued that the differences are so great as to constitute a qualitative difference. You can show me that the EEGs recorded when I am normally conscious and when I am hallucinating severely are only quantitatively different, but I have no trouble experiencing and recognizing their dramatic difference. Consider the near-death experience.

The Near-Death Experience

One of the driving forces behind religions, mysticism, spiritualism, the New Age movement, and belief in ESP and psychic powers is the desire to transcend the material world, to step beyond the here-and-now and pass through the invisible into another world beyond the senses. But where is this other world and how do we get there? What is the appeal of some place we know absolutely nothing about? Is death merely a transition to this other side?

Believers claim that we do know something about the other side through a phenomenon called the *perithanatic* or *near-death experience (NDE)*. The NDE, like its related partner the *out-of-body experience (OBE)*, is one of the most compelling phenomena in psychology. Apparently, upon a close encounter with death, some individuals' experiences are so similar as to lead many to believe that there is an afterlife or that death is a pleasant experience or both. The phenomenon was popularized in 1975 with the publication of Raymond Moody's book, *Life After Life*, and was substantiated by corroborative evidence from others. For example, cardiologist F. Schoonmaker (1979) reported that 50 percent of the more than two thousand patients he treated over an eighteen-year period had NDEs. A 1982 Gallup poll found that one out of twenty Americans had been through an NDE (Gallup 1982, p. 198). And Dean Sheils (1978) has studied the cross-cultural nature of the phenomenon.

When NDEs first came into prominence, they were perceived as isolated, unusual events and were dismissed by scientists and medical doctors as either exaggerations or flights of fantasy by highly stressed but very creative minds. In the 1980s, however, NDEs gained credibility through the work of Elisabeth Kübler-Ross, a medical doctor who publicized this now-classic example:

> Mrs. Schwartz came into the hospital and told us how she had had a near-death experience. She was a housewife from Indiana, a very simple and unsophisticated woman. She had advanced cancer, had hemorrhaged and was put into a private hospital, very close to death. The doctors attempted for 45 minutes to revive her, after which she had no vital signs and was declared dead. She told me later that while they were working on her, she had an experience of simply floating out of her physical body and hovering a few feet above the bed, watching the resuscitation team work very frantically. She described to me the designs of the doctors' ties, she repeated a joke one of the young doctors told, she remembered absolutely everything. And all she wanted to tell them was relax, take it easy, it is all right, don't struggle so hard. The more she tried to tell them, the more frantically they worked to revive her. Then, in her own language, she "gave up" on them and lost consciousness. After they declared her dead, she made a comeback and lived for another year and a half. (1981, p. 86)

This is a typical NDE, characterized by one of the three most commonly reported elements: (1) a floating OBE in which you look down and see your body; (2) passing through a tunnel or spiral chamber toward a bright light that represents transcendence to "the other side"; (3) emerging on the other side and seeing loved ones who have already passed away or a Godlike figure.

It seems obvious that these are hallucinatory wishful-thinking experiences, yet Kübler-Ross has gone out of her way to verify the stories. "We've had people who were in severe auto accidents, had no vital signs and told us how many blow torches were used to extricate them from the wreck" (1981, p. 86). Even more bizarre are stories of an imperfect or diseased body becoming whole again during an NDE. "Quadriplegics are no longer paralyzed, multiple-sclerosis patients who have been in wheelchairs for years say that when they were out of their bodies, they were able to sing and dance."

Memories from a previously whole body? Of course. A close friend of mine who became a paraplegic after an automobile accident often dreamed of being whole. It was not at all unusual for her to wake in the morning and fully expect to hop out of bed. But Kübler-Ross does not buy the prosaic explanation: "You take totally blind people who don't even have light perception, don't even see shades of gray. If they have a near-death experience, they can report exactly what the scene looked like at the accident or hospital room. They have described to me incredibly minute details. How do you explain that?" (1981, p. 90). Simple. Memories of verbal descriptions given by others during the NDE are converted into visual images of the scene and then rendered back into words. Further, quite frequently patients in trauma or surgery are not totally unconscious or under the anesthesia and are aware of what is happening around them. If the patient is in a teaching hospital, the attending physician or chief resident who performs the surgery would be describing the procedure for the other residents, thus enabling the NDE subject to give an accurate description of events.

Something is happening in the NDE that cries out for explanation, but what? Physician Michael Sabom, in his 1982 *Recollections of Death*, drew on the results of his correlational study of a large number of people who had had NDEs, noting age, sex, occupation, education, and religious affiliation, along with prior knowledge of NDEs, possible expectations as a result of religious or prior medical knowledge, the type of crisis (accident, arrest), location of crisis, method of resuscitation, estimated time of unconsciousness, description of the experience, and so on. Sabom followed these subjects for years, re-interviewing them as well as members of their families to see whether they altered their stories or found some other explanation for the experience. Even after years, every subject felt just as strongly about his or her experience and was convinced that the episode did occur. Almost all stated that the experience had a definite impact on their outlook on life and perception of death. They were no longer "afraid" of dying nor did they "mourn" the death of loved ones, as they were convinced that death is a pleasant experience. Each felt that he or she had been given a second chance and, although not every subject became "religious," they all felt a need to "do something with their lives."

Although Sabom notes that nonbelievers and believers had similar experiences, he fails to mention that we have all been exposed to the Judeo-Christian worldview. Whether or not we consciously believe, we have all heard similar ideas about God and the afterlife, heaven and hell. Sabom also does not point out that people of different religions see different religious figures during NDEs, an indication that the phenomenon occurs within the mind, not without.

What naturalistic explanations can be offered for NDEs? An early, speculative theory came from psychologist Stanislav Grof (1976; Grof and Halifax 1977), who argued that every human being has already experienced the characteristics of the NDE—the sensation of floating, the passage down a tunnel, the emergence into a bright light—birth. Perhaps the memory of such a traumatic event is permanently imprinted in our minds, to be triggered later by an equally traumatic event—death. Is it possible that recollection of perinatal memories accounts for what is experienced during an NDE? Not likely. There is no evidence for infantile memories of any kind. Furthermore, the birth canal does not look like a tunnel and besides the infant's head is normally down and its eyes are closed. And why do people who are born by cesarean section have NDEs? (Not to mention that Grof and his subjects were experimenting with LSD—not the most reliable method for retrieving memories, since it creates its own illusions.)

A more likely explanation looks to biochemical and neurophysiological causes. We know, for example, that the hallucination of flying is triggered by atropine and other belladonna alkaloids, some of which are found in mandrake and jimsonweed and were used by European witches and American Indian shamans. OBEs are easily induced by dissociative anesthetics such as the ketamines. DMT (dimethyltryptamine) causes the perception that the world is enlarging or shrinking. MDA (methylene-dioxyamphetamine) stimulates the feeling of age regression so that things we have long forgotten are brought back into memory. And, of course, LSD (lysergic acid diethylamide) triggers visual and auditory hallucinations and creates a feeling of oneness with the cosmos, among other effects (see Goodman and Gilman 1970; Grinspoon and Bakalar 1979; Ray 1972; Sagan 1979; Siegel 1977). The fact that there are receptor sites in the brain for such artificially processed chemicals means that there are naturally produced chemicals in the brain that, under certain conditions (the stress of trauma or an accident, for example), can induce any or all of the experiences typically associated with an NDE. Perhaps NDEs and OBEs are nothing more than wild "trips" induced by the extreme trauma of almost dying. Aldous Huxley's *Doors of Perception* (whence the rock group The Doors got its name) has a fascinating description, made by the author while under the influence of mescaline, of a flower in a vase. Huxley describes "seeing what Adam had seen on the morning of his creation—the miracle, moment by moment, of naked existence" (1954, p. 17).

Psychologist Susan Blackmore (1991, 1993, 1996) has taken the hallucination hypothesis one step further by demonstrating why different peo-

FIGURE 8:
Spiral chamber and striped tunneling effects of near-death experiences. Such effects are also produced by hallucinogenic drugs.

plc would experience similar effects, such as the tunnel. The visual cortex on the back of the brain is where information from the retina is processed. Hallucinogenic drugs and lack of oxygen to the brain (such as sometimes occurs near death) can interfere with the normal rate of firing by nerve cells in this area. When this occurs "stripes" of neuronal activity move across the visual cortex, which is interpreted by the brain as concentric rings or spirals. These spirals may be "seen" as a tunnel. Similarly, the OBE is a confusion between reality and fantasy, as dreams can be upon first awakening. The brain tries to reconstruct events and in the process visualizes them from above—a normal process we all do when "decentering" ourselves (when you picture yourself sitting on the beach or climbing a mountain, it is usually from above, looking down). Under the influence of hallucinogenic drugs, subjects saw images like those in figure 8; such images produce the tunneling effect of the NDE.

Finally, the "otherworldliness" of the NDE is produced by the dominance of the fantasy of imagining the other side, visualizing our loved ones who died before, seeing our personal God, and so on. But what happens to those who do not come back from an NDE? Blackmore gives this reconstruction of death: "Lack of oxygen first produces increased activity through disinhibition, but eventually it all stops. Since it is this activity that produces the mental models that give rise to consciousness, then all this will cease. There will be no more experience, no more self, and so that . . . is the end" (1991, p. 44). Cerebral anoxia (lack of oxygen), hypoxia (insufficient oxygen), or hypercardia (too much carbon dioxide) have all been proposed as triggers of NDEs (Saavedra-Aguilar and Gomez-Jeria 1989), but Blackmore points out that people with none of these conditions have had NDEs. She admits, "It is far from clear, as yet, how they are best to be explained. No amount of evidence is likely to settle, for good, the argument between the 'afterlife' and 'dying brain' hypotheses" (1996, p. 440). NDEs remain one of the great unsolved mysteries of psychology, leaving us once again with a Humean question: Which is more likely, that an NDE is an as-yet-to-be explained phenomenon of the brain or that it is evidence of what we have always wanted to be true—immortality?

The Quest for Immortality

Death, or at least the end of life, appears to be the outer limit of our consciousness and the frontier of the possible. Death is the ultimate altered state. Is it the end, or merely the end of the beginning? Job asked the same question: "If a man die, shall he live again?" Obviously no one knows for sure, but plenty of folks think they do know, and many of them are not shy about trying to convince the rest of us that their particular answer is the correct one. This question is one of the reasons that there are literally thousands of organized religions in the world, each claiming exclusive knowledge about what follows death. As humanist scholar Robert Ingersoll (1879) noted, "The only evidence, so far as I know, about another life is, first, that we have no evidence; and secondly, that we are rather sorry that we have not, and wish we had." Without some belief structure, however, many people find this world meaningless and without comfort. The philosopher George Berkeley (1713) penned this example of such sentiments: "I can easily overlook any present momentary sorrow when I reflect that it is in my power to be happy a thousand years hence. If it were not for this thought I had rather be an oyster than a man."

In one of Woody Allen's movies, his physician gives him one month to live. "Oh, no," he moans, "I only have thirty days to live?" "No," the doctor responds, "twenty-eight; this is February." Are we this bad? Sometimes. It might be splendid if we were all to adopt Socrates' reflectiveness just before his state-mandated suicide: "To fear death, gentlemen, is nothing other than to think oneself wise when one is not; for it is to think one knows what one does not know. No man knows whether death may not even turn out to be the greatest of blessings for a human being; and yet people fear it as if they knew for certain that it is the greatest of evils" (Plato 1952, p. 211). But most people feel more like Berkeley and his oyster, and thus, as Ingersoll was fond of pointing out, we have religion. But the quest for immortality is not restricted to the religious. Wouldn't we all like to live on in some capacity? We can, indirectly, and, if science can accomplish what some hope it will, perhaps even in reality.

Science and Immortality

Because purely religious theories of immortality—based on faith, not reason—are not testable, I will not discuss them here. Frank Tipler's *Physics of Immortality* is the subject of chapter 16 of this book, as Tipler's work requires extensive analysis. Suffice it to say that by "immortality" most people do not mean merely living on through one's legacy, whatever it may be. As Woody Allen said, "I don't want to gain immortality through my work; I want to gain immortality through not dying." Most people would not be content with the argument that parents are immortal in the sense that a significant part of their genetic make-up lives on in the genes of their offspring. From an evolutionary viewpoint, 50 percent of a person's genes live on in their offspring, 25 percent in their grandchildren, 12.5 percent in each great grandchild, and so on. What most of us think of as "real" immortality is living forever, or at least considerably longer than the norm. The rub is that it seems certain that the process of aging and death is a normal, genetically programmed part of the sequence of life. In evolutionary biologist Richard Dawkins's (1976) scenario, once we've passed reproductive age (or at least the period of intense and regular participation in sexual activity), then the genes have no more use for the body. Aging and death may be the species' way of eliminating those who are no longer genetically useful but are still competing for limited resources with those whose job it now is to pass along the genes.

To extend life significantly, we must understand the causes of death. Basically there are three: trauma, such as accidents; disease, such as cancer and arteriosclerosis; and entropy, or senescence (aging), which is a naturally occurring, progressive deterioration of various biochemical and cellular functions that begins early in adult life and ultimately results in an increased likelihood of dying from trauma or disease.

How long can we live? The *maximum life potential* is the age at death of the longest-lived member of the species. For humans, the record for the oldest documented age ever achieved is 120 years. It is held by Shigechiyo Izumi, a Japanese stevedore. There are many undocumented claims of people living beyond 150 years and even up to 200 years; these frequently involve such cultural oddities as adding the ages of father and son together. Data on documented centenarians (people who live to be 100 years old) reveal that only one person will live to be 115 years old for every 2,100 million (2.1 billion) people. Today's world population of slightly over five billion is likely to produce only two or three individuals who will reach 115 years old. *Life span* is the age at which the average individual would die if there were no premature deaths from accidents or disease. This age is approximately 85 to 95 years and has not changed for centuries, and probably millennia. Life span, like maximum life potential, is probably a fixed biological constant for each species. *Life expectancy* is the age at which the average individual would die when accidents and disease have been taken into consideration. In 1987, life expectancy for women in the West was 78.8 years and for men 71.8 years, for an overall expectancy of 75.3 years. Worldwide, in 1995 life expectancy was estimated at 62 years. The numbers are continually on the rise. In the United States, life expectancy was 47 years in 1900. By 1950 the figure had climbed to 68. In Japan, the life expectancy for girls born in 1984 is 80.18 years, making it the first country to pass the 80 mark. It is unlikely, however, that life expectancy will ever go higher than the life span of 85 to 95.

Though aging and death do appear to be certain, attempts to extend the biological functions of humans for as long as possible are slowly moving away from the lunatic fringe into the arena of legitimate science. Organ replacements, improved surgical techniques, immunizations against most major diseases, advanced nutritional knowledge, and the awareness of the salubrious effects of exercise have all contributed to the rapid rise in life expectancy.

Another futuristic possibility is *cloning*, the exact duplication of an organism from a body cell (which is diploid, or has a full set of genes, as opposed to a sex cell, which is haploid, or has only a half set of genes). Cloning lower organisms has been accomplished but the barriers to cloning humans are both scientific and ethical. If these barriers go down, cloning may play a significant role in life extension. One of the major problems with organ transplantation is the rejection of foreign tissue. This issue would not exist with duplicate organs from a clone—just raise your clone in a sterile environment to keep the organs healthy, and then replace your own aging parts with the clone's younger, healthier organs.

The ethical questions associated with this scenario are challenging, to say the least. Is the clone human? Does the clone have rights? Should there be a union for clones? (How about a new ACLU, the American Clone Liberties Union?) Is the clone a separate and independent individual? If no, then what about your individuality, since there is one of you living in two bodies? If yes, then are there two of "you"? For that matter, if you replace so many organs that all your original organs are gone, are you still "you"? If you believe in the Judeo-Christian form of immortality and you clone yourself, is there one soul or two?

Finally, there is the fascinating field of *cryonic suspension*, or what Alan Harrington calls the "freeze-wait-reanimate" process. The principles of the procedure are relatively simple, the application is not. When the heart stops and death is officially pronounced, all the blood is removed and replaced with a fluid that preserves the organs and tissues while they are in a frozen state. Then, no matter what kills us—accident or disease—sooner or later the technologies of the future should be equal to the task of reviving and curing us.

Cryonics is still so new and experimental that the ethical questions have yet to come to public attention. For now, cryonic suspension is considered by the government as a form of burial, and individuals are frozen after they are declared legally dead by natural means, never by choice. If cryonicists could succeed in reviving someone, the distinction between the living and the dead would blur. Life and death would become a continuum instead of the discrete states they have always been. Certainly, definitions of death would have to be rewritten. And what about the problem of the soul? If there is such a thing, where does it go while the body is in cryonic suspension? If an individual chooses to be put into cryonic suspension before he is actually dead, then is the technician committing murder? Would it be murder only if the reanimation procedure failed to revive this suspended individual?

If cryonic suspension technology ever matches the hopes and expectations of cryonicists, it may be feasible that someday one could choose to be frozen and reanimated at will, maybe even multiple times. Perhaps one could come back for ten-year stretches every century and essentially live a thousand years or more. Think of future historians able to write an oral history with someone who lived a thousand years before. But alas, as yet the entire field remains high-tech scientific speculation, or protoscience. Here are just a few of the problems:

1. We do not know whether anyone frozen to date or anyone who will be frozen in the foreseeable future will ever be successfully revived. No higher organism has ever been truly frozen and brought back alive.

2. The freezing technology appears to do considerable damage to brain cells, though the exact nature and extent of such damage have yet to be determined since no one has been revived to put it to the test. Even if the physical damage is slight, it still remains to be seen whether memory

and personal identity will be restored. Our scientific understanding of where and how memory and personal identity are stored is fairly unsophisticated. Neurophysiologists have come a long way toward an explanation of memory storage and retrieval, but the theory is by no means complete. It is possible, though seemingly unlikely, that complete restoration will still result in memory loss. We just do not know without an actual test case. If cryonic revival does not result in return of considerable personal memory and identity, then what's the point?

3. The entire science of cryonics presently depends on future technological developments. As cryonicists Mike Darwin and Brian Wowk explain, "Even the best known cryo-preservation methods still lead to brain injuries irreversible by present technology. Until brain cryo-preservation is perfected, cryonics will rely on future technologies, not just for tissue replacement, but also for repair of tissues essential to the patient's survival" (1989, p. 10). This is the biggest flaw in cryonics. Ubiquitous in the cryonics literature are reminders that the history of science and technology is replete with stories of misunderstood mavericks, surprise discoveries, and dogmatic closed-mindedness to revolutionary new ideas. The stories are all true, but cryonicists ignore all the revolutionary new ideas that were wrong. Unfortunately for cryonicists, past success does not guarantee future progress in any field. Cryonics presently depends on nanotechnology, the construction of tiny computer-driven machines. As Eric Drexler (1986) has shown, and Richard Feynman demonstrated as early as 1959, "There's plenty of room at the bottom" for molecular-size technologies. But theory and application are two different things, and a scientific conclusion cannot be based on what *might* be, no matter how logical it may seem or who endorses it. Until we have evidence, our judgment must remain, appropriately enough, suspended.

Historical Transcendence— Is It So Small a Thing?

Given these prospects, where can the nonreligious individual find meaning in an apparently meaningless universe? Can we transcend the banality of life without leaving the body? History is the one field of thought that deals with human action across time and beyond any one individual's personal story. History transcends the here-and-now through its fairly long past and near limitless future. History is a product of sequences of events that come together in their own unique ways. Those events are mostly human actions, so history is a product of the way individual human actions come together to produce the future, however constrained by certain previous conditions, such as laws of nature, economic forces, demographic trends, and cultural mores. We are free, but not to do just anything. And the significance of a human action is also restricted by *when* in the historical sequence the action is taken. The earlier the action is in a sequence, the more sensitive the sequence is to minor changes—the so-called butterfly effect.

The key to historical transcendence is that since you cannot know when in the sequence you are (since history is contiguous) and what effects present actions may have on future outcomes, positive change requires that you choose your actions wisely—all of them. What you do tomorrow could change the course of history, even if only long after you are gone. Think of all the famous people of the past who died relatively unknown. Today, they have transcended their own time because we perceive that some of their actions altered history, even if they were unaware that they were doing anything important. One may gain transcendence by affecting history, by actions whose influence extends well beyond one's biological existence. The alternatives to this scenario—apathy about one's effect on others and the world, or belief in the existence of another life for which science provides no proof—may lead one to miss something of profound importance in *this* life. We should heed Matthew Arnold's beautiful words from his *Empedocles on Etna* (1852):

> *Is it so small a thing, To have enjoyed the sun,*
> *To have lived light in the Spring,*
> *To have loved, to have thought, to have done;*
> *To have advanced true friends, and beat down baffling foes—*
> *That we must feign a bliss Of doubtful future date,*
> *And while we dream on this, Lose all our present state,*
> *And relegate to worlds . . . yet distant our repose?*

6

Abducted!

Encounters with Aliens

On Monday, August 8, 1983, I was abducted by aliens. It was late at night and I was traveling along a lonely rural highway approaching the small town of Haigler, Nebraska, when a large craft with bright lights hovered alongside me and forced me to stop. Alien creatures got out and cajoled me into their vehicle. I do not remember what happened inside but when I found myself traveling back down the road I had lost ninety minutes of time. Abductees call this "missing time," and my abduction a "close encounter of the third kind." I'll never forget the experience, and, like other abductees, I've recounted my abduction story numerous times on television and countless times to live audiences.

A Personal Abduction Experience

This may seem like a strange story for a skeptic to be telling, so let me fill in the details. As I explained in Chapter 1, for many years I competed as a professional ultra-marathon bicycle racer, primarily focusing on the 3,000-mile, nonstop, transcontinental Race Across America. "Nonstop" means racers go long stretches without sleep, riding an average of twenty-two out of every twenty-four hours. It is a rolling experiment on stress, sleep deprivation, and mental breakdown.

Under normal sleep conditions, most dream activity is immediately forgotten or fades fairly soon after waking into consciousness. Extreme sleep deprivation breaks down the wall between reality and fantasy. You have severe hallucinations that seem as real as the sensations and perceptions of daily life. The words you hear and speak are recalled like a normal memory. The people you see are as corporeal as those in real life.

During the inaugural 1982 race, I slept three hours on each of the first two nights and consequently fell behind the leader, who was proving that one could get by with considerably less sleep. By New Mexico, I began riding long stretches without sleep in order to catch up, but I was not prepared for the hallucinations that were to come. Mostly they were the garden-variety hallucinations often experienced by weary truck drivers, who call the phenomenon "white-line fever": bushes form into lifelike animals, cracks in the road make meaningful designs, and mailboxes look like people. I saw giraffes and lions. I waved to mailboxes. I even had an out-of-body experience near Tucumcari, New Mexico, where I saw myself riding on the shoulder of Interstate 40 from above.

Finishing third that year, I vowed to ride sleepless in 1983 until I got the lead or collapsed. Eighty-three hours away from the Santa Monica Pier, just shy of Haigler, Nebraska, and 1,259 miles into the race, I was falling asleep on the bike so my support crew (every rider has one) put me down for a forty-five-minute nap. When I awoke I got back on my bike, but I was still so sleepy that my crew tried to get me back into the motorhome. It was then that I slipped into some sort of altered state of consciousness and became convinced that my entire support crew were aliens from another planet and that they were going to kill me. So clever were these aliens that they even looked, dressed, and spoke like my crew. I began to quiz individual crew members about details from their personal lives and about the bike that no alien should know. I asked my mechanic if he had glued on my bike tires with spaghetti sauce. When he replied that he had glued them on with Clement glue (also red), I was quite impressed with the research the aliens had done. Other questions and correct answers followed. The context for this hallucination was a 1960s television program—*The Invaders*—in which the aliens looked exactly like humans with the exception of a stiff little finger. I looked for stiff pinkies on my crew members. The motorhome with its bright lights became their spacecraft. After the crew managed to bed me down for another forty-five minutes, I awoke clear-headed and the problem was solved. To this day, however, I recall the hallucination as vividly and clearly as any strong memory.

Now, I am not claiming that people who have had alien abduction experiences were sleep deprived or undergoing extreme physical and mental stress. However, I think it is fairly clear that if an alien abduction experience can happen under these conditions, it can happen under other conditions. Obviously I was not abducted by aliens, so what is more likely: that other people are having experiences similar to mine, triggered by other altered states and unusual circumstances, or that we really are being visited secretly by aliens from other worlds? By Hume's criterion of how to judge a miracle—"no testimony is sufficient to establish a miracle, unless the testimony be of such a kind, that its falsehood would be more miraculous than the fact which it endeavors to establish"—we would have to choose the first explanation. It is not impossible that aliens are traveling thousands of light years to Earth and dropping in undetected, but it is much more likely that humans are experiencing altered states of consciousness and interpreting them in the context of what is popular in our culture today, namely, space aliens.

Autopsy of an Alien

Humans have achieved space flight and even sent spacecraft out of the solar system, so why couldn't other intelligent beings have done the same thing? Perhaps they have learned to traverse the enormous distances between the stars by accelerating beyond the speed of light, even though all laws of nature known to us prohibit this. Perhaps they have solved the problem of collisions with space dust and particles which would shatter a spacecraft traveling at such enormous speeds. And somehow they have reached such technological sophistication without destroying themselves in their versions of war and genocide. These are very hard problems to solve, but look how much humans have accomplished since 1903 when the Wright brothers lofted their tiny craft into the air for twelve seconds. Should we be so arrogant as to think that only we exist and that only we could solve such problems?

This is a subject discussed at great length and in great detail by scientists, astronomers, biologists, and science fiction writers. Some, like astronomer Carl Sagan (1973, 1980), believe that the odds are good that the universe is teeming with life. Given the hundreds of billions of stars in our galaxy, and the hundreds of billions of galaxies in the known universe, what are the chances that ours is the only one that has evolved intelligent

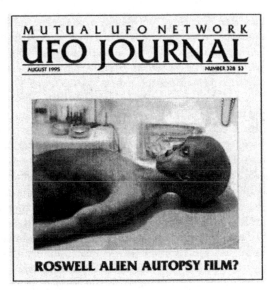

FIGURE 9:
Alien from alien autopsy film. [Courtesy Mutual UFO Network.]

sentients? Others, like cosmologist Frank Tipler (1981), are convinced that extraterrestrials do not exist because if they did they would be here by now. Given that there is nothing special about the timing of human evolution, it is fairly likely that if intelligent beings evolved elsewhere, at least half of them would be ahead of us in biological evolution, which should put them far, far ahead of us scientifically and technologically, which means they would have found Earth by now.

Some people claim that not only have aliens found Earth, they crash-landed near Roswell, New Mexico, in 1947, and we can see what they look like on film. On August 28, 1995, the Fox network aired what has come to be known as the "Roswell Incident," which featured footage of an autopsy of what appears to be an alien body (see figure 9). The footage came from Ray Santilli, a London-based video producer who claims to have come across the black-and-white film while he was searching the U.S. Army archives for footage of Elvis (who served eighteen months in the military) for a documentary on the singer. The individual who sold him the footage (reportedly for $100,000) remains anonymous, Santilli maintains, because it is illegal to sell U.S. government property. Santilli, in turn, sold use of the footage to Fox. The U.S. Air Force has stated that the wreckage at

Roswell came from a crashed top-secret surveillance balloon—"Project Mogul"—launched to monitor Soviet nuclear testing from the upper atmosphere. Given that the cold war was heating up in 1947, it is not surprising that at the time the Air Force was reluctant to discuss the crash, but this gave rise to decades of speculation by believers in UFOs, especially those with a bent for conspiracy theories. There are, however, numerous problems with the alien autopsy film as evidence of an alien encounter.

1. Santilli needs to give a significant sample of the original autopsy film to a credible institution equipped to date film footage. So far Kodak has been given a few inches of leader which could have come off of any film. If Santilli really wants to prove that the film was shot in 1947, why has he given Kodak only a small, entirely generic portion of the footage? Kodak routinely dates film for people who bring in old cameras.

2. According to the Fox documentary, the government ordered tiny coffins for the alien bodies. First of all, a bonfire would have been more efficient than burial if the government were intent on eradicating all traces of the aliens—no record of tiny coffins, no weird skeletons to explain later. Second, why would the government, no matter how paranoid, just bury the alien bodies a few days after the crash? As one of the most important discoveries in history, surely these bodies would be studied by experts from around the world for many years to come.

3. Given the number of people who were apparently involved in the discovery, isolation, transfer, handling, filming, autopsying, preservation, and burial of the bodies, there would have had to be a massive cover-up. How could the government have concealed from the public such a spectacular event? How do you keep all these people from talking?

4. In the Fox program, many people recalled that they were cautioned, threatened, and otherwise warned about talking or writing about the fact that some debris had been found. This is not unexpected, since we now know that a project involving the utmost secrecy was being carried out and that every effort was being made to keep it secret.

5. Can anyone seriously believe that arguably the most important event in human history was filmed using a hand-held Filmo camera, loaded with black-and-white film no less, and by a cameraman who was being jostled about so much that the camera was going in and out of focus?

6. We would not expect an alien from another planet (and thus another evolutionary sequence) to be humanoid in form. The enormous variety of life-forms here on Earth took many diverse shapes and configurations that might have displaced us, and might yet do so, but none are so nearly humanoid as this alleged alien from another planet. The chances against this happening are simply astronomical.

7. The alien in the film has six fingers and toes, yet the "original eye-witness accounts" recorded in 1947 reported aliens with four fingers and toes. Are we facing problems with the eyewitness accounts, problems with the film, problems with both, or *two* species of aliens?

8. The alien matches every detail called for by alien abductees, from short stature to bald head and large eyes. This look was created for a 1975 NBC movie called *The UFO Incident* and has been used by abductees ever since.

9. During the autopsy, the two guys in white suits show little interest in the organs. They make no attempt to measure or examine the organs and don't even turn them over. They just pull them out and plop them into a bowl, with no still-photographer or medical sketch artist present. Their suits are not radiation suits, and no radiation detectors or Geiger-Mueller counters are visible.

10. A vinyl alien would be easy to obtain from a prop warehouse, as would all the other items in the room.

11. Ed Uthman, a pathologist in Houston, Texas, made these observations (posted on the Internet, September 7, 1995):

> Any pathologist involved in such a case would be obsessed with documenting the findings. He would be systematically demonstrating findings every step of the way, such as showing how the joints worked, whether the eyelids closed, etc. He should be ordering the cameraman all over the place, but instead the cameraman was totally ignored, like he wasn't there at all. The pathologist acted more like an actor in front of a camera than someone who was cooperating in a photographic documentation session.
>
> The prosector used scissors like a tailor, not like a pathologist or surgeon. He held the scissors with thumb and forefinger, whereas pathologists and surgeons put the thumb in one scissors hole and the middle or ring finger in the other. The forefinger is used to steady the scissors further up toward the blades.
>
> The way the initial cuts in the skin were made was a little too Hollywood-like, too gingerly, like operating on a living patient. Autopsy cuts are deeper and faster.

12. Joachim Koch, a practicing surgeon in Germany who is a co-founder of the International Roswell Initiative, had this to say (posted on the Internet, September 12, 1995):

> If a preliminary autopsy in Roswell had been performed and the final dissection (in the Santilli film) was done in another place, then sutures placed during the first autopsy should have been visible during the second autopsy shown in the film, but they were not.
>
> Note the physical features of the "alien": extreme growth of the head, widespread eyes and deep eyesockets, a broad-based nose, increased growth of

the base of the skull, a crescent-shaped skin fold at the inner upper eyelid, mongoloid axis of the eyelids, no hair between the eyebrows, lowering of the outer ear, which is small, small lips, lower jaw underdeveloped, low birth weight, short length at birth, malformations of inner organs, unproportioned growth, and poly- and/or hexadactylism (six fingers and toes). This description is not that of an alien, but of a human being who suffers from "C-syndrome," or in the American medical literature, from "Opitz trigonocephaly syndrome." Only a few cases of C-syndrome have ever been described formally, and these few died very young.

It is interesting that this film, to date the best physical evidence ever presented for the alien encounter case, is discounted by most believers. Why? They, like the skeptics, suspect a hoax and don't want to hitch themselves to a soon-to-be-falling star. Yet if this is the best they've got, what does that say for this phenomenon? Unfortunately, the lack of physical evidence matters little to true believers. They have shared anecdotes and personal experiences, and for most this is good enough.

Encounters with Alien Abductees

In 1994 NBC began airing *The Other Side*, a New Age show that explored alien abduction claims as well as other mysteries, miracles, and unusual phenomena. I appeared numerous times on this show as the token skeptic, but most interesting for me was their two-part program on UFOs and alien abductions. The claims made by the alien abductees were quite remarkable indeed. They state that literally millions of people have been "beamed up" to alien spacecraft, some straight out of their bedrooms through walls and ceilings. One woman said the aliens took her eggs for use in a breeding experiment but could produce no evidence for how this was done. Another said that the aliens actually implanted a human-alien hybrid in her womb and that she gave birth to the child. Where is this child now? The aliens took it back, she explained. One man pulled up his pant leg to show me scars on his legs that he said were left by the aliens. They looked like normal scars to me. Another woman said the aliens had implanted a tracking device in her head, much as biologists do to track dolphins or birds. An MRI of her head proved negative. One man explained that the aliens took his sperm. I asked him how he knew that they took his sperm, since he had said he was asleep when he was abducted. He said he knew because he had had an orgasm. I responded, "Is it possible you simply had a wet dream?" He was not amused.

After the taping of this program, about a dozen of the "abductees" were going out to dinner. Since I tend to be a fairly friendly, nonconfrontational skeptic in these situations, disdaining the shouting so desired by talk-show producers, they invited me to join them. It was enlightening. I discovered that they were neither crazy nor ignorant, as one might suspect. They were perfectly sane, rational, intelligent folks who had in common an irrational experience. They were convinced of the reality of the experience—no rational explanation I could offer, from hallucinations to lucid dreams to false memories, could convince them otherwise. One man became teary-eyed while telling me how traumatic the abduction was for him. Another woman explained that the experience had cost her a happy marriage to a wealthy television producer. I thought, "What is wrong here? There isn't a shred of evidence that any of these claims is true, yet these are normal, rational folks whose lives have been deeply affected by these experiences."

In my opinion, the alien abduction phenomenon is the product of an unusual altered state of consciousness interpreted in a cultural context replete with films, television programs, and science fiction literature about aliens and UFOs. Add to this the fact that for the past four decades we have been exploring the solar system and searching for signs of extraterrestrial intelligence, and it is no wonder that people are seeing UFOs and experiencing alien encounters. Driven by mass media that revel in such tabloid-type stories, the alien abduction phenomenon is now in a positive feedback loop. The more people who have had these unusual mental experiences see and read about others who have interpreted similar incidents as abduction by aliens, the more likely it is that they will convert their own stories into their own alien abduction. The feedback loop was given a strong boost in late 1975 after millions watched NBC's *The UFO Incident*, a movie on Betty and Barney Hill's abduction dreams. The stereotypical alien with a large, bald head and big, elongated eyes, reported by so many abductees since 1975, was created by NBC artists for this program. The rate of information exchange took off as more and more alien abductions were reported on the news and recounted in popular books, newspapers, tabloids, and specialty publications dedicated solely to UFOs and alien abductions. As there seemed to be agreement on how the aliens looked and also on their preoccupation with human reproductive systems (usually women are sexually molested by the aliens), the feedback loop took off. Because of our fascination with the possibility of extraterrestrial life, and there is a real possibility that extraterrestrials might exist somewhere in the cosmos (a different question than their arrival here on Earth), this craze will probably wax and wane depending on what is hot in pop culture. Blockbuster films like *ET* and *Independence Day* and television shows like *Star Trek* and *The X-Files*, as well as best-selling books like Whitley Strieber's *Communion* and John Mack's *Abduction*, continue feeding the movement.

While dining with the abductees, I found out something very revealing: not one of them recalled being abducted immediately after the experience. In fact, for most of them, many years went by before they "remembered" the experience. How was this memory recalled? Under hypnosis. As we shall see in the next chapter, memories cannot simply be "recovered" like rewinding a videotape. Memory is a complex phenomenon involving distortions, deletions, additions, and sometimes complete fabrication. Psychologists call this *confabulation*—mixing fantasy with reality to such an extent that it is impossible to sort them out. Psychologist Elizabeth Loftus (Loftus and Ketcham 1994) has shown how easy it is to plant a false memory in a child's mind by merely repeating a suggestion until the child incorporates it as an actual memory. Similarly, Professor Alvin Lawson put students at California State University, Long Beach, into a hypnotic state and in their altered state told them over and over that they had been abducted by aliens. When asked to fill in the details of the abduction, the students elaborated in great detail, making it up as they went along in the story (in Sagan 1996). Every parent has stories about the fantasies their children create. My daughter once described to my wife a purple dragon we saw on our hike in the local hills that day.

True, not all abduction stories are recalled only under hypnosis, but almost all alien abductions occur late at night during sleep. In addition to normal fantasies and lucid dreams, there are rare mental states known as *hypnogogic hallucinations*, which occur soon after falling asleep, and *hypnopompic hallucinations*, which happen just before waking up. In these unusual states, subjects report a variety of experiences, including floating out of their bodies, feeling paralyzed, seeing loved ones who have passed away, witnessing ghosts and poltergeists, and, yes, being abducted by aliens. Psychologist Robert A. Baker presents as typical this subject's report: "I went to bed and went to sleep and then sometime near morning something woke me up. I opened my eyes and found myself wide awake but unable to move. There, standing at the foot of my bed was my mother, wearing her favorite dress—the one we buried her in" (1987/1988, p. 157). Baker also demonstrates that Whitley Strieber's encounter with aliens (one of the more famous in abduction lore) "is a classic, textbook description of a hypnopompic hallucination, complete with awakening from a sound sleep, the strong sense of reality and of being awake, the paralysis (due to the fact that the body's neural circuits keep our muscles relaxed and help preserve our sleep), and the encounter with strange beings" (p. 157).

Harvard psychiatrist John Mack, a Pulitzer Prize–winning author, gave the abduction movement a strong endorsement with his 1994 book, *Abduction: Human Encounters with Aliens.* Here at last was a mainstream scholar from a highly respectable institution lending credence (and his reputation) to the belief in the reality of these encounters. Mack was impressed by the commonalities of the stories told by abductees—the physical description of the aliens, the sexual abuse, the metallic probes, and so on. Yet I think we can expect consistencies in the stories since so many of the abductees go to the same hypnotist, read the same alien encounter books, watch the same science fiction movies, and in many cases even know one another and belong to "encounter" groups (in both senses of the word). Given the shared mental states and social contexts, it would be surprising if there was not a core set of characteristics of the abduction experience shared by the abductees. And what are we to do with the shared absence of convincing physical evidence?

Finally, the sexual component of alien abduction experiences demands comment. It is well known among anthropologists and biologists that humans are the most sexual of all primates, if not all mammals. Unlike most animals, when it comes to sex, humans are not constrained by biological rhythms and the cycle of the seasons. We like sex almost anytime or anywhere. We are stimulated by visual sexual cues, and sex is a significant component in advertising, films, television programs, and our culture in general. You might say we are obsessed with sex. Thus, the fact that alien abduction experiences often include a sexual encounter tells us more about humans than it does about aliens. As we shall see in the next chapter, women in the sixteenth and seventeenth centuries were often accused of (and even allegedly experienced or confessed to) having illicit sexual encounters with aliens—in this case the alien was usually Satan himself—and these women were burned as witches. In the nineteenth century, many people reported sexual encounters with ghosts and spirits at about the time that the spiritualism movement took off in England and America. And in the twentieth century, we have phenomena such as "Satanic ritual abuse," in which children and young adults are allegedly being sexually abused in cult rituals; "recovered memory syndrome," in which adult women and men are "recovering" memories of sexual abuse that allegedly occurred decades previously; and "facilitated communication," where autistic children are "communicating" through facilitators (teachers or parents) who hold the child's hand above a typewriter or computer keyboard reporting that they were sexually abused.

We can again apply Hume's maxim: is it more likely that demons, spirits, ghosts, and aliens have been and continue to sexually abuse humans or that humans are experiencing fantasies and interpreting them in the social context of their age and culture? I think it can reasonably be argued that such experiences are a very earthly phenomenon with a perfectly natural (albeit unusual) explanation. To me, the fact that humans have such experiences is at least as fascinating and mysterious as the possibility of the existence of extraterrestrial intelligence.

7

Epidemics of Accusations
Medieval and Modern Witch Crazes

n the small town of Máttoon, Illinois, a woman says that a stranger entered her bedroom late at night on Thursday, August 31, 1944, and anesthetized her legs with a spray gas. She reported the incident the next day, claiming she was temporarily paralyzed. The Saturday edition of the Mattoon *Daily Journal-Gazette* ran the headline "ANESTHETIC PROWLER ON LOOSE." In the days to come, several other cases were reported. The newspaper covered these new incidents under the headline "MAD ANESTHETIST STRIKES AGAIN." The perpetrator became known as the "Phantom Gasser of Mattoon." Soon cases were occurring all over Mattoon, the state police were brought in, husbands stood guard with loaded guns, and many firsthand sightings were recounted. In the course of thirteen days, a total of twenty-five cases were reported. After a fortnight, however, no one was caught, no chemical clues were discovered, the police spoke of "wild imaginations," and the newspapers began to characterize the story as a case of "mass hysteria" (see Johnson 1945; W. Smith 1994).

Where have we heard all this before? If this story sounds familiar, it might be because it has the same components as an alien abduction experience, only the paralysis is the work of a mad anesthetist rather than aliens. Strange things going bump in the night, interpreted in the context of the time and culture of the victims, whipped into a phenomenon through rumor and gossip—we are talking about modern versions of medieval witch crazes. Most people do not believe in witches anymore, and today no one is burned at the stake, yet the components of the early witch crazes are still alive in their many modern pseudoscientific descendants:

1. Victims tend to be women, the poor, the retarded, and others on the margins of society.

2. Sex or sexual abuse is typically involved.

3. Mere accusation of potential perpetrators makes them guilty.

4. Denial of guilt is regarded as further proof of guilt.

5. Once a claim of victimization becomes well known in a community, other similar claims suddenly appear.

6. The movement hits a critical peak of accusation, when virtually everyone is a potential suspect and almost no one is above suspicion.

7. Then the pendulum swings the other way. As the innocent begin to fight back against their accusers through legal and other means, the accusers sometimes become the accused and skeptics begin to demonstrate the falsity of the accusations.

8. Finally, the movement fades, the public loses interest, and proponents, while never completely disappearing, are shifted to the margins of belief.

So it went for the medieval witch crazes. So it will likely go for modern witch crazes such as the "Satanic panic" of the 1980s and the "recovered memory movement" of the 1990s. Is it really possible that thousands of Satanic cults have secretly infiltrated our society and that their members are torturing, mutilating, and sexually abusing tens of thousands of children and animals? No. Is it really possible that millions of adult women were sexually abused as children but have repressed all memory of the abuse? No. Like the alien abduction phenomenon, these are products of the mind, not reality. They are social follies and mental fantasies, driven by a curious phenomenon called the *feedback loop*.

A Witch Craze Feedback Loop

Why should there be such movements in the first place, and what makes these seemingly dissimilar movements play out in a similar manner? A helpful model comes from the emerging sciences of chaos and complexity theory. Many systems, including social systems like witch crazes, self-organize through *feedback loops*, in which outputs are connected to inputs, producing change in response to both (like a public-address system with feedback, or stock market booms and busts driven by flurries of buying and selling). The underlying mechanism driving a witch craze is the cycling of information through a closed system. Medieval witch crazes existed because the internal and external components of a feedback loop periodically occurred together, with deadly results. Internal components include the social control of one group of people by another, more powerful group, a prevalent feeling of loss of personal control and responsibility, and the need to place blame for misfortune elsewhere; external conditions include socioeconomic stresses, cultural and political crises, religious strife, and moral upheavals (see Macfarlane 1970; Trevor-Roper 1969). A conjuncture of such events and conditions can lead the system to self-organize, grow, reach a peak, and then collapse. A few claims of ritual abuse are fed into the system through word-of-mouth in the seventeenth century or the mass media in the twentieth. An individual is accused of being in league with the devil and denies the accusation. The denial serves as proof of

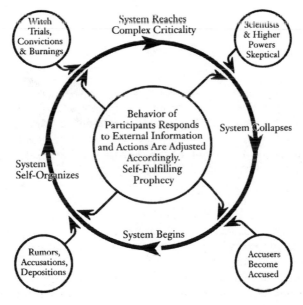

FIGURE 10:
Witch craze feedback loop.

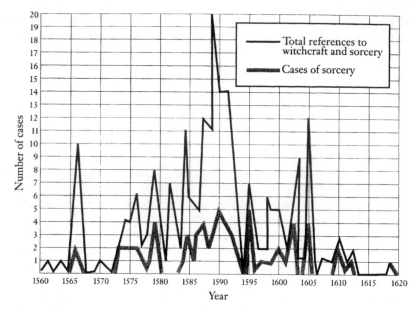

FIGURE 11:
Accusations of witchcraft at ecclesiastical courts, England, 1560–1620. [From Macfarlane 1970.]

guilt, as does silence or confession. Whether the defendant is being tried by the water test of the seventeenth century (if you float you are guilty, if you drown you are innocent) or in the court of public opinion today, accusation equals guilt (consider any well-publicized sexual abuse case). The feedback loop is now in place. The witch or Satanic ritual child abuser must name accomplices to the crime. The system grows in complexity as gossip or the media increase the amount and flow of information. Witch after witch is burned and abuser after abuser is jailed, until the system reaches criticality and finally collapses under changing social conditions and pressures (see figure 10). The "Phantom Gasser of Mattoon" is another classic example. The phenomenon self-organized, reached criticality, switched from a positive to a negative feedback loop, and collapsed— all in the span of two weeks.

Data supporting this model exist. For example, note in figure 11 the rise and fall of accusations of witchcraft brought before the ecclesiastical courts in England from 1560 to 1620, and trace through the various parts of figure 12 the pattern of accusations in the witch craze that began in 1645 in Manningtree, England. The density of accusation drives the feedback loop to self-organize and reach criticality.

Over the past century dozens of historians, sociologists, anthropologists, and theologians proffered theories to explain the medieval witch craze phenomenon. We can dismiss up-front the theological explanation that witches really existed and the church was simply reacting to a real threat. Belief in witches existed for centuries prior to the medieval witch craze without the church embarking on mass persecutions. Secular explanations are as varied as the writer's imagination would allow. Early in this historiography, Henry Lea (1888) speculated that the craze was caused by the active imaginations of theologians, coupled with the power of the ecclesiastical establishment. More recently, Marion Starkey (1963) and John Demos (1982) have offered psychoanalytic explanations. Alan Macfarlane (1970) used copious statistics to show that scapegoating was an important element of the craze, and Robin Briggs (1996) has recently reinforced this theory by showing how ordinary people used scapegoating as a means of resolving grievances. In one of the best books on the period, Keith Thomas (1971) argues that the craze was caused by the decline of magic and the rise of large-scale, formalized religion. H. C. E. Midelfort (1972) theorizes that it was caused by interpersonal conflict within and between various villages. Barbara Ehrenreich and Deirdre English (1973) correlated it with the suppression of midwives. Linnda Carporael (1976) attributed the craze in Salem to suggestible adolescents high on hallucinatory substances. More likely are the accounts of Wolfgang Lederer (1969), Joseph Klaits (1985), and Ann Barston (1994), which examine the hypothesis that the witch craze was a combination of misogyny and gender politics. Theories and books continue to be produced at a steady rate. Hans Sebald believes that this episode of medieval mass persecution "cannot be explained within a monocausal frame; rather the explanation most likely consists of a multivariable syndrome, in which important psychological and societal conditions are intermeshed" (1996, p. 817). I agree, but would add that these divers sociocultural theories can be taken to a deeper theoretical level by grafting them into the witch craze feedback loop. Theological imaginations, ecclesiastical power, scapegoating, the decline of magic, the rise of formal religion, interpersonal conflict, misogyny, gender politics, and possibly even psychedelic drugs were all, to lesser or greater degrees, components of the feedback loop. They all either fed into or out of the system, driving it forward.

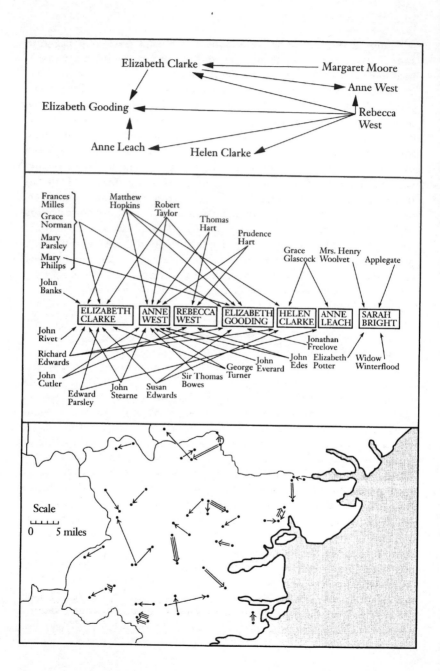

Hugh Trevor-Roper, in *The European Witch-Craze*, demonstrates how suspicions and accusations built upon one another as the scope and intensity of the feedback loop expanded. He provides this example from the county of Lorraine about the frequency of alleged witch meetings: "At first the interrogators . . . thought that they occurred only once a week, on Thursday; but, as always, the more evidence was pressed, the worse the conclusions that it yielded. Sabbats were found to take place on Monday, Wednesday, Friday, and Sunday, and soon Tuesday was found to be booked as a by-day. It was all very alarming and proved the need of ever greater vigilance by the spiritual police" (1969, p. 94). It is remarkable how quickly the feedback loop self-organizes into a full-blown witch craze, and interesting to discover what happens to skeptics who challenge the system. Trevor-Roper was appalled by what he read in the historical documents:

> To read these encyclopaedias of witchcraft is a horrible experience. Together they insist that every grotesque detail of demonology is true, that scepticism must be stifled, that sceptics and lawyers who defend witches are themselves witches, that all witches, "good" or "bad," must be burnt, that no excuse, no extenuation is allowable, that mere denunciation by one witch is sufficient evidence to burn another. All agree that witches are multiplying incredibly in Christendom, and that the reason for their increase is the indecent leniency of judges, the indecent immunity of Satan's accomplices, the sceptics. (p. 151)

What is especially curious about the medieval witch craze is that it occurred at the very time experimental science was gaining ground and popularity. This is curious because we often think that science displaces superstition and so one would expect belief in things like witches, demons, and spirits to have decreased as science grew. Not so. As modern examples show, believers in paranormal and other pseudoscientific phenomena try to wrap themselves in the mantle of science because science is a dominating force in our society but they still believe what they believe. Historically, as science grew in importance, the viability of all belief systems began to be directly attached to experimental evidence in favor of specific claims. Thus, scientists of the day found themselves investigating haunted houses

◄ **Figure 12:**
Witch craze that originated in Manningtree, England, 1645. (*top*) Accusations by suspected witches against other suspected witches; (*middle*) accusations against suspected witches (*central boxes*) by other villagers; (*bottom*) spread of craze—arrows point from village of the accused witch to village of the supposed victim. Modeled by the feedback loop of figure 10, these data show how a craze begins, spreads, and reaches criticality. [From Macfarlane 1970.]

and testing accused witches by using methods considered rigorous and scientific. Empirical data for the existence of witches would support belief in Satan which, in turn, would buttress belief in God. But the alliance between religion and science was uneasy. Atheism as a viable philosophical position was growing in popularity, and church authorities put themselves in a double-bind by looking to scientists and intellectuals to respond. As one observer at a seventeenth-century witch trial of an Englishman named Mr. Darrell noted, "Atheists abound in these days and witchcraft is called into question. If neither possession nor witchcraft [exists], why should we think that there are devils? If no devils, no God" (in Walker 1981, p. 71).

The Satanic Panic Witch Craze

The best modern example of a witch craze would have to be the "Satanic panic" of the 1980s. Thousands of Satanic cults were believed to be operating in secrecy throughout America, sacrificing and mutilating animals, sexually abusing children, and practicing Satanic rituals. In *The Satanism Scare*, James Richardson, Joel Best, and David Bromley argue persuasively that public discourse about sexual abuse, Satanism, serial murders, or child pornography is a barometer of larger social fears and anxieties. The Satanic panic was an instance of moral panic, where "a condition, episode, person or group of persons emerges to become defined as a threat to societal values and interests; its nature is presented in a stylized and stereotypical fashion by the mass media; the moral barricades are manned by editors, bishops, politicians and other right-thinking people; socially accredited experts pronounce their diagnoses and solutions; ways of coping are evolved or resorted to; the condition then disappears, submerges or deteriorates" (1991, p. 23). Such events are used as weapons "for various political groups in their campaigns" when someone stands to gain and someone stands to lose by the focus on such events and their outcome. According to these authors, the evidence for widespread Satanic cults, witches' covens, and ritualistic child abuse and animal killings is virtually nonexistent. Sure, there is a handful of colorful figures who are interviewed on talk shows or dress in black and burn incense or introduce late-night movies in a push-up bra, but these are hardly the brutal criminals supposedly disrupting society and corrupting the morals of humanity. Who says they are?

The key is in the answer to the question, "Who needs Satanic cults?" "Talk-show hosts, book publishers, anti-cult groups, fundamentalists, and certain religious groups" is the reply. All thrive from such claims. "Long a staple topic for religious broadcasters and 'trash TV' talk shows," the authors note, "satanism has crept into network news programs and prime-time programming, with news stories, documentaries, and made-for-TV movies about satanic cults. Growing numbers of police officers, child protection workers, and other public officials attend workshops supported by tax dollars to receive formal training in combating the satanist menace" (p. 3). Here is the information exchange fueling the feedback loop and driving the witch craze toward higher levels of complexity.

The motive, like the movement, is repeated historically from century to century as a shunt for personal responsibility—fob off your problems on the nearest enemy, the more evil the better. And who fits the bill better than Satan himself, along with his female co-conspirator, the witch? As sociologist Kai Erikson observed, "Perhaps no other form of crime in history has been a better index to social disruption and change, for outbreaks of witchcraft mania have generally taken place in societies which are experiencing a shift of religious focus—societies, we would say, confronting a relocation of boundaries" (1966, p. 153) Indeed, of the sixteenth- and seventeenth-century witch crazes, anthropologist Marvin Harris noted, "The principal result of the witch-hunt system was that the poor came to believe that they were being victimized by witches and devils instead of princes and popes. Did your roof leak, your cow abort, your oats wither, your wine go sour, your head ache, your baby die? It was the work of the witches. Preoccupied with the fantastic activities of these demons, the distraught, alienated, pauperized masses blamed the rampant Devil instead of the corrupt clergy and the rapacious nobility" (1974, p. 205).

Jeffrey Victor's book, *Satanic Panic: The Creation of a Contemporary Legend* (1993), is the best analysis to date on the subject, and the subtitle summarizes his thesis about the phenomenon. Victor traces the development of the Satanic cult legend by comparing it to other rumor-driven panics and mass hysterias and showing how individuals get caught up in such phenomena. Participation involves a variety of psychological factors and social forces, combined with information input from modern as well as historical sources. In the 1970s, there were rumors about dangerous religious cults, cattle mutilations, and Satanic cult ritual animal sacrifices; in the 1980s, we were bombarded by books, articles, and television programs about multiple personality disorder, Procter & Gamble's "Satanic" logo, ritual child abuse, the McMartin Preschool case, and devil worship; and the 1990s have given us the ritual child abuse scare in England, reports that the Mormon Church was infiltrated by secret Satanists who sexually abuse children in rituals, and the Satanic ritual abuse scare in San Diego (see Victor 1993, pp. 24–25). These cases, and many others, drove the feedback loop forward. But now it is reversing. In 1994, for example, Britain's Ministry of Health conducted a study that found no independent corroboration for eyewitness claims of Satanic abuse of children in Britain. According to Jean La Fontaine, a professor from the London School of Economics, "The alleged disclosures of satanic abuse by younger children were influenced by adults. A small minority involved children pressured or coached by their mothers." What was the driving force? Evangelical Christians, suggests La Fontaine: "The evangelical Christian campaign against new religious movements has been a powerful influence encouraging the identification of satanic abuse" (in Shermer 1994, p. 21).

The Recovered Memory Movement as a Witch Craze

A frightening parallel to the medieval witch crazes is what has come to be known as the "recovered memory movement." Recovered memories are alleged memories of childhood sexual abuse repressed by the victims but recalled decades later through use of special therapeutic techniques, including suggestive questioning, hypnosis, hypnotic age-regression, visualization, sodium amytal ("truth serum") injections, and dream interpretation. What makes this movement a feedback loop is the accelerating rate of information exchange. The therapist usually has the client read books about recovered memories, watch videotapes of talk shows on recovered memories, and participate in group counseling with other women with recovered memories. Absent at the beginning of therapy, memories of childhood sexual abuse are soon created through weeks and months of applying the special therapeutic techniques. Then names are named—father, mother, grandfather, uncle, brother, friends of father, and so on. Next is confrontation with the accused, who inevitably denies the charges, and termination of all relations with the accused. Shattered families are the result (see Hochman 1993).

Experts on both sides of this issue estimate that at least one million people have "recovered" memories of sexual abuse since 1988 alone, and this does not count those who really were sexually abused and never forgot it (Crews et al. 1995; Loftus and Ketcham 1994; Pendergrast 1995). Writer Richard Webster, in his fascinating *Why Freud Was Wrong* (1995), traces the movement to a group of psychotherapists in the Boston area who in the 1980s, after reading psychiatrist Judith Herman's 1981 book, *Father-Daughter Incest*, formed therapy groups for incest survivors. Since sexual abuse is a real and tragic phenomenon, this was an important step in bringing it to the attention of society. Unfortunately, the idea that the subconscious is the keeper of repressed memories was also proffered, based on Herman's description of one woman whose "previously repressed memories" of sexual abuse were reconstructed in therapy. In the beginning, membership mostly consisted of those who had always remembered their abuse. But gradually, Webster notes, the process of therapeutic memory reconstruction entered the sessions.

> In their pursuit of the hidden memories which supposedly accounted for the symptoms of these women, therapists sometimes used a form of time-limited group therapy. At the beginning of the ten or twelve weekly sessions, patients would be encouraged to set themselves goals. For many patients without memories of incest the goal was to recover such memories. Some of them actually defined their goal by saying "I just want to be in the group and feel I belong." After the fifth session the therapist would remind the group that they had reached the middle of their therapy, with the clear implication that time was running out. As pressure was increased in this way women with no memories would often begin to see images of sexual abuse involving father or other adults, and these images would then be construed as memories or "flashbacks." (1995, p. 519)

The feedback loop for the movement now began to self-organize, encouraged by psychotherapist Jeffrey Masson's 1984 book, *The Assault on Truth*, in which he rejected Freud's claim that childhood sexual abuse was fantasy and argued that Freud's initial position—that the sexual abuse so often recounted by his patients was actual, rampant, and responsible for adult women's neuroses—was the correct one. The movement became a full-blown witch craze when Ellen Bass and Laura Davis published *The Courage to Heal: A Guide for Women Survivors of Child Sexual Abuse* in 1988. One of its conclusions was "If you think you were abused and your life shows the symptoms, then you were" (p. 22). The book sold more than 750,000 copies and triggered a recovered memory industry that involved dozens of similar books, talk-show programs, and magazine and newspaper stories.

The controversy over recovered versus false memories still rages among psychologists, psychiatrists, lawyers, the media, and the general public. Because childhood sexual abuse does happen, and probably more frequently than any of us like to think, much is at stake when accusations

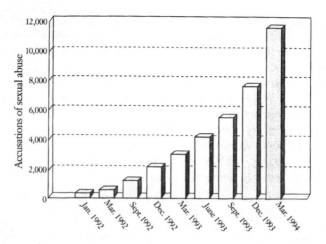

FIGURE 13:
Registered accusations of sexual abuse against parents, March 1992–March 1994.
[Courtesy False Memory Syndrome Foundation.]

made by the alleged victims themselves are discounted. But what we appear to be experiencing with the recovered memory movement is not an epidemic of childhood sexual abuse but an epidemic of accusations (see figure 13). It's a witch craze, not a sex craze. The supposed numbers alone should make us skeptical. Bass and Davis and others estimate that as many as one-third to one-half of all women were sexually abused as children. Using the conservative percentage, this means that in America alone 42.9 million women were sexually abused. Since they have to be abused by someone, this means about 42.9 million men are sex offenders, bringing us to a total of 85.8 million Americans. Additionally, many of these cases allegedly involve mothers who consent and friends and relatives who participate. This would push the figure to over 100 million Americans (about 38 percent of the entire population) involved in sexual abuse. Impossible. Impossible even if we cut that estimate in half. Something else is going on here.

This movement is made all the scarier by the fact that not only can anyone be accused, the consequences are extreme—incarceration. Many men and a number of women have been sent to jail, and some are still sitting there, after being convicted of sexual abuse on nothing more than a recovered memory. Given what is at stake, we must proceed with extreme caution. Fortunately, the tide seems to be turning in favor of the recovered memory movement being relegated to a bad chapter in the history of psychiatry. In 1994 Gary Ramona, father of his accuser, Holly Ramona, won his suit against her two therapists, Marche Isabella and Dr. Richard Rose, who had helped Holly "remember" such events as her father forcing her to perform oral sex on the family dog. The jury awarded Gary Ramona $500,000 of the $8 million he sought mainly because he had lost his $400,000-a-year job at the Robert Mondavi winery as a result of the fiasco.

Not only are the accused taking action but accusers are suing their therapists for planting false memories. And they are winning. Laura Pasley (1993), who once believed she was a victim of sexual abuse during her childhood, has since recanted her recovered memory, sued and won a settlement from her therapist, and her story has made the rounds in the mass media. Many other women are now reversing their original claims and filing lawsuits against their therapists. These women have become known as "retractors," and there is now even a therapist retractor (Pendergrast 1996). Lawyers are helping to reverse the feedback loop by holding therapists accountable through the legal system. The positive feedback loop is now becoming a negative one, and thanks to people like Pasley and organizations like the False Memory Syndrome Foundation, the direction of information exchange is reversing.

The reversal of the feedback loop was given another boost in October 1995, when a six-member jury in Ramsey County, Minnesota, awarded $2.7 million to Vynnette Hamanne and her husband after a six-week trial about charges that Hamanne's St. Paul psychiatrist, Dr. Diane Bay Humenansky, planted false memories of childhood sexual abuse. Hamanne went to Humenansky in 1988 with general anxiety and no memories whatsoever of childhood sexual abuse. After a year of therapy with Humenansky, however, Hamanne was diagnosed with multiple personality disorder— Humenansky "discovered" no less than 100 different personalities. What had caused Hamanne to become so many different people? According to Humenansky, Hamanne was sexually abused by her mother, father, grandmother, uncles, neighbors, and many others. Because of the trauma, Hamanne allegedly repressed the memories. Through therapy, Humenansky reconstructed a past for Hamanne that even included Satanic ritual abuse featuring dead babies being served as meals "buffet style." The jury didn't buy it. Neither did another jury, which on January 24, 1996, awarded another one of Humenansky's clients, E. Carlson, $2.5 million (Grinfeld 1995, p. 1).

Finally, one of the most famous cases involving repressed memories was recently dismissed and the accused released from jail. In 1989 George

Franklin's daughter, Eileen Franklin-Lipsker, told police that her father had killed her childhood friend Susan Nason in 1969. Her evidence? A twenty-year-old recovered memory upon which (and without further evidence) Franklin was found guilty of first-degree murder and sentenced to life in prison in January 1991. Franklin-Lipsker claimed that the memory of the murder returned to her while she was playing with her own daughter, who was close to the age of her murdered childhood friend. But in April 1995, U.S. District Court Judge Lowell Jensen ruled that Franklin had not received a fair trial because the original judge refused to let the defense introduce newspaper articles about the murder that could have provided Franklin-Lipsker with the details of the crime. In other words, her memory may have been constructed, not recovered. Additionally, Franklin-Lipsker's sister, Janice Franklin, in sworn testimony revealed that she and her sister had been hypnotized before their father's trial to "enhance" their memories. The final straw was when Franklin-Lipsker told investigators that she remembered her father committing two more murders but investigators were unable to link Franklin to either of them. One of the memories was so general that they could not even find an actual murder to go with it. In the other, Franklin allegedly raped and murdered an eighteen-year-old girl in 1976, but investigators placed Franklin at a union meeting at the time of the murder, and DNA and semen tests confirmed Franklin's innocence. Franklin's wife, Leah, who testified against him in the 1990 trial, has now recanted and says she no longer believes in the concept of repressed memories. Franklin's attorney concluded, "George has been in prison or jail for six years, seven months, and four days. It is an absolute travesty and a tragedy. This has been a Kafkaesque experience for him" (Curtius 1996). Indeed, the entire recovered memory movement is a Kafkaesque experience.

The parallels with Trevor-Roper's description of how a medieval witch craze worked can be eerie. Consider the case of East Wenatchee, Washington, in 1995. Detective Robert Perez, a sex-crimes investigator who took as his mission the rescue of the children of his city from what he believed was an epidemic of sexual abuse. Perez accused, charged, convicted, and terrorized citizens of this rural community with literally unbelievable claims. One woman was charged with over 3,200 acts of sexual abuse. One elderly gentleman was charged with having sexual intercourse twelve times in one day, which he admitted was impossible even when he was a teenager. And who were the accused? As in the medieval witch crazes, they were mainly poor men and women unable to hire adequate legal counsel. And who was doing the accusing? Young girls with active imaginations who had spent a lot of time with Detective Perez. And who was Perez? According to a police department evaluation, Perez had a history of petty crime and domestic strife, and it described him as "pompous," with an "arrogant approach." The report also stated that Perez appeared "to pick out people and target them." Soon after he was hired, Perez began interrogating vulnerable, dysfunctional girls without their parents being present. Not surprising, he did not tape the interviews; instead, he wrote out statements of accusation for the girls, who then signed them, usually after hours of relentless questioning (Carlson 1995, pp. 89–90).

While no one was burned in East Wenatchee, these young girls (the most prolific accuser was ten years old), because of Perez's influence and powers as a police officer, put more than twenty adults in jail. Over half of the incarcerated were poor women. Interestingly, anyone who hired a private attorney was not imprisoned. The message was clear—fight back. In the case of the ten-year-old accuser, Perez pulled her out of school, questioned her for four hours, then threatened to arrest her mom unless she admitted to being the victim of sex orgies that included her mom. "You have ten minutes to tell the truth," Perez insisted, promising that he would let her go home if she did. The girl signed the paper and Perez promptly arrested and jailed the mother. The girl did not see her mother again for six months. When the mother finally hired a lawyer, all 168 charges were dropped. East Wenatchee was firmly locked in a witch craze feedback loop that reached criticality when this epidemic of accusation was reported in the mass media (including a one-hour special on ABC and a *Time* magazine article). Now that Perez has been exposed, the accused are turning on him, the girls are retracting their accusations, lawsuits are being filed by the victims and their destroyed families, and the feedback loop has reversed itself.

The troubling aspect of this particular craze and of the sexual abuse hysteria sweeping across America over the last few years is that some genuine sex offenders may well go free in the inevitable backlash against the panic. Childhood sexual abuse is real. Now that it has been turned into a witch craze, it may be some time before society finds its balance in dealing with it.

8

The Unlikeliest Cult

Ayn Rand, Objectivism, and the Cult of Personality

Accroding to psychoanalysts, *projection* is the process of attributing one's own ideas, feelings, or attitudes to other people or objects—the guilt-laden adulterer accuses his spouse of adultery, the homophobe actually harbors latent homosexual tendencies. A subtle form of projection is at work when fundamentalists make the accusation that secular humanism and evolution are "religions" or announce that skeptics are themselves a cult and that reason and science have cultic properties, a claim that sounds absurd given that a cult is by definition 180 degrees out of phase with reason. And while it should be obvious to the reader by now that I am strongly pro-science and pro-reason, a recent historical phenomenon has convinced me that the seductiveness of facts, theory, evidence, and logic may mask some flaws in the system. The phenomenon provides a lesson about what happens when a truth becomes more important than the *search* for truth, when the final results of inquiry become more important than the *process* of inquiry, when reason leads to so absolute a certainty about one's beliefs that anyone who is not for them is anathematized as against them, and when supposedly intellectual inquiry becomes the basis of a personality cult.

The story begins in the United States in 1943 when an obscure Russian immigrant published her first successful novel after two consecutive failures. It was not an instant success. In fact, the reviews were harsh and initial sales sluggish. But slowly a following grew around the novel, not because it was well written (which it wasn't) but because of the power of its ideas. Word of mouth became its most effective marketing tool, and the author began to develop a large following. The initial print-run of 7,500 copies was followed by print-runs in multiples of 5,000 and 10,000 until by 1950 a half-million copies were circulating in the country.

The book was *The Fountainhead*, and the author was Ayn Rand. Her commercial success allowed her the time and freedom to write her magnum opus, *Atlas Shrugged*, published in 1957. *Atlas Shrugged* is a murder mystery about the murder not of a human body but of a human spirit. It is a sweeping story of a man who said he would stop the ideological motor of the world. When he did, there was a panoramic collapse of civilization, but its flame was kept burning by a handful of heroic individuals whose reason and morals directed both the collapse and the subsequent return of culture.

As with *The Fountainhead*, reviewers panned *Atlas Shrugged* with a sarcastic brutality that only seemed to reinforce followers' belief in the book, its author, and her ideas. And, also like *The Fountainhead*, sales of *Atlas Shrugged* have sputtered and clawed forward, to the point where the book has regularly sold over 300,000 copies a year. "In all my years of publishing," recalled Random House's head, Bennett Cerf, "I've never seen anything like it. To break through against such enormous opposition!" (in Branden 1986, p. 298). Such is the power of an individual hero . . . and a cult-like following.

What is it about Rand's philosophy as presented in these novels that so emotionally stimulates proponents and opponents alike? At a sales conference at Random House before *Atlas Shrugged* was published, a salesman asked Rand if she could summarize the essence of her philosophy, called *Objectivism*, while standing on one foot. She did so as follows (Rand 1962, p. 35):

1. *Metaphysics:* Objective Reality
2. *Epistemology:* Reason
3. *Ethics:* Self-interest
4. *Politics:* Capitalism

In other words, reality exists independent of human thought. Reason is the only viable method for understanding reality. Every human should seek personal happiness and exist for his own sake, and no one should sacrifice himself for or be sacrificed by others. And laissez-faire capitalism is the political-economic system in which the first three flourish best. This combination, said Rand, allows people to "deal with one another, not as victims and executioners, nor as masters and slaves, but as traders, by free, voluntary exchange to mutual benefit." This is not to say, however, that "anything goes." In these free exchanges, "no man may initiate the use of physical force against others" (Rand 1962, p. 1). Ringing through Rand's works is the philosophy of individualism, personal responsibility, the power of reason, and the importance of morality. One should think for oneself and never allow any authority to dictate truth, especially the authority of government, religion, and other such groups. Those who use reason to act in the highest moral fashion, and who never demand favors or handouts, are much more likely to find success and happiness than the irrational and unreasonable. Objectivism is the ultimate philosophy of unsullied reason and unadulterated individualism, as expressed by Rand through the primary character in *Atlas Shrugged*, John Galt:

> Man cannot survive except by gaining knowledge, and reason is his only means to gain it. Reason is the faculty that perceives, identifies and integrates the material provided by his senses. The task of his senses is to give him the evidence of existence, but the task of identifying it belongs to his reason, his senses tell him only that something is, but what it is must be learned by his mind. (1957, p. 1016)
>
> In the name of the best within you, do not sacrifice this world to those who are its worst. In the name of the values that keep you alive, do not let your vision of man be distorted by the ugly, the cowardly, the mindless in those who have never achieved his title. Do not lose your knowledge that man's proper estate is an upright posture, an intransigent mind and a step that travels unlimited roads. Do not let your fire go out, spark by irreplace-able spark, in the hopeless swamps of the approximate, the not-quite, the not-yet, the not-at-all. Do not let the hero in your soul perish, in lonely frus-tration for the life you deserved, but have never been able to reach. Check your road and the nature of your battle. The world you desired can be won, it exists, it is real, it is possible, it's yours. (1957, p. 1069)

How could such a highly individualistic philosophy become the basis of a cult, an organization that thrives on group thinking, intolerance of dissent, and the power of the leader? The last thing a cult leader wants is for follow-ers to think for themselves and exist as individuals apart from the group.

The 1960s were years of anti-establishment, anti-government, find-yourself individualism. Rand's philosophy exploded across the nation, par-ticularly on college campuses. *Atlas Shrugged* became *the* book to read. Though it is 1,168 pages long, readers devoured the characters, plot, and philosophy. The book stirred emotions and provoked action. Ayn Rand clubs were founded at hundreds of colleges. Professors taught courses on the philosophy of Objectivism and the literary works of Rand. Rand's inner circle of friends grew, and one of this circle, Nathaniel Branden, founded the Nathaniel Branden Institute (NBI) in 1958, which sponsored lectures and courses on Objectivism, first in New York and then nationally.

As Rand's popularity shot skyward, so too did confidence in her philosophy, both Rand's and her followers'. Thousands of people attended classes, thousands of letters poured into the offices of the NBI, and millions of books were sold. By 1948, *The Fountainhead* had been made into a successful film starring Gary Cooper and Patricia Neal, and the movie rights for *Atlas Shrugged* were being negotiated. Rand's ascent to power and influence was nothing short of miraculous. Readers of her novels, especially *Atlas Shrugged*, told Rand they had changed their lives and their way of thinking. Their comments include (Branden 1986, pp. 407–415 passim):

- A twenty-four-year-old "traditional housewife" (her own label) read *Atlas Shrugged* and said, "Dagny Taggart [the book's principle heroine] was an inspiration to me; she is a great feminist role model. Ayn Rand's works gave me the courage to be and to do what I had dreamed of."

- A law school graduate said of Objectivism, "Dealing with Ayn Rand was like taking a post-doctoral course in mental functioning. The universe she created in her work holds out hope, and appeals to the best in man. Her lucidity and brilliance was a light so strong I don't think anything will ever be able to put it out."

- A philosophy professor concluded, "Ayn Rand was one of the most original thinkers I have ever met. There is no escape from facing the issues she raised. At a time in my life when I thought I had learned at least the essentials of most philosophical views, being confronted with her . . . suddenly changed the entire direction of my intellectual life, and placed every other thinker in a new perspective."

The November 20, 1991 issue of *Library of Congress News* reported the results of a survey conducted by the Library of Congress and the Book of the Month Club of readers' "lifetime reading habits," indicating that *Atlas Shrugged* was ranked second only to the *Bible* in its significance to their lives. But to those in the inner circle surrounding and protecting Rand (in a fit of irony, they named themselves "the Collective"), their leader soon was more than just extremely influential—she was venerated. Her seemingly omniscient ideas were inerrant. The power of her personality made her

so persuasive that no one dared to challenge her. And Objectivism, since
it was derived through pure reason, revealed final Truth and dictated
absolute morality.

The cultic flaw in Rand's philosophy of Objectivism is not its use of
reason, emphasis on individuality, view that humans ought to be motivated
by rational self-interest, or conviction that capitalism is the ideal system.
The fallacy in Objectivism is its belief that absolute knowledge and final
Truth are attainable through reason, and therefore that there are absolutes
of right and wrong knowledge and of moral and immoral thought and
action. For Objectivists, once a principle has been discovered by (the
Objectivists' version of) reason to be True, the discussion is at an end. If
you disagree with the principle, then your reasoning is flawed. If your reasoning is flawed, it can be corrected, but if you don't correct your reasoning (i.e., learn to accept the principle), you are flawed and do not belong in
the group. Excommunication is the final solution for such unreformed
heretics.

One of those closest to Rand was Nathaniel Branden, a young philosophy student who joined the Collective in the early days, before *Atlas
Shrugged* was published. In his autobiographical memoirs, entitled *Judgment
Day*, he recalled, "There were implicit premises in our world to which
everyone in our circle subscribed, and which we transmitted to our students
at NBI." Incredibly, and here is where a philosophical movement mutated
into a cult of personality, their creed became, in Nathaniel Branden's words:

- Ayn Rand is the greatest human being who has ever lived.

- *Atlas Shrugged* is the greatest human achievement in the history
  of the world.

- Ayn Rand, by virtue of her philosophical genius, is the supreme
  arbiter in any issue pertaining to what is rational, moral, or
  appropriate to man's life on earth.

- Once one is acquainted with Ayn Rand and/or her work, the measure of one's virtue is intrinsically tied to the position one takes
  regarding her and/or it.

- No one can be a good Objectivist who does not admire what Ayn
  Rand admires and condemn what Ayn Rand condemns.

- No one can be a fully consistent individualist who disagrees with Ayn
  Rand on any fundamental issue.

- Since Ayn Rand has designated Nathaniel Branden as her "intellectual
  heir," and has repeatedly proclaimed him to be an ideal exponent of her
  philosophy, he is to be accorded only marginally less reverence than Ayn
  Rand herself.

- But it is best not to say most of these things explicitly (excepting, perhaps, the first two items). One must always maintain that one arrives at one's beliefs solely by reason. (1989, pp. 255–256)

Rand and her followers were, in their own time, accused of being a cult, a charge that, of course, they denied. "My following is not a cult. I am not a cult figure," Rand once told an interviewer. Barbara Branden, in her biography, *The Passion of Ayn Rand*, stated, "Although the Objectivist movement clearly had many of the trappings of a cult—the aggrandizement of the person of Ayn Rand, the too ready acceptance of her personal opinions on a host of subjects, the incessant moralizing—it is nevertheless significant that the fundamental attraction of Objectivism . . . was the precise opposite of religious worship" (1986, p. 371). And Nathaniel Branden addressed the issue this way: "We were not a cult in the literal, dictionary sense of the word, but certainly there was a cultish aspect to our world. We were a group organized around a powerful and charismatic leader, whose members judged one another's character chiefly by loyalty to that leader and to her ideas" (1989, p. 256).

But when you leave the "religious" component out of the definition of *cult*, thus broadening the word's usage, it becomes clear that Objectivism was (and is) a type of cult—a cult of personality—as are many other, nonreligious groups. A cult is characterized by

Veneration of the leader: Glorification of the leader to the point of virtual sainthood or divinity.

Inerrancy of the leader: Belief that the leader cannot be wrong.

Omniscience of the leader: Acceptance of the leader's beliefs and pronouncements on all subjects, from the philosophical to the trivial.

Persuasive techniques: Methods, from benign to coercive, used to recruit new followers and reinforce current beliefs.

Hidden agendas: The true nature of the group's beliefs and plans is obscured from or not fully disclosed to potential recruits and the general public.

Deceit: Recruits and followers are not told everything they should know about the leader and the group's inner circle, and particularly disconcerting flaws or potentially embarrassing events or circumstances are covered up.

Financial and/or sexual exploitation: Recruits and followers are persuaded to invest money and other assets in the group, and the leader may develop sexual relations with one or more of the followers.

Absolute truth: Belief that the leader and/or the group has discovered final knowledge on any number of subjects.

Absolute morality: Belief that the leader and/or the group has developed a system of right and wrong thought and action applicable to members and nonmembers alike. Those who strictly follow the moral code become and remain members; those who do not are dismissed or punished.

The ultimate statement of Rand's moral absolutism heads the title page of Nathaniel Branden's book. Says Rand,

> The precept: "Judge not, that ye be not judged" . . . is an abdication of moral responsibility: it is a moral blank check one gives to others in exchange for a moral blank check one expects for oneself. There is no escape from the fact that men have to make choices; so long as men have to make choices, there is no escape from moral values; so long as moral values are at stake, no moral neutrality is possible. To abstain from condemning a torturer, is to become an accessory to the torture and murder of his victims. The moral principle to adopt . . . is: "Judge, and be prepared to be judged."

The absurd lengths to which such thinking can go are demonstrated by Rand's judgments on her followers for even the most trivial things. Rand had argued, for example, that musical taste could not be objectively defined, yet, as Barbara Branden observed, "if one of her young friends responded as she did to Rachmaninoff . . . she attached deep significance to their affinity." By contrast, Barbara tells of a friend of Rand's who remarked that he enjoyed the music of Richard Strauss: "When he left at the end of the evening, Ayn said, in a reaction becoming increasingly typical, 'Now I understand why he and I can never be real soul mates. The distance in our sense of life is too great.' Often, she did not wait until a friend had left to make such remarks" (1986, p. 268).

In both Barbara and Nathaniel Branden's assessments, we see all the characteristics of a cult. Deceit and sexual exploitation? In this case, exploitation may be too strong, but the act was present nonetheless, and deceit was rampant. In what has become the most scandalous (and now oft-told) story in the brief history of the Objectivist movement, starting in 1953 and lasting until 1958 (and on and off for another decade after), Ayn Rand and Nathaniel Branden, twenty-five years her junior, carried on a love affair and kept it secret from everyone *except* their respective spouses.

By their reckoning, the affair was ultimately "reasonable" since the two of them were, de facto, the two greatest intellects on the planet. "By the total logic of who we are—by the total logic of what love and sex mean—we *had* to love each other," Rand rationalized to Barbara Branden and her own husband, Frank O'Connor. "Whatever the two of you may be feeling I know your intelligence, I know you recognize the rationality of what we feel for each other, and that you hold no value higher than reason" (Branden 1986, p. 258). Amazingly, both spouses bought this line and agreed to allow Rand and Nathaniel an afternoon and evening of sex and love once a week. "And so," Barbara said later, "we all careened toward disaster."

The disaster came in 1968, when Rand found out that Nathaniel had not only fallen in love with yet another woman but begun an affair with her. Even though the affair between Rand and Nathaniel had long since dwindled, the master of the absolute moral double standard would not tolerate such a breach of conduct by anyone else. "Get that bastard down here," Rand screamed upon hearing the news, "or I'll drag him here myself!" Nathaniel, according to Barbara, slunk into Rand's apartment to face judgment day. "It's finished, your whole act!" she told him. "I'll tear down your facade as I built it up! I'll denounce you publicly, I'll destroy you as I created you! I don't even care what it does to me. You won't have the career I gave you, or the name, or the wealth, or the prestige. You'll have nothing." The barrage continued for several minutes until she pronounced her final curse: "If you have an ounce of morality left in you, an ounce of psychological health—you'll be impotent for the next twenty years!" (1986, pp. 345–347).

Rand followed up with a six-page open letter to her followers in which she explained that she had completely broken with the Brandens and extended the pattern of deceit through lies of omission: "About two months ago . . . Mr. Branden presented me with a written statement which was so irrational and so offensive to me that I had to break my personal association with him." Without so much as a hint of the nature of the offense, Rand continued, "About two months later Mrs. Branden suddenly confessed that Mr. Branden had been concealing from me certain ugly actions and irrational behavior in his private life, which was grossly contradictory to Objectivist morality." Nathaniel's second affair was judged immoral, his first was not. This excommunication was followed by a barrage from NBI's associate lecturers, fired in complete ignorance of what really happened, that sounds all too ecclesiastical: "Because Nathaniel Branden and Barbara Branden, in a series of actions, have betrayed fundamental principles of Objectivism, we condemn and repudiate these two persons irrevocably, and have terminated all association with them" (Branden 1986, pp. 353–354).

Confusion reigned in the Collective and among the rank-and-file. What were they to think about such a formidable condemnation for unnamed sins? The logical extreme of cultish thinking was articulated several months later. In the words of Barbara Branden, "A half-demented former student of NBI . . . raised the question of whether or not it would be morally appropriate to assassinate Nathaniel because of the suffering he had caused Ayn; the man concluded that it should not be done on practical grounds, but would be morally legitimate. Fortunately, he was shouted down at once by a group of appalled students" (1986, p. 356n).

It was the beginning of Rand's long decline and fall, of the slow loosening of her tight grip on the Collective. One by one, they sinned, the condemnations growing in ferocity as the transgressions became more minor. And, one by one, they left or were asked to leave. When Rand died in 1982, there remained only a handful of friends. Today, the designated executor of her estate, Leonard Peikoff, carries on the cause at the Center for the Advancement of Objectivism, the southern California–based Ayn Rand Institute. While the cultic qualities of the group sabotaged the inner circle, there remained (and remains) a huge following of those who ignore the indiscretions, infidelities, and moral inconsistencies of the founder and focus instead on the positive aspects of her philosophy. There is much in it to admire, if you do not have to accept the whole package.

This analysis, then, suggests two important caveats about cults, skepticism, and reason. One, *criticism of the founder or followers of a philosophy does not, by itself, constitute a negation of any part of the philosophy.* The fact that some religious sects have been some of the worst violators of their own moral codes does not mean that such ethical axioms as "Thou shalt not murder" or "Do unto others as you would have them do unto you" are negated. The components of a philosophy must stand or fall on their own internal consistency or empirical support, regardless of the founder's or followers' personality quirks or moral inconsistencies. By most accounts Newton was a cantankerous and relatively unpleasant person to be around. This fact has nothing at all to do with the truth or falsity of his principles of natural philosophy. When founders or adherents proffer moral principles, as in the case of Rand, this caveat is more difficult to apply because one would hope that they would live by their own standards, but it is true nonetheless. Two, *criticism of part of a philosophy does not gainsay the whole.* Likewise, one may reject some parts of the Christian philosophy of moral behavior while embracing other parts. I might, for example, attempt to treat others as I would have them treat me but at the same time renounce the belief that women should remain silent in church and be obedient to their husbands. One may disavow Rand's absolute morality, while accepting her metaphysics of objective reality, her epistemology of reason, and her political philosophy of capitalism (though Objectivists would say they all follow inexorably from her metaphysics).

Rand critics come from all political positions—left, right, and center. Professional novelists generally disdain her style. Professional philosophers generally refuse to take her work seriously (both because she wrote for popular audiences and because her work is not considered a complete philosophy). There are more Rand critics than followers, although some of them have attacked *Atlas Shrugged* without reading it and rejected Objectivism without knowing anything about it. The conservative intellectual William F. Buckley, Jr., spoke of the "desiccated philosophy" and tone of "over-riding arrogance" of *Atlas Shrugged* and derided the "essential aridity of Miss Rand's philosophy," yet later confessed, "I never read the book. When I read the review of it and saw the length of the book, I never picked it up" (Branden 1986, p. 298).

I have read *Atlas Shrugged*, as well as *The Fountainhead* and all of Rand's nonfiction works. I accept much of Rand's philosophy, but not all of it. Certainly the commitment to reason is admirable (although clearly this is a philosophy, not a science); wouldn't most of us on the face of it, agree that individuals need to take personal responsibility for their actions? The great flaw in her philosophy is the belief that morals can be held to some absolute standard or criteria. This is not scientifically tenable. Morals do not exist in nature and thus cannot be discovered. In nature there are only actions—physical actions, biological actions, human actions. Humans act to increase their happiness, however they personally define it. Their actions become moral or immoral only when someone else judges them as such. Thus, morality is strictly a human creation, subject to all sorts of cultural influences and social constructions, just as other human creations are. Since virtually every person and every group claims they know what constitutes right versus wrong human action, and since virtually all of these moralities differ from all others to a greater or lesser extent, reason alone tells us they cannot all be correct. Just as there is no absolute right type of human music, there is no absolute right type of human action. The broad range of human action is a rich continuum that precludes pigeonholing into the unambiguous rights and wrongs that political laws and moral codes tend to require.

Does this mean that all human actions are morally equal? Of course not, any more than all human music is equal. We create hierarchies of what we like or dislike, desire or reject, and make judgments based on those standards. But the standards are themselves human creations and cannot be discovered in nature. One group prefers classical music over rock, and so judges Mozart to be superior to the Moody Blues. Similarly, one group prefers patriarchal dominance, and so judges male privilege to be morally honorable. Neither Mozart nor males are absolutely better, but only so when judged by a particular group's standards. Male ownership of females, for example, was once thought to be moral and is now thought immoral. The change happened not because we have discovered this as immoral but because our society (thanks primarily to the efforts of women) has realized that women should have rights and opportunities denied to them when they are in bondage to males. And having half of society happier raises the overall happiness of the group significantly.

Morality is relative to the moral frame of reference. As long as it is understood that morality is a human construction influenced by human cultures, one can be more tolerant of other human belief systems, and thus other humans. But as soon as a group sets itself up as the final moral arbiter of other people's actions, especially when its members believe they have discovered absolute standards of right and wrong, it marks the beginning of the end of tolerance, and thus reason and rationality. It is this characteristic more than any other that makes a cult, a religion, a nation, or any other group dangerous to individual freedom. Its absolutism was the biggest flaw in Ayn Rand's Objectivism, the unlikeliest cult in history. The historical development and ultimate destruction of her group and philosophy is the empirical evidence that documents this assessment.

What separates science from all other human activities (and morality has never been successfully placed on a scientific basis) is its commitment to the tentative nature of all its conclusions. There are no final answers in science, only varying degrees of probability. Even scientific "facts" are just conclusions confirmed to such an extent that it would be reasonable to offer temporary agreement, but that assent is never final. Science is not the affirmation of a set of beliefs but a process of inquiry aimed at building a testable body of knowledge constantly open to rejection or confirmation. In science, knowledge is fluid and certainty fleeting. That is at the heart of its limitations. It is also its greatest strength.

PART 3

EVOLUTION

AND

CREATIONISM

I have given the evidence to the best of my ability. We must, however, acknowledge, as it seems to me, that man with all his noble qualities, with sympathy which feels for the most debased, with benevolence which extends not only to other men but to the humblest living creature, with his god-like intellect which has penetrated into the movements and constitution of the solar system—with all these exalted powers—Man still bears in his bodily frame the indelible stamp of his lowly origin.

—Charles Darwin, *The Descent of Man*, 1871

9

In the Beginning

An Evening with Duane T. Gish

On the the evening of March 10, 1995, I entered a 400-seat lecture hall at the University of California, Los Angeles, five minutes before the debate was to begin. There wasn't an empty seat in the house, and the aisles were beginning to fill. Fortunately, I had a seat on the dais, as I was the latest in a long line of challengers to Duane T. Gish, creationist laureate and one of the directors of the Institute for Creation Research, the "research" arm of Christian Heritage College in San Diego. This was my first debate with a creationist. It was Gish's 300th-plus debate against an evolutionist. Las Vegas was not even giving odds. What could I say that hundreds of others had not already said?

In preparation, I read much of the creationist literature and reread the Bible. Twenty years ago, I had read the Bible very carefully as a theology student at Pepperdine University (before I switched to psychology), and, like many in the early 1970s, I had been a born-again Christian, taking up the cause with considerable enthusiasm, including "witnessing" to nonbelievers. Then, during my graduate training in experimental psychology and ethology (the study of animal behavior) at California State University, Fullerton, I ran into the brilliant but eccentric Bayard Brattstrom and the insightful and wise Meg White. Brattstrom was far more than one of the world's leading experts in behavioral herpetology (the study of reptilian behavior). He was well versed in the philosophical debates of modern biology and science, and regularly regaled us for hours with philosophical musings over beer and wine at the 301 Club (named for the nightclub's address) after the Tuesday night class. Somewhere between Brattstrom's 301 Club discussions of God and evolution and White's ethological explanations about the evolution of animal behavior, my Christian icthus (the fish with Greek symbols that Christians wore in the 1970s to publicly indicate their faith) got away, and with it my religion. Science became my belief system, and evolution my doctrine. Since that time the Bible had taken on less importance for me, so it was refreshing to read it again.

As additional preparation, I interviewed others who had debated Gish successfully, including my colleague at Occidental College, Don Prothero, and watched videotapes of earlier debates with Gish. I noticed that regardless of his opponent, his opponent's strategy, or even what his opponent said, Gish delivered the same automated presentation—same opening, same assumptions about his opponent's position, same outdated slides, and even the same jokes. I made a note to steal his jokes if I went first. A toss of the coin determined that I would start.

Rather than go toe-to-toe with a man so seasoned in the ways of debate, I had decided to try a version of Muhammed Ali's rope-a-dope strategy by refusing to engage in debate. That is, I turned it into a meta-debate about the difference between religion and science. I began by explaining that the goal of skeptics is not just to debunk claims; it is also to examine belief systems and understand how people are affected by them. I quoted Baruch Spinoza—"I have made a ceaseless effort not to ridicule, not to bewail, not to scorn human actions, but to understand them"—and explained that my real purpose was to understand Gish and the creationists so that I could understand how they can reject the well-confirmed theory called evolution.

I then read parts of the biblical creation story (Gen. 1) to the audience.

In the beginning God created the heavens and the earth.

And the earth was without form, and void; and darkness was upon the face of the deep. And the Spirit of God moved upon the face of the waters.

And God said, "Let there be light"; and there was light. . . . And God called the light Day, and the darkness He called Night. And the evening and the morning were the first day.

And God said, "Let the waters under the heaven be gathered together unto one place, and let the dry land appear"; and it was so.

And God said, "Let the earth bring forth grass, the herb yielding seed, and the fruit tree yielding fruit after his kind, whose seed is in itself, upon the earth"; and it was so.

And God created great whales, and every living creature that moveth, which the waters brought forth abundantly, after their kind, and every winged fowl after his kind; and God saw that it was good.

And God said, "Let the earth bring forth the living creature after his kind, cattle, and creeping thing, and beast of the earth after his kind"; and it was so.

And God said, "Let Us make man in Our image, after Our likeness: and let them have dominion over the fish of the sea, and over the fowl of the air, and over the cattle, and over all the earth, and over every living thing that creepeth upon the earth."

The Bible follows the story of creation with a re-creation story (Gen. 7–8).

And Noah went in, and his sons, and his wife, and his sons' wives with him, into the ark, because of the waters of the flood.

And the rain was upon the earth forty days and forty nights.

And all flesh died that moved upon the earth, both of fowl, and of cattle, and of beast, and of every creeping thing that creepeth upon the earth, and every man.

And the waters returned from off the earth continually; and after the end of the hundred and fifty days the waters were abated.

These stories of creation and re-creation, birth and rebirth, are among the most sublime myths in the history of Western thought. Such myths and stories play an important role in every culture, including ours. Around the world and across the millennia, the details vary but the types converge.

No Creation Story: "The world has always existed as it is now, unchanging from eternity." (Jainists of India)

Slain Monster Creation Story: "The world was created from the parts of a slain monster." (Gilbert Islanders, Greeks, Indochinese, Kabyles of Africa, Koreans, Sumero-Babylonians)

Primordial Parents Creation Story: "The world was created by the interaction of primordial parents." (Cook Islanders, Egyptians, Greeks, Luiseño Indians, Tahitians, Zuñi Indians)

Cosmic Egg Creation Story: "The world was generated from an egg." (Chinese, Finns, Greeks, Hindus, Japanese, Persians, Samoans)

Spoken Edict Creation Story: "The world sprang into being at the command of a god." (Egyptians, Greeks, Hebrews, Maidu Indians, Mayans, Sumerians)

Sea Creation Story: "The world was created from out of the sea."
(Burmese, Choctaw Indians, Egyptians, Icelanders, Maui
Hawaiians, Sumerians)

The Noachian flood story, in fact, is but one variation on the Sea
Creation Story, except that it is a myth of re-creation. The earliest version
we have is ancient, predating the biblical story by over a thousand years.
Around 2800 B.C.E., a Sumerian myth presents the flood hero as the priest-
king Ziusudra, who built a boat to survive a great deluge. Around 2000 to
1800 B.C.E., the hero of the famous Babylonian *Epic of Gilgamesh* learns of
the flood from an ancestor named Utnapishtim. Warned by the Earth-god
Ea that the gods were about to destroy all life by a flood, Utnapishtim was
instructed to build an ark in the form of a cube 120 cubits (180 feet) to a
side, with seven floors, each divided into nine compartments, and to take
aboard one pair of each living creature. The Gilgamesh flood story floated
(pardon the pun) for centuries throughout the Near East and was known
in Palestine before the arrival of the Hebrews. Literary comparison makes
its influence on the Noachian flood story obvious.

We know that a culture's geography influences its myths. For example,
cultures whose major rivers flooded and destroyed the surrounding villages
and cities told flood stories, as in Sumeria and Babylonia where the Tigris
and Euphrates rivers periodically flood. Even cultures in arid regions have
flood stories if they are subject to the whims of flash flooding. By contrast,
cultures not on major bodies of water typically have no flood stories.

Does all this mean that the biblical creation and re-creation stories
are false? To even ask the question is to miss the point of the myths, as
Joseph Campbell (1949, 1988) spent a lifetime making clear. These flood
myths have deeper meanings tied to re-creation and renewal. Myths are
not about truth. Myths are about the human struggle to deal with the
great passages of time and life—birth, death, marriage, the transitions
from childhood to adulthood to old age. They meet a need in the psychological or spiritual nature of humans that has absolutely nothing to do
with science. To try to turn a myth into a science, or a science into a
myth, is an insult to myths, an insult to religion, and an insult to science.
In attempting to do this, creationists have missed the significance, meaning, and sublime nature of myths. They took a beautiful story of creation
and re-creation and ruined it.

To show the absurdity of trying to turn a myth into a science, one has
only to consider the realities of fitting two each of millions of species, let
alone their food, into a boat 450 by 75 by 45 feet. Consider the logistics of

FIGURE 14:
Painting of Noah's Ark at the Institute of Creation Research Museum, San Diego, California.
Note the *Stegosaurus* plates peeking over the stall in the foreground. [Photograph courtesy
Bernard Leikind.]

feeding and watering and cleaning up after all those animals. How do you
keep them from preying on one another? Do you have a predators-only
deck? One might also ask why fish and water-based dinosaurs would
drown in a flood. Creationists are undaunted. The Ark carried "only"
30,000 species, the rest "developing" from this initial stock. The Ark did
indeed have separate decks for predators and prey. It even had a special
deck for dinosaurs (see figure 14). Fish? They died from the silt churned
up by the violent storms of the flood clogging their gills. With faith one
can believe anything because God can accomplish anything.

It would be difficult to find a supposedly scientific belief system more
extraordinary than creationism, whose claims deny not only evolutionary
biology but most of cosmology, physics, paleontology, archeology, historical geology, zoology, botany, and biogeography, not to mention much of
early human history. Of all the claims we have investigated at *Skeptic*, I
have found only one that I could compare to creationism for the ease and
certainty with which it asks us to ignore or dismiss so much existing
knowledge. That is Holocaust denial. Further, the similarities between the
two in their methods of reasoning are startling:

1. Holocaust deniers find errors in the scholarship of historians and then imply that therefore their conclusions are wrong, as if historians never make mistakes. Evolution deniers (a more appropriate title than creationists) find errors in science and imply that all of science is wrong, as if scientists never make mistakes.

2. Holocaust deniers are fond of quoting, usually out of context, leading Nazis, Jews, and Holocaust scholars to make it sound like they are supporting Holocaust deniers' claims. Evolution deniers are fond of quoting leading scientists like Stephen Jay Gould and Ernst Mayr out of context and implying that they are cagily denying the reality of evolution.

3. Holocaust deniers contend that genuine and honest debate between Holocaust scholars means they themselves doubt the Holocaust or cannot get their stories straight. Evolution deniers argue that genuine and honest debate between scientists means even they doubt evolution or cannot get their science straight.

The irony of this analogy is that the Holocaust deniers can at least be partially right (the best estimate of the number of Jews killed at Auschwitz, for example, has changed), whereas the evolution deniers cannot even be partially right—once you allow divine intervention into the scientific process, all assumptions about natural law go out the window, and with them science.

It is also important to understand that what may appear to be "warfare" between science and religion, especially when this debate is promoted as "evolution v. creationism," or in this case "Shermer v. Gish," is not a war in most people's minds. Even Charles Darwin saw no problem with integrating his theory with the prevailing doctrines of his age, as he wrote in a letter late in his life: "It seems to me absurd to doubt that a man can be an ardent Theist and an Evolutionist. Whether a man deserves to be called a Theist depends upon the definition of the term, which is much too large a subject for a note. In my most extreme fluctuations I have never been an Atheist in the sense of denying the existence of a God. I think that generally (and more and more as I grow older, but not always), that an Agnostic would be the more correct description of my state of mind" (1883, p. 107).

Many creationists would be surprised to learn that some prominent skeptics either harbor no animosity against religion or are themselves believers. Stephen Jay Gould once wrote, "Unless at least half my colleagues are dunces, there can be—on the most raw and empirical grounds—no conflict between science and religion" (1987a, p. 68). Steve Allen explained, "My present position as to the existence of God is that though it seems utterly fantastic, I accept it because the alternative seems even more fantastic" (1993, p. 40). Martin Gardner (1996), the skeptics' skeptic, calls himself a *fideist*, a philosophical theist who says *credo consolans*—I believe because it is consoling. Given a metaphysical problem impossible to resolve through science or reason (like the existence of God), says Gardner, it is acceptable to make a leap of faith. These are hardly fighting words.

Even Pope John Paul II, on October 27, 1996, in an address to the Pontifical Academy of Sciences in Rome, declared his acceptance of evolution as a fact of nature and noted that there is no war between science and religion: "Consideration of the method used in diverse orders of knowledge allows for the concordance of two points of view which seem irreconcilable. The sciences of observation describe and measure with ever greater precision the multiple manifestations of life . . . while theology extracts . . . the final meaning according to the Creator's designs." Pushing the warfare model, creationists and the Christian right reacted angrily. Henry Morris, emeritus president of the Institute for Creation Research, responded that "the pope is just an influential person; he's not a scientist. There is no scientific evidence for evolution. All the real solid evidence supports creation." Cal Thomas, the conservative right-wing author, stated in his *Los Angeles Times* column that despite the pope's stand against communism, "he has accepted a philosophy that stands at the core of communism." Thomas explained away this error in the pope's thinking by concluding that he "has succumbed in his declining years to the tyranny of evolutionary scientists who claim we are related to monkeys." (All cited in *Skeptic*, Vol. 4, No. 4, 1996.)

For some believers, the warfare model forces an either-or choice between science and religion to account for the woes of civilization. Since a benevolent and omnipotent God could not cause such evil as we see around us, the explanation is obvious, as Judge Braswell Dean of the Georgia Court of Appeals noted in his opinion on whether creationism should be taught in public schools: "This monkey mythology of Darwin is the cause of permissiveness, promiscuity, pills, prophylactics, perversions, pregnancies, abortions, pornography, pollution, poisoning, and proliferation of crimes of all types" (*Time*, March 16, 1981, p. 82). The alliteration is lovely. The sentiment is not.

Nell Segraves, of the Creation-Science Research Center, was no less adamant: "The research conducted by CSRC has demonstrated that the results of evolutionary interpretations of science data result in a widespread breakdown of law and order. This cause and effect relationship

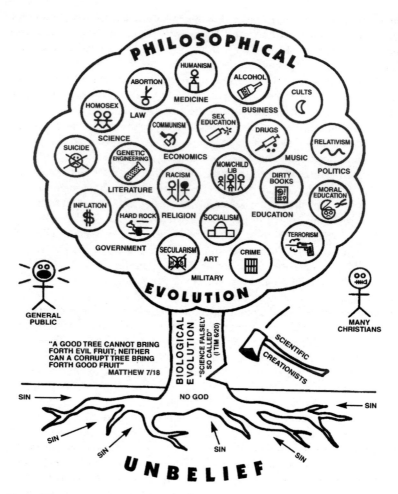

FIGURE 15:
Evolution as a tree routed in unbelief and bearing evil fruit. [From flier distributed by the Pittsburgh Creation Society, Bairdford, Pennsylvania. Redrawn from Toumey 1994.]

stems from the moral decay of mental health and loss of a sense of well being on the part of those involved with this belief system, i.e., divorce, abortion, and rampant venereal disease" (1977, p. 17). The evolution tree from the Pittsburgh Creation Society (figure 15) sums up this warfare model—evolution must fall, along with the evils of humanism, alcohol, abortion, cults, sex education, communism, homosexuality, suicide, racism, dirty books, relativism, drugs, moral education, terrorism, socialism, crime, inflation, secularism, that evil of all evils, hard rock, and, God forbid, women's and children's liberation.

The perceived implications of evolution for ethics and religion are what really disturb Gish and the creationists; for them, all other arguments about evolution are secondary. They are convinced that somehow belief in evolution leads to loss of faith and all sorts of social evils. How do we deal with these fears? Here are four brief replies.

- The use or misuse of a theory does not negate the validity of the theory itself. Marx once claimed that he was not a Marxist. Darwin would undoubtedly be spinning in his grave if he knew how the twentieth century has used his theory to justify all manner of ideologies, from Marxism to capitalism to Fascism. The fact that Hitler implemented a eugenics program does not negate the theory of genetics. Similarly, any correlation between loss of faith and belief in evolution cannot touch the theory of evolution. Scientific theories are neutral; the use of theories is not. They are two different things.

- The creationists' list of social problems—promiscuity, pornography, abortion, infanticide, racism, and so on—obviously existed long before Darwin and the theory of evolution. In the several thousand years before Darwin came along, Judaism, Christianity, and other organized religions failed to resolve these social problems. There is no evidence that the fall of evolution-science will either mitigate or eradicate social ills. To blame Darwin, evolutionary theory, and science for our own social and moral problems is to distract us from a deeper analysis and better understanding of these complex social issues.

- Evolution theory cannot replace faith and religion, and science has no interest in pretending that it can. The theory of evolution is a scientific theory, not a religious doctrine. It stands or falls on evidence alone. Religious faith, by definition, depends on belief when evidence is absent or unimportant. They fill different niches in the human psyche.

- To fear the theory of evolution is an indication of a shortcoming in one's faith, as is looking to scientific proof for justification of one's religious beliefs. If creationists have true faith in their religion, it should not matter what scientists think or say and scientific proof of God or biblical stories should be of no interest.

I concluded my meta-debate analysis with a show of goodwill by offering Gish an honorary membership in the Skeptics Society. I was later forced to retract the offer, however, when Gish refused to retract his characterization of me as an atheist. As Darwin said, "An Agnostic would be the more correct description of my state of mind." I knew Gish had a lengthy section in his presentation on the evils of atheism as a technique to destroy his opponents (who typically are atheists), so I made a point of stating in my introduction, loud and clear, that I am not an atheist. I even called the audience's attention to the man passing out anti-Christian literature, who was now sitting in the front row, and I told him that I thought he was doing more harm than good. Nonetheless, in his opening statement Gish called me an atheist and then proceeded with his automated diatribe against atheism.

The rest of Gish's presentation was his stock litany of jokes and jabs against evolution. He demanded one transitional fossil (I provided several), argued that the bombardier beetle could not have evolved its noxious spray (it could), claimed that evolution violates the second law of thermodynamics (it doesn't because the Earth is in an open system with the Sun as a continuing source of energy), stated that neither evolution-science nor creation-science is scientific (odd for someone calling himself a creation-scientist), and so on. I rebutted all of his points, and in the next chapter I summarize them one by one and provide evolutionists' answers to them.

Who won the debate? Who knows? A more important question to address is whether skeptics and scientists should participate in such debates. Deciding how to respond to fringe groups and extraordinary claims is always a tough call. It is our job at *Skeptic* to investigate claims to discover if they are bogus, but we do not want to dignify them in the process. The principle we use at *Skeptic* is this: when a fringe group or extraordinary claim has gained widespread public exposure, a proper rebuttal deserves equal public exposure. Whether my meta-debate tactic worked with Gish, I have no way of knowing, although a number of people who had come to root for Gish thanked me afterward for at least trying to understand them. It is for these folks, and for those in the middle who are uncertain as to which direction to lean, that I think debates such as this can make a difference. If we can offer a natural explanation for apparently supernatural phenomena and make three or four simple points about science and critical thinking so that listeners can learn *how* to think instead of *what* to think, then I believe it is well worth the effort.

10

Confronting Creationists

Twenty-five Creationist Arguments, Twenty-five Evolutionist Answers

---

Late in his life, Charles Darwin received many letters asking for his views on God and religion. On October 13, 1880, for example, he answered a letter from the editor of a book on evolution and free thought who was hoping to dedicate it to him. Knowing that the book had an antireligious slant, Darwin dissembled: "Moreover though I am a strong advocate for free thought on all subjects, yet it appears to me (whether rightly or wrongly) that direct arguments against christianity & theism produce hardly any effect on the public; & freedom of thought is best promoted by the gradual illumination of men's minds, which follow from the advance of science. It has, therefore, been always my object to avoid writing on religion, & I have confined myself to science" (in Desmond and Moore 1991, p. 645).

In classifying the relationship of science and religion, I would like to suggest a three-tiered taxonomy:

> The *same-worlds model:* Science and religion deal with the same subjects and not only is there overlap and conciliation but someday science may subsume religion completely. Frank Tipler's cosmology (1994), based on the anthropic principle and the eventual resurrection of all humans through a supercomputer's virtual reality in the far future of the universe, is one example. Many humanists and evolutionary psychologists foresee a time when science not only can explain the purpose of religion, it will replace it with a viable secular morality and ethics.

The *separate-worlds model:* Science and religion deal with different subjects, do not conflict or overlap, and the two should coexist peacefully with one another. Charles Darwin, Stephen Jay Gould, and many other scientists hold this model.

The *conflicting-worlds model:* One is right and the other is wrong, and there can be no reconciliation between the two viewpoints. This model is predominantly held by atheists and creationists, who are often at odds with one another.

This taxonomy allows us to see that Darwin's advice is as applicable today as it was a century ago. Thus, let us be clear that refuting creationists' arguments is not an attack on religion. Let us also be clear that creationism is an attack on science—all of science, not just evolutionary biology—so the counterarguments presented in this chapter are a response to the antiscience of creationism and have nothing whatsoever to do with antireligion. If creationists are right, then there are serious problems with physics, astronomy, cosmology, geology, paleontology, botany, zoology, and all the life sciences. Can all these sciences be wrong in the same direction? Of course not, but creationists think they are, and, worse, they want their antiscience taught in public schools.

Creationists and religious fundamentalists will go to absurd lengths to protect their beliefs from science. The Summer 1996 issue of the National Center for Science Education's *Reports* notes that in Marshall County, Kentucky, elementary school superintendent Kenneth Shadowen found a rather unique solution to a vexing problem with his fifth- and sixth-graders' science textbooks. It seems that the heretical textbook *Discovery Works* claimed that the universe began with the Big Bang but did not present any "alternatives" to this theory. Since the Big Bang was explained on a two-page spread, Shadowen recalled all the textbooks and glued together the offending pages. Shadowen told the *Louisville Courier-Journal,* "We're not going to teach one theory and not teach another theory" and that the textbook's recall "had nothing to do with censorship or anything like that" (August 23, 1996, A1, p. 1). It seems doubtful that Shadowen was lobbying for equal time for the Steady State theory or Inflationary Cosmology. Perhaps Shadowen found his solution by consulting librarian Ray Martin's "Reviewing and Correcting Encyclopedias," a guide for Christians on how to doctor books:

> Encyclopedias are a vital part of many school libraries. . . . [They] represent the philosophies of present day humanists. This is obvious by the bold display

of pictures that are used to illustrate painting, art, and sculpture. . . . One of the areas that needs correction is immodesty due to nakedness and posture. This can be corrected by drawing clothes on the figures or blotting out entire pictures with a magic marker. This needs to be done with care or the magic marker can be erased from the glossy paper used in printing encyclopedias. You can overcome this by taking a razor blade and lightly scraping the surface until it loses its glaze. . . . [Regarding evolution] cutting out the sections is practical if the portions removed are not thick enough to cause damage to the spine of the book as it is opened and closed in normal use. When the sections needing correction are too thick, paste the pages together being careful not to smear portions of the book not intended for correction. (*Christian School Builder*, April 1983, pp. 205–207)

Fortunately, creationists have failed in their top-down strategy of passing antievolution, pro-creationism laws (Ohio, Tennessee, and Georgia recently rejected creationist legislation), but their bottom-up grassroots campaign bent on injecting Genesis into the public school curriculum has met with success. In March 1996, for example, Governor Fob James used a discretionary fund of taxpayers' money to purchase and send a copy of Phillip Johnson's antievolution book, *Darwin on Trial*, to every high school biology teacher in Alabama. Their success should not be surprising. Politically, the United States has taken a sharp turn to the right, and the political strength of the religious right has grown. What can we do? We can counter with our own literature. For example, the National Center for Science Education, Eugenie Scott's Berkeley-based group specializing in tracking creationist activities, countered Governor James's mailing with a mailing that included a critical review of Johnson's book. We can also try to understand the issue thoroughly so that we are prepared to counter pro-creationist arguments wherever we meet them.

The following is a list of arguments put forth by creationists and answers put forth by evolutionists. The arguments are primarily attacks on evolutionary theory and secondarily (in a minor way) positive statements of creationists' own beliefs. The arguments and answers are simplified due to space constraints; nonetheless, they provide an overview of the principal points of the debate. This list is not meant to substitute for critical reading, however. While these answers might be adequate for casual conversation, they would not be adequate for a formal debate with a well-prepared creationist. Numerous books offer fuller discussions (e.g., Berra 1990; Bowler 1989; Eve and Harrold 1991; Futuyma 1983; Gilkey 1985; Godfrey 1983; Gould 1983a, 1991; Lindberg and Numbers 1986; Numbers 1992; Ruse 1982; and, especially, Strahler 1987).

What Is Evolution?

Before reviewing creationists' arguments against evolution, a brief summary of the theory itself might be useful. Darwin's theory, outlined in his 1859 *On the Origin of Species by Means of Natural Selection*, can be summarized as follows (Gould 1987a; Mayr 1982, 1988):

> *Evolution:* Organisms change through time. Both the fossil record and nature today make this obvious.
>
> *Descent with modification:* Evolution proceeds via branching through common descent. Offspring are similar to but not exact replicas of their parents. This produces the necessary variation to allow for adaptation to an ever-changing environment.
>
> *Gradualism:* Change is slow, steady, stately. *Natura non facit saltum*—Nature does not make leaps. Given enough time, evolution accounts for species change.
>
> *Multiplication of speciation:* Evolution does not just produce new species; it produces an increasing number of new species.
>
> *Natural selection:* The mechanism of evolutionary change, co-discovered by Darwin and Alfred Russel Wallace, operates as follows:
>
> A. Populations tend to increase indefinitely in a geometric ratio: 2, 4, 8, 16, 32, 64, 128, 256, 512, ...
>
> B. In a natural environment, however, population numbers stabilize at a certain level.
>
> C. Therefore, there must be a "struggle for existence" because not all of the organisms produced can survive.
>
> D. There is variation in every species.
>
> E. In the struggle for existence, those individuals with variations that are better adapted to the environment leave behind more offspring than individuals that are less well adapted. This is known in the jargon of the trade as *differential reproductive success.*

Point E is crucial. Natural selection, and thus evolutionary change, operate primarily at the local level. It is just a game of who can leave behind the most offspring, that is, who can most successfully propagate their genes into the next generation. Natural selection has nothing to say about evolutionary direction, species progress, or any of the other teleological goals, such as human inevitability or the necessary evolution of intelligence, which are commonly attributed to it. There is no ladder of evolutionary progress with humans at the top, only a richly branching bush with humans as one tiny twig among millions. There is nothing special about humans; we just happen to be extremely good at differential reproductive success—we leave behind lots of offspring and are good at getting them into adulthood—a trait that could eventually cause our demise.

Of the five points of Darwin's theory, the most controversial today are gradualism, with Niles Eldredge (1971, 1985; Eldredge and Gould 1972) and Stephen Jay Gould (1985, 1989, 1991) and their supporters pushing for a theory called *punctuated equilibrium*, which involves rapid change and stasis, to replace gradualism; and the exclusivity of natural selection, with Eldredge, Gould, and others arguing for change at the level of genes, groups, and populations in addition to individual natural selection (Somit and Peterson 1992). Ranged against Eldredge, Gould, and their supporters are Daniel Dennett (1995), Richard Dawkins (1995), and those who opt for a strict Darwinian model of gradualism and natural selection. The debate rages, while creationists sit on the sidelines hoping for a double knockout. They will not get it. These scientists are not arguing about *whether* evolution happened; they are debating the *rate* and *mechanism* of evolutionary change. When it all shakes down, the theory of evolution will be stronger than ever. It is sad that while science moves ahead in exciting new areas of research, fine-tuning our knowledge of how life originated and evolved, creationists remain mired in medieval debates about angels on the head of a pin and animals in the belly of an Ark.

Philosophically Based Arguments and Answers

1. Creation-science is scientific and therefore should be taught in public school science courses.

Creation-science is scientific in name only. It is a thinly disguised religious position rather than a theory to be tested using scientific methods, and therefore it is not appropriate for public school science courses, just as calling something Muslim-science or Buddha-science or Christian-science would not mean that it requires equal time. The following statement from the Institute for Creation Research, which must be adhered to by all faculty members and researchers, is a powerful illumination of creationist beliefs: "The scriptures, both Old and New Testaments, are inerrant in relation to any subject with which they deal, and are to be accepted in their natural and intended sense . . . all things in the universe were created and made by God in the six days of special creation described in Genesis. The creationist account is accepted as factual, historical and perspicuous and is thus fundamental in the understanding of every fact and phenomenon in the created universe" (in Rohr 1986, p. 176).

Science is subject to disproof and is ever-changing as new facts and theories reshape our views. Creationism prefers faith in the authority of the Bible no matter what contradictory empirical evidence might exist: "The main reason for insisting on the universal Flood as a fact of history and as the primary vehicle for geological interpretation is that God's Word plainly teaches it! No geological difficulties, real or imagined, can be allowed to take precedence over the clear statements and necessary inferences of Scripture" (in Rohr 1986, p. 190). Here is an analogy. Professors at Caltech declare Darwin's *Origin of Species* dogma, the authority of this book and its author absolute, and any further empirical evidence for or against evolution irrelevant.

2. Science only deals with the here-and-now and thus cannot answer historical questions about the creation of the universe and the origins of life and the human species.

Science does deal with past phenomena, particularly in historical sciences such as cosmology, geology, paleontology, paleoanthropology, and archeology. There are experimental sciences and historical sciences. They use different methodologies but are equally able to track causality. Evolutionary biology is a valid and legitimate historical science.

3. Education is a process of learning all sides of an issue, so it is appropriate for creationism and evolution to be taught side-by-side in public school science courses. Not to do so is a violation of the principles of education and of the civil liberties of creationists. We have a right to be heard, and, besides, what is the harm in hearing both sides?

Exposure to the many facets of issues is indeed a part of the general educational process, and it might be appropriate to discuss creationism in courses on religion, history, or even philosophy but most certainly not science; similarly, biology courses should not include lectures on American Indian creation myths. There is considerable harm in teaching creation-science as science because the consequent blurring of the line between religion and science means that students will not understand what the scientific paradigm is and how to apply it properly. Moreover, the assumptions behind creationism comprise a two-pronged attack on all the sciences, not just on evolutionary biology. One, if the universe and Earth are only about ten thousand years old, then the modern sciences of cosmology, astronomy, physics, chemistry, geology, paleontology, paleoanthropology, and early human history are all invalid. Two, as soon as the creation of even one species is attributed to supernatural intervention, natural laws and inferences about the workings of nature become void. In each case, all science becomes meaningless.

4. There is an amazing correlation between the facts of nature and the acts of the Bible. It is therefore appropriate to use creation-science books and the Bible as reference tools in public school science courses and to study the Bible as a book of science alongside the book of nature.

There is also an amazing correlation between acts in the Bible for which there are no facts in nature and between facts in nature for which there are no acts in the Bible. If a group of Shakespeare scholars believe that the universe is explained in the bard's plays, does that mean science courses should include readings of Shakespeare? Shakespeare's plays are literature, the Bible contains scriptures sacred to several religions, and neither has any pretensions to being a book of science or a scientific authority.

5. The theory of natural selection is tautological, or a form of circular reasoning. Those who survive are the best adapted. Who are the best adapted? Those who survive. Likewise, rocks are used to date fossils, and fossils are used to date rocks. Tautologies do not make a science.

Sometimes tautologies are the beginning of science, but they are never the end. Gravity can be tautological, but its inference is justified by the way this theory allows scientists to accurately predict physical effects and phenomena. Likewise, natural selection and the theory of evolution are testable and falsifiable by looking at their predictive power. For example, population genetics demonstrates quite clearly, and with mathematical prediction, when natural selection will and will not effect change on a population. Scientists can make predictions based on the theory of natural selection and then test them, as the geneticist does in the example just given or the paleontologist does in interpreting the fossil record. Finding hominid fossils in the same geological strata as trilobites, for instance, would be evidence against the theory. The dating of fossils with rocks, and vice versa, could only be done *after* the geological column was established. The geological column exists nowhere in its entirety because strata are disrupted, convoluted, and always incomplete for a variety of reasons. But strata order is unmistakably *nonrandom*, and chronological order can be accurately pieced together using a variety of techniques, only one of which is fossils.

6. There are only two explanations for the origins of life and existence of humans, plants, and animals: either it was the work of a creator or it was not. Since evolution theory is unsupported by the evidence (i.e., it is wrong), creationism must be correct. Any evidence that does not support the theory of evolution is necessarily scientific evidence in support of creationism.

Beware of the either-or fallacy, or the fallacy of false alternatives. If A is false, B must be true. Oh? Why? Plus, shouldn't B stand on its own regardless of A? Of course. So even if evolutionary theory turns out to be completely wrong, that does not mean that, ergo, creationism is right. There may be alternatives C, D, and E we have yet to consider. There is, however, a true dichotomy in the case of natural versus supernatural explanations. Either life was created and changed by natural means, or it was created and changed by supernatural intervention and according to a supernatural design. Scientists assume natural causation, and evolutionists debate the various natural causal agents involved. They are not arguing about whether it happened by natural or supernatural means. And, again, once you assume supernatural intervention, science goes out the window—so there can be no scientific evidence in support of creationism because natural laws no longer hold and scientific methodology has no meaning in the world of creationists.

7. Evolutionary theory is the basis of Marxism, communism, atheism, immorality, and the general decline of the morals and culture of America, and therefore is bad for our children.

This partakes of the *reductio ad absurdum* fallacy. Neither the theory of evolution in particular nor science in general is no more the basis of these "isms" and Americans' so-called declining morals and culture than the printing press is responsible for Hitler's *Mein Kampf* or *Mein Kampf* is responsible for what people did with Hitler's ideology. The fact that the atomic bomb, the hydrogen bomb, and many even more destructive weapons have been invented does not mean we should abandon the study of the atom. Moreover, there may well be Marxist, communist, atheistic, and even immoral evolutionists, but there are probably just as many capitalist, theist, agnostic, and moral evolutionists. As for the theory itself, it can be used to support Marxist, communist, and atheistic ideologies, and it has; but so has it been used (especially in America) to lend credence to laissez-faire capitalism. The point is that linking scientific theories to political ideologies is tricky business, and we must be cautious of making connections that do not necessarily follow or that serve particular agendas (e.g., one person's cultural and moral decline is another person's cultural and moral progress).

8. Evolutionary theory, along with its bedfellow, secular humanism, is really a religion, so it is not appropriate to teach it in public schools.

To call the science of evolutionary biology a religion is to so broaden the definition of religion as to make it totally meaningless. In other words, religion becomes any lens that we look through to interpret the world. But that is not what religion is. Religion has something to do with the service and worship of God or the supernatural, whereas science has to do with physical phenomena. Religion has to do with faith and the unseen, science focuses on empirical evidence and testable knowledge. Science is a set of methods designed to describe and interpret observed or inferred phenomena, past or present, and aimed at building a testable body of knowledge open to rejection or confirmation. Religion—whatever it is—is certainly neither testable nor open to rejection or confirmation. In their methodologies, science and religion are 180 degrees out of phase with each other.

9. Many leading evolutionists are skeptical of the theory and find it problematic. For example, Eldredge and Gould's theory of punctuated equilibrium proves Darwin wrong. If the world's leading evolutionists cannot agree on the theory, the whole thing must be a wash.

It is particularly ironic that the creationists would quote a leading spokesman against creationism—Gould—in their attempts to marshal the forces of science on their side. Creationists have misunderstood, either naively or intentionally, the healthy scientific debate among evolutionists about the causal agents of organic change. They apparently take this normal exchange of ideas and the self-correcting nature of science as evidence that the field is coming apart at the seams and about to implode. Of the many things evolutionists argue and debate within the field, one thing they are certain of and all agree upon is that evolution has occurred. Exactly how it happened, and what the relative strengths of the various causal mechanisms are, continue to be discussed. Eldredge and Gould's theory of punctuated equilibrium is a refinement of and improvement upon Darwin's theory of evolution. It no more proves Darwin wrong than Einsteinian relativity proves Newton wrong.

10. "The Bible is the written Word of God . . . all of its assertions are historically and scientifically true. The great Flood described in Genesis was an historical event, worldwide in its extent and effect. We are an organization of Christian men of science, who accept Jesus Christ as our Lord and Savior.

The account of the special creation of Adam and Eve as one man and one woman, and their subsequent Fall into sin, is the basis for our belief in the necessity of a Savior for all mankind" (in Eve and Harrold 1991, p. 55).

Such a statement of belief is clearly religious. This does not make it wrong, but it does mean that creation-science is really creation-religion and to this extent breaches the wall separating church and state. In private schools funded or controlled by creationists, they are free to teach whatever they like to their children. But one cannot make the events in any text historically and scientifically true by fiat, only by testing the evidence, and to ask the state to direct teachers to teach a particular religious doctrine as science is unreasonable and onerous.

11. All causes have effects. The cause of "X" must be "X-like." The cause of intelligence must be intelligent—God. Regress all causes in time and you must come to the first cause—God. Because all things are in motion, there must have been a prime mover, a mover who needs no other mover to be moved—God. All things in the universe have a purpose, therefore there must be a purposeful designer—God.

If this were true, should not nature then have a natural cause, not a supernatural cause? But causes of "X" do not have to be "X-like." The "cause" of green paint is blue paint mixed with yellow paint, neither one of which is green-like. Animal manure causes fruit trees to grow better. Fruit is delicious to eat and is, therefore, very unmanure-like! The first-cause and prime-mover argument, brilliantly proffered by St. Thomas Aquinas in the fourteenth century (and more brilliantly refuted by David Hume in the eighteenth century), is easily turned aside with just one more question: Who or what caused and moved God? Finally, as Hume demonstrated, purposefulness of design is often illusory and subjective. "The early bird gets the worm" is a clever design if you are the bird, not so good if you are the worm. Two eyes may seem like the ideal number, but, as psychologist Richard Hardison notes cheerfully, "Wouldn't it be desirable to have an additional eye in the back of one's head, and certainly an eye attached to our forefinger would be helpful when we're working behind the instrument panels of automobiles" (1988, p. 123). Purpose is, in part, what we are accustomed to perceiving. Finally, not everything is so purposeful and beautifully designed. In addition to problems like evil, disease, deformities, and human stupidity which creationists conveniently overlook, nature is filled with the bizarre and seemingly unpurposeful. Male nipples and the panda's thumb are just two examples flaunted by Gould as purposeless and poorly designed structures. If God designed life to fit neatly together like a jigsaw puzzle, then what do you do with such oddities and problems?

12. Something cannot be created out of nothing, say scientists. Therefore, from where did the material for the Big Bang come? From where did the first life forms that provided the raw material for evolution originate? Stanley Miller's creation of amino acids out of an inorganic "soup" and other biogenic molecules is not the creation of life.

Science may not be equipped to answer certain "ultimate"-type questions, such as what there was before the beginning of the universe or what time it was before time began or where the matter for the Big Bang came from. So far these have been philosophical or religious questions, not scientific ones, and therefore have not been a part of science. (Recently, Stephen Hawking and other cosmologists have made some attempts at scientific speculations on these questions.) Evolutionary theory attempts to understand the causality of change *after* time and matter were "created" (whatever that means). As for the origin of life, biochemists do have a very rational and scientific explanation for the evolution from inorganic to organic compounds, the creation of amino acids and the construction of protein chains, the first crude cells, the creation of photosynthesis, the invention of sexual reproduction, and so on. Stanley Miller never claimed to have created life, just some of its building blocks. While these theories are by no means robust and are still subject to lively scientific debate, there is a reasonable explanation for how you get from the Big Bang to the Big Brain in the known universe using the known laws of nature.

Scientifically Based Arguments and Answers

13. Population statistics demonstrate that if we extrapolate backward from the present population using the current rate of population growth, there were only two people living approximately 6,300 years before the present (4300 B.C.E.). This proves that humans and civilization are quite young. If the Earth were old—say, one million years—over the course of 25,000 generations at a 0.5 percent rate of population growth and an average of 2.5 children per family, the present population would be 10 to the power of 2,100 people, which is impossible since there are only 10 to the power of 130 electrons in the known universe.

If you want to play the numbers game, how about this? Applying their model, we find that in 2600 B.C.E. the total population on Earth would have been around 600 people. We know with a high degree of certainty that in 2600 B.C.E. there were flourishing civilizations in Egypt, Mesopotamia, the Indus River Valley, and China. If we give Egypt an extremely generous one-sixth of the world's population, then 100 people built the pyramids, not to mention all the other architectural monuments—they most certainly needed a miracle or two . . . or perhaps the assistance of ancient astronauts!

The fact is that populations do not grow in a steady manner. There are booms and busts, and the history of the human population before the Industrial Revolution is one of prosperity and growth, followed by famine and decline, and punctuated by disaster. In Europe, for instance, about half of the population was killed by a plague during the sixth century, and in the fourteenth century the bubonic plague wiped out about one-third of the population in three years. As humans struggled for millennia to fend off extinction, the population curve was one of peaks and valleys as it climbed uncertainly but steadily upward. It is only since the nineteenth century that the rate of increase has been steadily accelerating.

14. Natural selection can never account for anything other than minor changes within species—microevolution. Mutations used by evolutionists to explain macroevolution are always harmful, rare, and random, and cannot be the driving force of evolutionary change.

I shall never forget the four words pounded into the brains of the students of evolutionary biologist Bayard Brattstrom at California State University, Fullerton: "Mutants are not monsters." His point was that the public perception of mutants—two-headed cows and the like at the county fair—is not the sort of mutants evolutionists are discussing. Most mutations are small genetic or chromosomal aberrations that have small effects—slightly keener hearing, a new shade of fur. Some of these small effects may provide benefits to an organism in an ever-changing environment.

Moreover, Ernst Mayr's (1970) theory of *allopatric speciation* seems to demonstrate precisely how natural selection, in conjunction with other forces and contingencies of nature, can and does produce new species. Whether they agree or disagree with the theory of allopatric speciation and punctuated equilibrium, scientists all agree that natural selection can produce significant change. The debate is over how much change, how rapid a change, and what other forces of nature act in conjunction with or contrary to natural selection. No one, and I mean *no one*, working in the field is debating whether natural selection is the driving force behind evolution, much less whether evolution happened or not.

15. There are no transitional forms in the fossil record, anywhere, including and especially humans. The whole fossil record is an embarrassment to evolutionists. Neanderthal specimens, for example, are diseased skeletons distorted by arthritis, rickets, and other diseases that create the bowed legs,

brow ridge, and larger skeletal structure. *Homo erectus* and *Australopithecus* are just apes.

Creationists always quote Darwin's famous passage in the *Origin of Species* in which he asks, "Why then is not every geological formation and every stratum full of such intermediate links? Geology assuredly does not reveal any such finely graduated organic chain; and this, perhaps, is the gravest objection which can be urged against my theory" (1859, p. 310). Creationists end the quote there and ignore the rest of Darwin's chapter, in which he addresses the problem.

One answer is that plenty of examples of transitional forms have been discovered since Darwin's time. Just look in any paleontology text. The fossil *Archeopteryx*—part reptile, part bird—is a classic example of a transitional form. In my debate with Duane Gish, I presented a slide of the newly discovered *Ambulocetus natans*—a beautiful example of a transitional form from land mammal to whale (see *Science*, January 14, 1994, p. 180). And the charges about the Neanderthals and *Homo erectus* are simply absurd. We now have a treasure trove of human transitional forms.

A second answer is a rhetorical one. Creationists demand just one transitional fossil. When you give it to them, they then claim there is a gap between these two fossils and ask you to present a transitional form between these two. If you do, there are now two more gaps in the fossil record, and so on ad infinitum. Simply pointing this out refutes the argument. You can do it with cups on a table, showing how each time the gap is filled with a cup it creates two gaps, which when each is filled with a cup creates four gaps, and so on. The absurdity of the argument is visually striking.

A third answer was provided in 1972 by Eldredge and Gould, when they argued that gaps in the fossil record do not indicate missing data of slow and stately change; rather, "missing" fossils are evidence of rapid and episodic change (punctuated equilibrium). Using Mayr's allopatric speciation, where small and unstable "founder" populations are isolated at the periphery of the larger population's range, Eldredge and Gould showed that the relatively rapid change in this smaller gene pool creates new species but leaves behind few, if any, fossils. The process of fossilization is rare and infrequent anyway, but it is almost nonexistent during these times of rapid speciation because the number of individuals is small and the change is swift. A lack of fossils may be evidence for rapid change, not missing evidence for gradual evolution.

16. The Second Law of Thermodynamics proves that evolution cannot be true since evolutionists state that the universe and life move from chaos to

order and simple to complex, the exact opposite of the entropy predicted by the Second Law.

First of all, on any scale other than the grandest of all—the 600-million-year history of life on Earth—species do not evolve from simple to complex, and nature does not simply move from chaos to order. The history of life is checkered with false starts, failed experiments, local and mass extinctions, and chaotic restarts. It is anything but a neat Time/Life-book fold-out from single cells to humans. Even in the big picture, the Second Law allows for such change because the Earth is in a system that has a constant input of energy from the Sun. As long as the Sun is burning, life may continue thriving and evolving, automobiles may be prevented from rusting, burgers can be heated in ovens, and all manner of other things in apparent violation of the Second Law may continue. But as soon as the Sun burns out, entropy will take over and life will cease and chaos come again. The Second Law of Thermodynamics applies to closed, isolated systems. Since the Earth receives a constant input of energy from the Sun, entropy may decrease and order increase (although the Sun itself is running down in the process). Thus, because the Earth is not strictly a closed system, life may evolve without violating natural laws. In addition, recent research in chaos theory suggests that order can and does spontaneously generate out of apparent chaos, all without violating the Second Law of Thermodynamics (see Kauffman 1993). Evolution no more breaks the Second Law of Thermodynamics than one breaks the law of gravity by jumping up.

17. Even the simplest of life forms are too complex to have come together by random chance. Take a simple organism consisting of merely 100 parts. Mathematically there are 10 to the power of 158 possible ways for the parts to link up. There are not enough molecules in the universe, or time since the beginning, to allow for these possible ways to come together in even this simple life form, let alone to produce human beings. The human eye alone defies explanation by the randomness of evolution. It is the equivalent of the monkey typing *Hamlet*, or even "To be or not to be." It will not happen by random chance.

Natural selection is not random, nor does it operate by chance. Natural selection preserves the gains and eradicates the mistakes. The eye evolved from a single, light-sensitive cell into the complex eye of today through hundreds if not thousands of intermediate steps, many of which still exist in nature (see Dawkins 1986). In order for the monkey to type the thirteen letters opening Hamlet's soliloquy by chance, it would take 26 to the power of 13 trials for success. This is sixteen times as great as the total number of seconds that have elapsed in the lifetime of our solar system. But if each correct letter is preserved and each incorrect letter eradicated, the process operates much faster. How much faster? Richard Hardison (1988) wrote a computer program in which letters were "selected" for or against, and it took an average of only 335.2 trials to produce the sequence of letters TOBEORNOTTOBE. It takes the computer less than ninety seconds. The entire play can be done in about 4.5 days.

18. Hydrodynamic sorting during the Flood explains the apparent progression of fossils in geological strata. The simple, ignorant organisms died in the sea and are on the bottom layers, while more complex, smarter, and faster organisms died higher up.

Not one trilobite floated upward to a higher stratum? Not one dumb horse was on the beach and drowned in a lower stratum? Not one flying pterodactyl made it above the Cretaceous layer? Not one moronic human did not come in out of the rain? And what about the evidence provided by other dating techniques such as radiometry?

19. The dating techniques of evolutionists are inconsistent, unreliable, and wrong. They give false impressions of an old Earth, when in fact it is no older than ten thousand years, as proven by Dr. Thomas Barnes from the University of Texas at El Paso when he demonstrated that the half-life of the Earth's magnetic field is 1,400 years.

First of all, Barnes's magnetic field argument assumes that the decay of the magnetic field is linear when geophysics has demonstrated that it fluctuates through time. He is working from a false premise. Second, not only are the various dating techniques quite reliable on their own but there is considerable independent corroboration between them. For example, radiometric dates for different elements from the same rock will all converge on the same date. Finally, how can creationists dismiss all dating techniques with a sweep of the hand except those that purportedly support their position?

20. Classification of organisms above the species level is arbitrary and man-made. Taxonomy proves nothing, especially because so many of the links between species are missing.

The science of classification is indeed man-made, like all sciences, and of course it cannot prove anything about the evolution of organisms absolutely. But its grouping of organisms is anything but arbitrary, even though there is an element of subjectivity to it. An interesting cross-cultural test of taxonomy is the fact that Western-trained biologists and native peoples from New Guinea identify the same types of birds as separate species (see Mayr 1988). Such groupings really do exist in nature. Moreover, the goal of modern cladistics—the science of classification through nested hierarchies of similarities—is to make taxonomy less subjective, and it successfully uses inferred evolutionary relationships to arrange taxa in a branching hierarchy such that all members of a given taxon have the same ancestors.

21. If evolution is gradual, there should be no gaps between species.

Evolution is not always gradual. It is often quite sporadic. And evolutionists never said there should not be gaps. Finally, gaps do not prove creation any more than blank spots in human history prove that all civilizations were spontaneously created.

22. "Living fossils" like the coelacanth and horseshoe crab prove that all life was created at once.

The existence of living fossils (organisms that have not changed for millions of years) simply means that they evolved a structure adequate for their relatively static and unchanging environment, so they stopped once they could maintain their ecological niche. Sharks and many other sea creatures are relatively unchanged over millions of years, while other sea creatures, such as marine mammals, have obviously changed rapidly and dramatically. Evolutionary change or lack of change, as the case may be, all depends on how and when a species' immediate environment changes.

23. The incipient structure problem refutes natural selection. A new structure that evolves slowly over time would not provide an advantage to the organism in its beginning or intermediate stages, only when it is completely developed, which can only happen by special creation. What good is 5 percent of a wing, or 55 percent? You need all or nothing.

A poorly developed wing may have been a well-developed something else, like a thermoregulator for ectothermic reptiles (who depend on external sources of heat). And it is not true that incipient stages are completely useless. As Richard Dawkins argues in *The Blind Watchmaker* (1986) and *Climbing Mount Improbable* (1996), 5 percent vision is significantly better than none and being able to get airborne for any length of time can provide an adaptive advantage.

24. Homologous structures (the wing of a bat, the flipper of a whale, the arm of a human) are proof of intelligent design.

By invoking miracles and special providence, the creationist can pick and choose anything in nature as proof of God's work and then ignore the rest. Homologous structures actually make no sense in a special creation paradigm. Why should a whale have the same bones in its flipper as a human has in its arm and a bat has in its wing? God has a limited imagination? God was testing out the possibilities of His designs? God just wanted to do things that way? Surely an omnipotent intelligent designer could have done better. Homologous structures are indicative of descent with modification, not divine creation.

25. The whole history of evolutionary theory in particular and science in general is the history of mistaken theories and overthrown ideas. Nebraska Man, Piltdown Man, Calaveras Man, and *Hesperopithecus* are just a few of the blunders scientists have made. Clearly science cannot be trusted and modern theories are no better than past ones.

Again, it is paradoxical for creationists to simultaneously draw on the authority of science and attack the basic workings of science. Furthermore, this argument reveals a gross misunderstanding of the nature of science. Science does not just change. It constantly builds upon the ideas of the past, and it is cumulative toward the future. Scientists do make mistakes aplenty and, in fact, this is how science progresses. The self-correcting feature of the scientific method is one of its most beautiful features. Hoaxes like Piltdown Man and honest mistakes like *Hesperopithecus* are, in time, exposed. Science picks itself up, shakes itself off, and moves on.

Debates and Truth

These twenty-five answers only scratch the surface of the science and philosophy supporting evolutionary theory. If confronted by a creationist, we would be wise to heed the words of Stephen Jay Gould, who has encountered creationists on many an occasion:

> Debate is an art form. It is about the winning of arguments. It is not about the discovery of truth. There are certain rules and procedures to debate that really have nothing to do with establishing fact—which they are very good at. Some of those rules are: never say anything positive about your own position because it can be attacked, but chip away at what appear to be the weaknesses in your opponent's position. They are good at that. I don't think I could beat the creationists at debate. I can tie them. But in courtrooms they are terrible, because in courtrooms you cannot give speeches. In a courtroom you have to answer direct questions about the positive status of your belief. We destroyed them in Arkansas. On the second day of the two-week trial, we had our victory party! (Caltech lecture, 1985)

11

Science Defended, Science Defined

Evolution and Creationism at the Supreme Court

On August 18, 1986, a press conference was held at the National Press Club in Washington, D.C., to announce the filing of an *amicus curiae* brief on behalf of seventy-two Nobel laureates, seventeen state academies of science, and seven other scientific organizations. This brief supported the appellees in *Edwards v. Aguillard*, the Supreme Court case testing the constitutionality of Louisiana's Balanced Treatment for Creation-Science and Evolution-Science Act, an equal-time law passed in 1982 requiring, essentially, that the Genesis version of creation be taught side-by-side with the theory of evolution in public school classrooms in Louisiana. Attorneys Jeffrey Lehman and Beth Shapiro Kaufman from the firm of Caplin and Drysdale, Nobel laureate Christian Anfinsen, biologist Francisco Ayala from the University of California, Davis, and paleontologist Stephen Jay Gould from Harvard University faced a room filled with television, radio, and newspaper reporters from across the country.

Gould and Ayala made opening statements, and a statement by Nobel laureate Murray Gell-Mann was read in absentia. The emotional commitment of these representatives from the scientific community was clear from the outset and baldly disclosed in their statements. Gould noted, "As a term, creation-science is an oxymoron—a self-contradictory and meaningless phrase—a whitewash for a specific, particular, and minority religious view in America—Biblical literalism." Ayala added, "To claim that the statements of Genesis are scientific truths is to deny all the evidence. To teach such statements in the schools as if they were science would do untold harm to the education of American students, who need scientific

FIGURE 16:
Putting the creationist in his place. [Editorial cartoon by Bill Day, *Detroit Free Press.*]

literacy to prosper in a nation that depends on scientific progress for national security and for individual health and economic gain." Gell-Mann concurred with Ayala on the broad, national scope of the problem but went further, saying, in no uncertain terms, that this was an assault on all science:

> I should like to emphasize that the portion of science that is attacked by the statute is far more extensive than many people realize, embracing very important parts of physics, chemistry, astronomy, and geology as well as many of the central ideas of biology and anthropology. In particular, the notion of reducing the age of the earth by a factor of nearly a million, and that of the visible expanding universe by an even larger factor, conflicts in the most basic way with numerous robust conclusions of physical science. For example, fundamental and well-established principles of nuclear physics are challenged, for no sound reason, when "creation-scientists" attack the validity of the radioactive clocks that provide the most reliable methods used to date the earth.

Reviews of the brief appeared in a broad range of publications, including *Scientific American, Nature, Science, Omni, The Chronicle of Higher Education, Science Teacher,* and *California Science Teacher's Journal.* The *Detroit Free Press* even published an editorial cartoon in which a creationist joins the famous evolutionary "march of human progress" (figure 16).

Equal Time or All the Time?

In general, creationists are Christian fundamentalists who read the Bible literally—when Genesis speaks of the six days of creation, for example, it means six 24-hour days. In particular, of course, there are many different types of creationists, including young-Earth creationists, who hold to the 24-hour-day interpretation; old-Earth creationists, who are willing to take the biblical days as figurative speech representing geological epochs; and gap-creationists, who allow for a gap of time between the initial creation and the rise of humans and civilization (thus adapting to scientific notions of deep time, dating back billions of years).

Card-carrying creationists are small in number. But what they lack in numbers they make up in volume. And they have been able to touch the nerve that somewhere deep in the national psyche connects many Americans to our country's religious roots. We may be a pluralistic society—melting pots, salad bowls, and all that—but Genesis remains at our beginning. A 1991 Gallup poll found that 47 percent of Americans believed that "God created man pretty much in his present form at one time within the last ten thousand years." A centrist view, that "Man has developed over millions of years from less advanced forms of life, but God guided this process, including man's creation," was held by 40 percent of Americans. Only 9 percent believed that "Man has developed over millions of years from less advanced forms of life. God had no part in this process." The remaining 4 percent answered, "I don't know" (Gallop and Newport 1991, p. 140).

Why, then, is there a controversy? Because 99 percent of scientists take the strict naturalist view shared by only 9 percent of Americans. This is a startling difference. It would be hard to imagine any other belief for which there is such a wide disparity between the person on the street and the expert in the ivory tower. Yet science is the dominant force in our culture, so in order to gain respectability and, what is more important for creationists, access to public school science classrooms, creationists have been forced to deal with this powerful minority. Over the past eighty years, creationists have used three basic strategies to press their religious beliefs. The Louisiana case was the culmination of a series of legal battles that began in the 1920s and may be grouped into the following three approaches.

Banning Evolution

In the 1920s, a perceived degeneration in the moral fiber of America was linked to Darwin's theory of evolution. For example, a supporter of fundamentalist orator William Jennings Bryan commented in 1923, "Ramming poison down the throats of our children is nothing compared with damning their souls with the teaching of evolution" (in Cowen 1986, p. 8). Fundamentalists rallied to check the moral decline by removing evolution from the public schools. In 1923, Oklahoma passed a bill offering free textbooks to public schools on the condition that neither the teachers nor the textbooks mentioned evolution, and Florida went even further by passing an antievolution law. In 1925, the Butler Act, which made it "unlawful for any teacher in any of the Universities, Normals and all other public schools of the state . . . to teach any theory that denies the story of the Divine Creation of man as taught in the Bible, and to teach instead that man has descended from a lower order of animals" (in Gould 1983a, p. 264), was passed by the Tennessee legislature. This act was viewed as an obvious violation of civil liberties and resulted in the famous 1925 Scopes "Monkey Trial," which has been well documented by Douglas Futuyma (1983), Gould (1983a), Dorothy Nelkin (1982), and Michael Ruse (1982).

John T. Scopes was a substitute teacher who volunteered to provide the test case by which the American Civil Liberties Union (ACLU) could challenge Tennessee's antievolution law. The ACLU intended to take the case all the way to the U.S. Supreme Court, if necessary. Clarence Darrow, the most famous defense attorney of the day, provided legal counsel for Scopes, and William Jennings Bryan, three-time presidential candidate and known voice of biblical fundamentalism, served as defender of the faith for the prosecution. The trial was labeled the "trial of the century," and the hoopla surrounding it was intense; it was, for example, the first trial in history for which daily updates were broadcast by radio. The two giants pontificated for days, but in the end Scopes was found guilty and fined $100 by Judge Raulston (Scopes did, indeed, break the law). Because of a little-known catch in Tennessee law, which required all fines above $50 to be set by a jury, not a judge, the court overturned Scopes's conviction, leaving the defense nothing to appeal. It never was taken to the U.S. Supreme Court, and the law stood on the books until 1967.

Most people think that Scopes, Darrow, and the scientific community scored a great victory in Tennessee. H. L. Mencken, covering the trial for the *Baltimore Sun*, summarized it and Bryan this way: "Once he had one leg in the White House and the nation trembled under his roars. Now he is a tinpot pope in the Coca-Cola belt and a brother to the forlorn pastors who belabor half-wits in galvanized iron tabernacles behind the railroad yards. . . . It is a tragedy, indeed, to begin life as a hero and to end it as a buffoon" (in Gould 1983a, p. 277). But, in fact, there was no victory for evolution. Bryan died a few days after the trial ended, but he had the last laugh, as the controversy stirred by the trial made others, particularly textbook publishers and state boards of education, reluctant to deal with the theory of evolution in any manner. Judith Grabiner and Peter Miller (1974) compared high school textbooks before and after the trial: "Believing that they had won in the forum of public opinion, the evolutionists of the late 1920s in fact lost on their original battleground— teaching of evolution in the high schools—as judged by the content of the average high school biology textbooks [which] declined after the Scopes trial." A trial that seems comical in retrospect was really a tragedy, as Mencken concluded: "Let no one mistake it for comedy, farcical though it may be in all its details. It serves notice on the country that Neanderthal man is organizing in these forlorn backwaters of the land, led by a fanatic, rid of sense and devoid of conscience. Tennessee, challenging him too timorously and too late, now sees its courts converted into camp meetings and its Bill of Rights made a mock of by its sworn officers of the law" (in Gould 1983a, pp. 277–278).

So matters stood for over thirty years, until October 4, 1957, when the Soviet Union launched Sputnik I, the first orbiting artificial satellite, thereby announcing to America that, unlike political secrets, nature's secrets cannot be concealed—no nation can hold a monopoly on the laws of nature. The Sputnik scare prompted a renaissance in American science education, during which evolution worked its way back into the mainstream of public education. In 1961, the National Science Foundation, in conjunction with the Biological Science Curriculum Study, outlined a basic program for teaching the theory of evolution and published a series of biology books in which the organizing principle was evolution.

Equal Time for Genesis and Darwin

The next generation of fundamentalists and biblical literalists responded with a new approach. In the late 1960s and early 1970s, they demanded equal time for the Genesis story and the theory of evolution, and insisted that evolution was "only" a theory, not a fact, and should be designated as such. The flash point for this new fire was the 1961 publication of John Whitcomb and Henry Morris's *The Genesis Flood: The Biblical Record and Its Scientific Implications*. Whitcomb and Morris were not interested in the origins of species, as the authors themselves explained: "The geologic record may provide much valuable information concerning earth history subsequent to the finished Creation . . . but it can give no information as to the processes or sequences employed by God during the Creation, since God has plainly said that those processes no longer operate" (p. 224). The book presented classic Flood geology in a new light, and it was promoted by new creationist organizations, like the Creation Research Society, founded in 1963. These organizations helped push through creationist legislation. For example, in 1963 the state senate of Tennessee passed by a vote of 69 to 16 a bill that required all textbooks to carry a disclaimer that any idea about "the origin and creation of man and his world . . . is not represented to be scientific fact" (in Bennetta 1986, p. 21). The Bible, designated as a reference book instead of a textbook, was exempt from the disclaimer.

The bill was appealed by the National Association of Biology Teachers on First Amendment arguments. At about the same time, Susan Epperson, a high school biology teacher in Little Rock, Arkansas, filed suit against the state on the grounds that an antievolution bill passed in 1929 violated her rights to free speech. She won, but the case was overturned by the Arkansas Supreme Court in 1967 and later appealed to the U.S. Supreme Court. In 1967, Tennessee repealed its antievolution law, and in 1968, the U.S. Supreme Court found Epperson in the right. The Court viewed the 1929 Arkansas law as "an attempt to blot out a particular theory because of its supposed conflict with the biblical account" (in Cowen 1986, p. 9) and interpreted it as an attempt to establish a religious position in a public classroom. On the basis of the Establishment Clause, the Arkansas law was overturned and the Court ruled all such antievolution laws unconstitutional. This series of legal contingencies led directly to a third course of action on the part of the creationists.

Equal Time for Creation-Science and Evolution-Science

If evolution could not be excluded from the classroom, and if the teaching of religious tenets was unconstitutional, creationists needed a new strategy to gain access to public school classrooms. Enter "creation-science." In 1972, Henry Morris organized the Creation-Science Research Center as an arm of the San Diego–based Christian Heritage College. Morris and his colleagues focused on the production and distribution of *Science and Creation* booklets designed for grades 1 through 8, which they managed to introduce in twenty-eight states in 1973 and 1974, along with other tracts such as Robert Kofahl's *Handy Dandy Evolution Refuter* (1977) and Kelly Segraves's *The Creation Explanation: A Scientific Alternative to Evolution* (1975).

The argument was that since academic honesty calls for a balanced treatment of competing ideas, creation-science should be taught side-by-side with evolution-science. Backers made a clear distinction between biblical creationism, with its openly fundamentalist religious basis, and scientific creationism, which emphasized the nonreligious scientific evidence against evolution and in favor of creation. Throughout the late 1970s and 1980s, the Creation-Science Research Center, the Institute for Creation Research, the Bible Science Association, and other such organizations pressed state boards of education and textbook publishers to include the science of creation alongside the science of evolution. Their goal was clearly stated: "to reach the 63 million children of the United States with the scientific teaching of Biblical creationism" (in Overton 1985, p. 273).

On the legal end of this third strategy, in 1981 Act 590 was enacted, requiring "balanced treatment of creation-science and evolution-science in public schools. Its purposes were to protect academic freedom by providing student choice; to ensure freedom of religious exercise; to guarantee freedom of speech; . . . [and] to bar discrimination on the basis of creationist or evolutionist belief" (in Overton 1985, p. 260). According to the *California Science Teacher's Journal*, "The Statute was introduced by a Senator who hadn't written a word of it, and didn't know who had. It was debated for 15 minutes in the State Senate, there was no floor debate in the House of Representatives, and the Governor signed it without reading it" (in Cowen 1986, p. 9). Nonetheless, it was law, and a year later the state of Louisiana passed a similar bill.

The constitutionality of Act 590 was challenged on May 27, 1981, with the filing of a suit by Reverend Bill McLean and others. The case was brought to trial in Little Rock on December 7, 1981, as *McLean v. Arkansas*. The contestants were, on one side, established science, scholarly religion, and liberal teachers (backed by the ACLU) and, on the other, the Arkansas Board of Education and various creationists. Federal Judge William R. Overton of Arkansas ruled against the state on the following grounds: First, creation-science conveys "an inescapable religiosity" and is therefore unconstitutional. "Every theologian who testified," Overton explained, "including defense witnesses, expressed the opinion that the statement referred to a supernatural creation which was performed by God." Second, the creationists employed a "contrived dualism" that "assumes only two explanations for the origins of life and existence of man, plants and animals: It was either the work of a creator or it was not." Given this either-or paradigm, the creationists claim that any evidence "which fails to support the theory of evolution is necessarily scientific evidence in support of creationism." But, as Overton clarified, "Although the subject of origins of life is within the province of biology, the scientific community does not consider origins of life a part of evolutionary theory." Furthermore, he noted, "Evolution does not presuppose the absence of a creator or God and the plain inference conveyed by Section 4 [of Act 590] is erroneous." Finally, Overton summarized the arguments of expert witnesses (including Gould, Ayala, and Michael Ruse) that creation-science is not science, as the scientific enterprise is usually defined: "science is what is 'accepted by the scientific community' and is 'what scientists do.'" Overton then listed the "essential characteristics" of science as outlined by the expert witnesses: "(1) It is guided by natural law; (2) It has to be explanatory by reference to natural law; (3) It is testable against the empirical world; (4) Its conclusions are tentative . . . ; and (5) It is falsifiable." Overton concluded, "Creation-science . . . fails to meet these essential characteristics." Moreover, Overton noted, "Knowledge does not require the imprimatur of legislation in order to become science" (1985, pp. 280–283).

To the Supreme Court

Despite this decision, creationists continued their lobbying for equal-time laws and revised textbooks. But this top-down strategy of passing laws and pressuring textbook publishers was hampered by the outcome of the case against the Louisiana law. In 1985, the Louisiana law was struck down by summary judgment (i.e., without trial) in the Federal Court of Louisiana when U.S. District Judge Adrian Duplantier ruled in concurrence with Overton that creation-science was actually religious dogma. Judge Duplantier's decision ignored the characteristics of science, centering instead on a religious argument—that teaching creation-science requires teaching the existence of a divine creator, which is in violation of the Establishment Clause. Despite the fact that over a thousand pages dealing with the characteristics of science were filed, Judge Duplantier declined "the invitation to judge that debate" (in Thomas 1986, p. 50). The decision was appealed to the U.S. Court of Appeals for the Fifth Circuit, where the value of that debate was argued. That court, initially with a panel of three judges and subsequently en banc with all fifteen judges voting, agreed with the district court that the statute was unconstitutional.

But when a federal court holds a state statute unconstitutional, by "mandatory jurisdiction," the U.S. Supreme Court must hear the case. And since the vote was only 8 to 7, Louisiana submitted a "jurisdictional statement," thus establishing a substantial federal question. At least four of the nine Supreme Court justices concurred that it was substantial, and by the "rule of four" agreed they would hear the case. The initial oral arguments in *Edwards v. Aguillard* were made on December 10, 1986, with Wendell Bird representing the appellants, and Jay Topkis and the ACLU the appellees. Bird first argued that because of some confusion about what the Louisiana statute means, "a trial, with factual development, ought to occur to enable expert witnesses on both sides to give definitions" (*Official Transcript Proceedings* 1986 [hereafter *OTP*], p. 8). After lengthy discussion of the "actual" intent of the Louisiana statute, Bird pushed the "academic freedom concern"—the "rights" of students to a balanced treatment of evolution and creation (p. 14).

Using a minimalist approach, and responding to the focus of Duplantier's decision, Topkis argued that creation-science was merely religion posing as science and was therefore unconstitutional. In this instance, however, the argument failed on the grounds that if the science were valid, it should have a place in the curriculum of public school science classes, no matter what its relation to religion. The justices' historical analogies brilliantly countered Topkis's arguments. For example, Chief Justice William Rehnquist demonstrated to Topkis that it is possible to believe in the creation of life by God with no religious intent (*OTP*, pp. 35–36).

Rehnquist: My next question is going to be whether you considered Aristotelianism a religion?

Topkis: Of course not.

Rehnquist: Well, then, you could believe in a first cause, an unmoved mover, that may be impersonal, and has no obligation of obedience or veneration from men, and in fact, doesn't care what's happening to mankind.

Topkis: Right.

Rehnquist: And believe in creation.

Topkis: Not when creation means creation by a divine creator.

Rehnquist: And I ask you, it depends on what you mean by divine. If all you mean is a first cause, an impersonal mover—

Topkis: Divine, Your Honor, has connotations beyond, I respectfully submit.

Rehnquist: But the statute doesn't say "divine."

Topkis: No.

Rehnquist: All it says is "creation."

Later in the arguments, Justice Antonin Scalia became "concerned about whether purpose alone would invalidate a State action, if a State action has a perfectly valid secular purpose," and drove home the issue with an even more enlightening historical argument about the irrelevancy of intent:

> Let's assume that there is an ancient history professor in a State high school who has been teaching that the Roman Empire did not extend to the southern shore of the Mediterranean in the first century A.D. And let's assume a group of Protestants who are concerned about that fact, inasmuch as it makes it seem that the Biblical story of the crucifixion has things a bit wrong—because of that concern, and really, no other reason—I mean, this fellow's also teaching other things that are wrong. He's teaching that the Parthians came out of Egypt. They don't care about that. They do care that Romans were in Jerusalem in the first century A.D. So they go to the principal of the school, and say, this history professor is teaching what is just falsehood. I mean, everybody knows that Rome was there. And the principal says, gee, you're right. And he goes in and directs the teacher to teach that Rome was on the southern shore of the Mediterranean in the first century A.D. Clearly a religious motivation. The only reason the people were concerned about that, as opposed to the Parthians, was the fact that it contradicted their religious view. Now, would it be unconstitutional for the principal to listen to them, and on the basis of that religious motivation, to make the change in the high school? (pp. 40–41)

Justice Lewis Powell followed with still another historical example about a hypothetical school presenting "only the Protestant view of the Reformation in their medieval history classes," with Catholics demanding equal time on religious grounds. The Catholics' demands would be historically tenable, so Powell inquired whether their demands would "raise any problems." Topkis responded, "So long as the purpose of the school authorities, in taking this position, was an historical purpose rather than a religious one, I couldn't quarrel with it" (pp. 47–48).

After Powell joined Rehnquist and Scalia in questioning whether the religious motives of the appellants were sufficient to call into question the legitimacy of their claims on behalf of creation-science, it seemed that Topkis's minimalist strategy of establishing religious intent was about to backfire and that there was a real possibility that the Louisiana statute would be upheld.

Science Defended

One of the appellees' witnesses in the trial, Stephen Jay Gould, in a letter to Jack Novik of the ACLU dated December 15, 1986, noted that Topkis was "nailed, absolutely nailed, by both Scalia and Rehnquist (the last two men in America I thought I'd ever be praising, but they were spot on in this)." Gould continued, "I entered with the conviction that we had four votes for sure (Brennan, Marshall, Blackmun, and Stevens), they had two (Rehnquist and Scalia), and that we probably had our key fifth vote in Powell, and probably a sixth and maybe even a seventh in O'Connor and White. I am no longer so sure that I know where the fifth vote will come from. Am I unduly pessimistic?" At the time, possibly not. After all, Topkis and the ACLU were using the very strategy preferred by creationists whenever they debate evolutionists: go on the offensive and say nothing about your own position so that you do not need to be defensive. Gould expressed his extreme frustration when he wrote to Novik: "It would have been sad enough if we had only argued badly. But I feel especially down-hearted because I think that we also argued indecently as well. We did the very thing that we have always accused the creationists of promoting—argument by innuendo rather than content. I never thought it could happen. We were not honorable. I feel like the little boy tugging on Shoeless Joe Jackson's sleeve—'say it ain't so, Jack.' Am I wrong?" If the key fifth vote could not be swung, the Louisiana appeal would be successful, negating Judge Overton's decision in the Arkansas trial and setting a precedent for other states to pass their own equal-time laws.

Since the argument attacking the religious motivations of the creationists was not valid in the view of the Court, another tack was needed. Denying the scientific content of creation-science seemed to be the only hope for the appellees. What was needed was a clear-cut and succinct definition of science so that the Court could see that the scientific content of creation-science failed to meet criteria that would legitimize its claim to "scientific" standing.

In spite of centuries of attention by scientists and philosophers of science, no concise definition of science has ever been accepted by the community of scientists and scholars. This situation changed temporarily with the *amicus curiae* brief submitted on August 18, 1986, to the Supreme Court. For this brief, the *amici* managed to define and agree upon the nature and scope of science. The brief was instigated by Murray Gell-Mann, Paul MacCready, and other members of the Southern California Skeptics Society after they read in the *Los Angeles Times* that the U.S.

Supreme Court had agreed to hear the Louisiana case. Worried, they contacted attorney Jeffrey Lehman, who had recently clerked for Justice John Paul Stevens. Lehman told them that "an *amicus* brief is the proper way for independent outsiders to present their views to the Supreme Court" (Lehman 1989).

The idea was born in March 1986. The brief would have to be submitted in five months. Time was of the essence. Lehman enlisted the help of Beth Kaufman, a colleague with expertise on the Establishment Clause. William Bennetta, a historian of the creationist movement, flew to Washington, D.C., to brief Lehman and Kaufman. Gell-Mann sent letters to state academies of science and to Nobel laureates in science and medicine in which he outlined the goals of the brief—which included showing that the language of the statute "displays and propagates misconceptions about the processes and vocabulary of science, that enforcement of the statute would promote the confusion of science with religion, and that such enforcement would subvert and distort efforts to teach well-established scientific conclusions about cosmic, planetary, and organic evolution." As a result, Gell-Mann noted, the statute "can be explained only as an attempt to misrepresent science for the sake of promoting fundamentalist religion" (letter to Nobel laureates, June 25, 1986).

The scientific community responded thoroughly and positively. For example, the Iowa Academy of Science joined the *amici* and sent Gell-Mann a copy of their position statement on "creationism as a scientific explanation of natural phenomena." Nobel laureate Leon N. Cooper accepted the invitation and sent Gell-Mann a copy of a lecture he had given on creation-science. The president of the Institute of Medicine, Samuel O. Thier, offered Gell-Mann his best wishes but declined to join only because the institute was filing its own *amicus* brief.

As it turned out, because the oral arguments went so badly, the briefs were significantly more important than anyone had anticipated. In a letter sent the same day as the one to Novik, Gould expressed his disappointment and concern to Gell-Mann (and revealed the level of his emotional commitment to the defense of science against the creationists): "God, I never thought those bozos could ever possibly come off better than our side in a high-level argument where it really mattered. But there is another side to all this. Our oral argument was so bad that our only hope now resides in the briefs. This makes what you did in securing the Nobelist brief all the more important, indeed probably crucial. And so I write, on behalf of the entire company of evolutionary biologists, to thank you for taking so much time for such important service in the truly common defense." Gell-Mann recalled that "we were very upset about the oral presentation. It wasn't that creationists are religious. Lots of scientists are religious. It's that they are claiming to be presenting science when it is really just total nonsense. It would be like the Flat Earth Society insisting their theory be taught in the public schools" (1990).

Science Defined

The *amicus curiae* brief was written primarily by Jeffrey Lehman, with input from Kaufman, Gell-Mann, Bennetta, and others. Lehman said that the "difficulty in writing this brief from a lawyer's point of view was to clarify what makes science different from religion, and why creationism isn't scientific. When I talked with scientists they weren't at all clear in trying to briefly define what they do" (1989). The brief is concise (twenty-seven pages), well-documented (thirty-two lengthy footnotes), and argues that creation-science, on the one hand, is just a new label for the old religious doctrines of decades past and, on the other, does not meet the criteria of "science" as defined in the brief by the *amici*.

The first argument is stated directly: "The term 'creation-science' in the act embodies religious dogma, not the sterilized 'abrupt-appearance' construct propounded by appellants in this litigation" (*Amicus curiae* brief 1986 [hereafter *AC*], p. 5). In the repackaging of their position, the creationists removed God from their arguments by "sterilizing" the creation act as "origin through abrupt appearance in complex form of biological life, life itself, and the physical universe" (p. 6). Kaufman explained, "We argued that the 'abrupt-appearance' construct is not a sufficiently well defined alternative to orthodox 'creation-science.' It fails to define a concrete alternative to evolution; accordingly, it is implausible that the Louisiana legislature intended the Act to embody it.... Therefore, the sterilized 'abrupt-appearance' construct can only be understood as a post hoc explanation, erected for the purpose of defending this unconstitutional Act" (1986, p. 5). A review of the creationist literature reveals that the creationists have merely substituted words, not belief. For example, members of the Creation Research Society must subscribe to the following "statement of belief" (in *AC*, p. 10):

> (1) The Bible is the written Word of God ... all of its assertions are historically and scientifically true in all of the original autographs. ... This means that the account of origins in Genesis is a factual presentation of simple historical truths. (2) All basic types of living things, including man, were made

by direct creative acts of God during Creation Week as described in Genesis. Whatever biological changes have occurred since Creation have accomplished only changes within the original created kinds. (3) The great Flood described in Genesis, commonly referred to as the Noachian Deluge, was an historical event, worldwide in its extent and effect. (4) Finally, we are an organization of Christian men of science, who accept Jesus Christ as our Lord and Savior. The account of the special creation of Adam and Eve as one man and one woman, and their subsequent Fall into sin, is the basis for our belief in the necessity of a Savior for all mankind. Therefore, salvation can come only thru accepting Jesus Christ as our Savior.

Similar statements issued by the Institute for Creation Research and other creationists make it clear that creationists prefer the authority of the Bible over any possibly contradictory empirical evidence. This lack of interest in empirical data is outlined in the brief to demonstrate that creation-science is not "scientific," as the *amici* would insist in the second section, in which a definition of science would have to be established and agreed upon. This second section begins by offering a very general definition: "Science is devoted to formulating and testing naturalistic explanations for natural phenomena. It is a process for systematically collecting and recording data about the physical world, then categorizing and studying the collected data in an effort to infer the principles of nature that best explain the observed phenomena." Next, the scientific method is discussed, beginning with the collection of "facts," the data of the world. "The grist for the mill of scientific inquiry is an ever increasing body of observations that give information about underlying 'facts.' Facts are the properties of natural phenomena. The scientific method involves the rigorous, methodical testing of principles that might present a naturalistic explanation for those facts" (p. 23).

Based on well-established facts, testable hypotheses are formed. The process of testing "leads scientists to accord a special dignity to those hypotheses that accumulate substantial observational or experimental support." This "special dignity" is called a "theory." When a theory "explains a large and diverse body of facts," it is considered "robust"; if it "consistently predicts new phenomena that are subsequently observed," then it is considered "reliable." Facts and theories are not to be used interchangeably. Facts are the world's data; theories are explanatory ideas about those facts. "An explanatory principle is not to be confused with the data it seeks to explain." Constructs and other nontestable statements are not a part of science. "An explanatory principle that by its nature cannot be tested is outside the realm of science." Thus, science seeks only naturalistic explanations for phenomena. "Science is not equipped to evaluate supernatural explanations for our observations; without passing judgment on the truth or falsity of supernatural explanations, science leaves their consideration to the domain of religious faith" (pp. 23–24).

It follows from the nature of the scientific method that no explanatory principles in science are final. "Even the most robust and reliable theory . . . is tentative. A scientific theory is forever subject to reexamination and—as in the case of Ptolemaic astronomy—may ultimately be rejected after centuries of viability." The creationists' *certainty* stands in sharp contrast with the *uncertainty* scientists encounter as a regular and natural part of their work. "In an ideal world, every science course would include repeated reminders that each theory presented to explain our observations of the universe carries this qualification: 'as far as we know now, from examining the evidence available to us today'" (p. 24). But, as Gell-Mann remarked, the creationists have an obsession "with the inerrancy of the Bible. It doesn't matter what the evidence is, they will continue to believe their doctrines to the end." Thus, Gell-Mann noted, the creationists "aren't doing science. They just insert the word":

> It reminds me of a Monty Python routine where a guy goes into a pet store to get his fish a license. He is told they don't make fish licenses. He replies that he has a cat license, so why can't he get a fish license? but is told they don't make cat licenses either. So he shows the pet store owner his cat license. "That's not a cat license," the owner responds. "That's a dog license. You just scratched out the word 'dog' and wrote in 'cat.'" That's all the creationists are doing. They've just scratched out "religion" and in its place put "science." (1990)

According to the *amici*, any body of knowledge accumulated within the guidelines they described is considered "scientific" and suitable for public school education; and any body of knowledge not accumulated within these guidelines is not considered scientific. "Because the scope of scientific inquiry is consciously limited to the search for naturalistic principles, science remains free of religious dogma and is thus an appropriate subject for public-school instruction" (*AC*, p. 23). By this line of reasoning, in singling out evolutionary theory as "speculative and baseless" compared to other "proven scientific facts" the Louisiana law is not consistent. Rather, even though the theory of evolution is considered by virtually all biologists to be as robust and reliable as any in science, it has attracted the attention of the creationists because they perceive it as directly opposing their static and inflexible religious beliefs. The *amici* thus conclude, "The Act, however construed, is structured to 'convey a message that religion or a particular religious belief is favored or preferred,'" and is thus unconstitutional (p. 26).

Creationists Respond

Calling the scientific community "scared," and the brief "the last hurrah on behalf of the dominance the teaching of evolutionism has had in our public schools," the Creation Research Legal Defense Fund immediately took up a collection to support its stand against the *amicus* brief. Noting that the brief had struck a "significant blow," a fund-raising letter requested creationists to "please pray about sending us the best possible gift you can." It told readers that this was a "David vs. Goliath battle" and reminded them that in the original confrontation "Goliath died and David became King of Israel." Finally, the letter noted the Nobelists' "atheistic orientation" and stated that the Nobelists "realize this is the most important court case they have ever faced—even more important than the original Scopes Trial" because their own "religion of secular humanism" was at stake.

After calling the press conference "media propaganda," and the brief a "clever ploy by the evolutionary establishment," Henry Morris was no less vitriolic in an issue of *Acts and Facts*, a publication of the Institute for Creation Research. "To keep this prestigious 'brief' in proper perspective . . . it should be remembered that Nobel scientists are probably no better informed on the creation/evolution question than any other group of people," Morris contended, leaving us to wonder what other group of people Morris had in mind to compare with seventy-two Nobel laureates. Morris did admit that the brief would "no doubt have much influence" but hoped "that most fair-minded people will see through it." In arguing for the scientific basis of creationism, Morris stated that not only are there "thousands of fully qualified scientists today who are creationists" but the "founding fathers of science," such as "Newton, Kepler, Pascal, and others," were also creationists and were "at least as knowledgeable in science as these modern Nobelists" (in Kaufman 1986, pp. 5–6).

Finally, an emotional commitment to their position by the creationists that matched that of the evolutionists was revealed in personal letters sent by rank-and-file creationists to some of the Nobelists. One letter sent to Gell-Mann said, "The blood of Jesus Christ cleanses us from all sin. Whosoever is not found written in the book of life will be cast into the lake of fire. The wages of sin is death, but the gift of God is eternal life through Jesus Christ our Lord. Ask the Lord Jesus to save you now! The second law of thermodynamics proves evolution is impossible. Why are you so afraid of the truth of creation-science?"

The U.S. Supreme Court Justices Respond

The case from the U.S. Court of Appeals for the Fifth Circuit, No. 85-1513, was argued before the U.S. Supreme Court on December 10, 1986, and decided June 19, 1987. The Supreme Court voted 7 to 2 in favor of the appellees. The Court held that "the Act is facially invalid as violative of the Establishment Clause of the First Amendment, because it lacks a clear secular purpose" and that "[t]he Act impermissibly endorses religion by advancing the religious belief that a supernatural being created human-kind" (*Syllabus* 1987, p. 1). Did the brief swing votes? It is hard to say. The key fifth vote that the brief probably swung was Justice Byron White's, whose short, two-page concurring opinion closely parallels section D, page 21, of the brief. Lehman noted that "insiders have told me that 'loose lips' in the court say that the brief mattered in the Justices' decision" (1989).

Justice William Brennan delivered the opinion of the Court, joined by Justices Thurgood Marshall, Harry Blackmun, Powell, Stevens, and Sandra Day O'Connor. White filed a separate but concurring opinion, as did Powell and O'Connor, who wanted "to emphasize that nothing in the Court's opinion diminishes the traditionally broad discretion accorded state and local school officials in the selection of the public school curriculum" (*Syllabus* 1987, p. 25). Scalia and Rehnquist filed a dissenting opinion, in which they argued (as in the oral arguments of December 10) that "so long as there was a genuine secular purpose" the Christian fundamentalist intent "would not suffice to invalidate the Act." Recalling the academic freedom issue as argued in the Scopes trial, Scalia and Rehnquist noted, "The people of Louisiana, including those who are Christian fundamental-ists, are quite entitled, as a secular matter, to have whatever scientific evidence there may be against evolution presented in their schools, just as Mr. Scopes was entitled to present whatever scientific evidence there was for it" (p. 25).

The creationists' "secular" integrity becomes questionable, however, under the weight of the following, progressively bolder statements, which scientists would argue are completely fallacious: "The body of scientific evidence supporting creation-science is as strong as that supporting evolution. In fact, it may be *stronger*"; "The evidence for evolution is far less compelling than we have been led to believe. Evolution is not a scientific 'fact,' since it cannot actually be observed in a laboratory. Rather, evolution is merely a scientific theory or guess"; "It is a very bad guess at that. The scientific problems with evolution are so serious that it could accurately be termed a 'myth'" (*Syllabus* 1987, p. 14).

Science Unified

The Louisiana trial in general, and the *amicus* brief in particular, had the effect of temporarily galvanizing the scientific community into not only *defending* science as a way of understanding the world that is different from religion but *defining* science as a body of knowledge accumulated through a particular method—the scientific method. Calling the case "the single biggest thrill of my practicing career as a lawyer," Lehman observed that "this issue more than anything else crystallizes what it means to be a scientist" (1989).

The event has significance in the history of science in that it unified a diverse group of individuals perhaps best characterized by their fierce independence. Nobel laureate Arno Penzias said the commonality among the Nobel laureates on the creationism case was unusual and that he could not imagine another issue receiving such support. Among the other Nobel Prize winning signers of the brief were individuals with whom Penzias "often had violent arguments on other issues" (Kaufman 1986, p. 6).

It would seem that there are two possible explanations for this unity. First, the scientific community felt itself directly under attack from the outside and, as social psychologists have demonstrated, in such conditions almost *any* group will respond by circling the wagons. A social psychologist might find this a most enlightening and instructive study of the process of "deindividuation," in which individuals temporarily suppress conflicts within a group in order to defend themselves from a perceived common enemy. As Nobel laureate Val Fitch observed, "When scientific method and education are attacked, the laureates close ranks and speak with one voice" (Kaufman 1986, p. 6).

Yet scientists have encountered "outside forces" before and have not responded quite so collectively and emotionally. A second factor in explaining the unification in the Louisiana case may be the scientists' nearly unanimous perception that the creationists' position lacked any validity whatsoever. As Fitch noted, the Louisiana creationism attack was turned back with unprecedented collective force because "it defies all scientific reason." Gell-Mann agrees: "That's right. It's not so much that we were being attacked from the outside, since outsiders can make worthwhile contributions. It's that these people were talking utter nonsense" (1990).

These two components explain why the defense and definition of science was an interim one—lasting for the duration of the case and left there to be recalled should similar circumstances again arise. Certainly philosophers of science have not suspended their research into the nature of science and the scientific method with the publication of the brief. This agreement was made politically, not philosophically. In our democratic society such conflicts are solved (if only for a while) by a vote. In the Louisiana case, the vote was taken and the Court followed the advice of the defenders and definers of science—the scientists themselves.

PART 4

HISTORY

AND

PSEUDOHISTORY

We believe we can construct a past that is veritable, that is accurate in terms of actual past events, since the past has left its mark in the present. The message of this book has been that, while there are many different possibilities, not all of these constructed pasts—not all of the possibilities—are equally plausible. Ultimately, then, we get the past we deserve. In every generation, thinkers, writers, scholars, charlatans, and kooks (these are not necessarily mutually exclusive categories) attempt to cast the past in an image either they or the public desire or find comforting. We deserve better and can do better than weave a past from the whole cloth of fantasy and fiction.

—Kenneth L. Feder, *Frauds, Myths, and Mysteries:*
Science and Pseudoscience in Archaeology, 1986

12

Doing Donahue

History, Censorship, and Free Speech

On March 14, 1994, Phil Donahue became the first of the talk-show hosts to address the Holocaust deniers, who claim that this event was radically different from what we have all come to accept. Many of the major talk shows had considered doing something on the subject, yet for a variety of reasons had not done so before. Montel Williams had taped a program on April 30, 1992, but it was pulled from major markets because, according to deniers, they looked too good and the Holocaust scholar offered nothing better than *ad hominem* attacks. I saw the show, and the deniers were correct. If it had been a fight, they would have stopped it.

The *Donahue* producer promised us that there would be no skinheads or neo-Nazis, nor would the show be allowed to erupt into violence or degenerate into mere shouting. The deniers—Bradley Smith, who places advertisements in college newspapers, and David Cole, the young Jewish video producer who primarily focuses on denying that gas chambers and crematoria were used for mass murder—were promised that they would be allowed to make their claims. I, in turn, was promised that I could properly answer their arguments. Edith Glueck, who had been in Auschwitz, albeit for only a few weeks, also appeared on the show, and her close friend, Judith Berg, who had been in Auschwitz for seven months, was seated in the studio audience. What was promised was quite different from what actually unfolded on the air.

Five minutes before the show, the producer came into the Green Room, panic-stricken. "Phil is very concerned about this show. He is in over his head and is worried it might not come off well." In the weeks prior to the show, I had prepared a list of denier claims and constructed sound-bite replies, so I assured the producer that I was ready to answer all the deniers' claims and told him not to worry.

Donahue opened the show with these words: "How do we know the Holocaust really happened? And what proof do we have that even one Jew was killed in a gas chamber?" As the producers rolled stock footage from Nazi concentration camps, Donahue continued:

> In just the last six months, fifteen college newspapers across the country have run advertisements that call for an open debate of the Holocaust. The ad claims that the U.S. Holocaust Memorial Museum in Washington, D.C., has no proof whatever of homicidal gassing chambers, and no proof that even one individual was gassed in a German program of genocide. The ads have caused an uproar everywhere, sparking protests from students and boycotts of the papers. The man who placed all the ads, Bradley Smith, has been called anti-semitic and a neo-Nazi because of the challenges of the Holocaust. Smith claims he simply wants the truth to be told—that Jews were never placed in gas chambers and that the figure of six million Jewish deaths is an irresponsible exaggeration. And he is not alone in his beliefs. A recent poll by the Roper organization found that 22 percent of all Americans believe it's possible the Holocaust never happened. Another 12 percent say they don't know. So in a time when over five thousand visitors are crowding the new Holocaust museum every day, and the film *Schindler's List* is reducing jaded movie-goers to tears, the question should be asked, How can anyone claim the Holocaust was a hoax?

It was obvious from the start that Donahue was, indeed, in over his head. He knew little about the Holocaust and even less about the debating style of the deniers. He immediately tried to reduce the discussion to accusations of antisemitism.

Donahue: You do not deny that antisemitism in Europe in the '30s, most especially Germany, Poland, and environs, was visceral and that Hitler . . .

Smith: We're not talking about any of that. Listen . . .

Donahue: Please don't be upset with my questions.

Smith: I'm not upset. But the question is outside the parameter of the issue. I'm running an advertisement that says the museum . . .

Donahue: We're three minutes into this program and you don't like my question.

Smith: The question has nothing to do with what I'm doing.

Donahue: Do you believe that there was engineered by Hitler and the Third Reich a strategy of eliminating Jews called the Final Solution? Do you believe that?

With this question, it looked like Phil was going to zero in on one of the deniers' major points—the moral equivalence argument that in times of war all people are treated badly and that the Nazis were no worse than the other major combatants in this and other wars. But Smith moved Donahue right by this issue.

Smith: I don't believe it anymore. I used to. But that's not what I'm talking about. If you don't understand what I'm talking about you won't ask the right question. The question is this. We have a $200 million museum in Washington, D.C. It's in America. It's not in Europe. And the whole museum is dedicated to the proposition that Jews were killed in gas chambers. They don't have any proof in the museum that Jews were killed in gas chambers. As a matter of fact, they are so sure that guys like you will never ask them the question . . .

Donahue: Guys like me? [Audience laughter.]

This sort of patter went on for another fifteen minutes, with Donahue continually returning to the issue of antisemitism, and Smith and Cole desperately trying to make their points that the Holocaust is debatable and that the camp gas chambers and crematoria were not used to kill prisoners. David Cole showed some of his footage from Auschwitz and Majdanek, and began discussing Zyklon-B trace deposits and other technical matters. Assuming that this was over the heads of his audience, Donahue switched to trying to associate Cole with the noted neo-Nazi, Ernst Zündel.

Donahue: David, you are familiar, and know, and have traveled with Ernst Zündel. Is that so?

Cole: No, I have not traveled with Ernst Zündel.

Donahue: Did you meet him in Poland?

Cole: I met him in Poland. I met him twice in my entire life.

Donahue: All right, what did you do, have a beer? I mean, what's travel mean? [Audience laughter.] You met him in Poland. He is a neo-Nazi. You don't deny that?

Cole: No, I'm sorry Phil. This is not about who I've met in my life. I just met you. Does that mean I'm Marlo Thomas? [Loud audience laughter.] This is about physical evidence. This is about Zyklon-B residue. This is about windows in a gas chamber . . .

Donahue: Were you bar mitzvahed David?

Cole: I'm an atheist. I made that clear to your production staff.

This meaningless chatter went on for several more minutes until a commercial break. The producer, page, make-up artist, and microphone technician now escorted me into the studio. My entrance had the look and feel of a prizefighter going into the ring. The producer told me to stay away from the technical matters and stick to analyzing their methods. In the days prior to the show, he had interviewed me extensively and I had told him everything I would say. There should have been no surprises.

I launched into my presentation, knowing that I only had a few minutes. After summarizing the methods of deniers, I began to move into their specific claims. Now was the time to put up on the screen the photographs and blueprints of gas chambers and crematoria and the short quotes about "elimination" and "extermination" of Jews that I had provided. Instead, Donahue showed film footage from Dachau, now known *not* to have been an extermination camp. Unfortunately, no one had told Donahue where the footage was taken or anything else about it. Cole promptly nailed him.

Cole: I'd like to ask Dr. Shermer a question. They just showed the Dachau gas chamber in that footage. Is that gas chamber ever claimed to have killed people?

Shermer: No. And in fact, the important point here . . .

Donahue: There is a sign at Dachau notifying tourists of that fact.

Cole: That it was not used to kill people. So why did you just show it in the clip?

Donahue: I'm not at all sure that was Dachau.

> **Cole:** Oh, that was Dachau. Now wait a minute. You're not sure that was Dachau? *You* show a clip on *your* show and you're not sure it was Dachau?

I jumped in to try to redirect the discussion back to the point: "History is knowledge and like all knowledge it progresses and changes. We continually refine our certainty about claims. . . . And that's what historical revisionism is all about." Meanwhile, David Cole left the studio, disgusted that he had not been allowed to make his points. Donahue said, "Let him walk!"

Thinking that I had done fairly well in analyzing the methodologies of the deniers, I was comfortably awaiting the next segment when the producer came running over to me. "Shermer, what are you doing? *What are you doing?* You need to be more aggressive. My boss is furious. Come on!" I was shocked. Apparently either Donahue thought the Holocaust deniers could be refuted in a matter of minutes or he was hoping I would just call them antisemites as he did and be done with it. It was suddenly obvious that Donahue was not privy to the briefing I had given the producer. As I anxiously tried to think of new things to say, the studio audience and callers started asking questions, resulting in talk-show chaos.

One caller wanted to know why Smith was doing this to the Jews. The ensuing exchange demonstrated the problem of having a host and guests who are not prepared to deal with the specific claims and tactics of the deniers.

> **Smith:** One of the problems here is we have a feeling that if we talk about this issue nobody is involved but Jews. Germans are involved. For instance, if we tell, there is something vulgar about lying about Germans and thinking that it's proper. For example, it was a lie that Germans cooked Jews to make soap from them. It was a lie . . .
>
> **Shermer:** No, not a lie. It's a mistake . . .
>
> **Judith Berg** [from the front row]: It was true. They made lampshades and they cooked soap. That's true.
>
> **Smith:** Ask the professor.
>
> **Shermer:** Excuse me, historians make mistakes. Everybody makes mistakes. We're always refining our knowledge, and some of these things come down and they don't turn out to be true. But let me tell you what I think is going on here . . .

Smith: Ask why they're doing that to this woman. Why have they taught this woman to believe that the Germans cooked and skinned . . .

Berg [jumps out of seat, screaming]: I was seven months in Auschwitz. I lived near the crematorium as far as I am from you. I smelled. . . . You would never eat roast chicken if you had been there. Because I smelled . . .

Smith: Let's get to the bottom of one thing. She says soap and lampshades. The professor says you're mistaken.

Berg: Even the Germans admit it. They admit it that they had lampshades . . .

Donahue [to Smith]: Do you have any empathy at all? . . . Are you concerned about the pain that you cause this woman?

Smith: Sure, but why should we ignore the Germans who are accused of this despicable story?

Berg [in an emotion-filled voice, pointing finger at Smith]: I was seven months there. If you are blind someone else can see it. I was seven months there . . .

Smith: What does that have to do with soap? No soap, no lampshades. The professor says you're wrong, that's all.

Berg: He wasn't there. The people there told me not to use that [soap] because it could be your mother.

Smith: A doctor of history, Occidental College. He says you're mistaken.

Because Mrs. Berg had told me that she had seen Nazis burning large numbers of bodies in an open field, I began to explain: "They burnt bodies in mass graves . . . " but I was cut off when Donahue broke for a commercial.

Before the show, I had told both Mrs. Berg and Mrs. Glueck not to exaggerate or embellish anything, to just tell the audience exactly what they remembered. Most survivors know little about the Holocaust outside of what happened to them half a century ago, and deniers are good at tripping them up when they get dates wrong or, worse, claim they saw someone or something they could not have seen. By turning her actual experience of seeing burning bodies into evidence for human soap, Mrs. Berg provided a perfect setup, and Smith capitalized on it. He not only avoided the issue of burning bodies and undermined the credibility of what Mrs. Berg *did* see but also managed to make it look as if I and other Holocaust historians were on his side. Donahue, having exhausted his knowledge of the Holocaust, returned to the free-speech issues and, once again, antisemitism and *ad hominem* attacks on Smith's character and credentials. During each of the subsequent segments, the producer stood on the sidelines pointing at me and mouthing, "Say something! Say something!"

Because of the chaos during the commercials and stimulation overload during the show, it was difficult for me to know how the program was perceived by viewers. I thought that it was a total disaster and the deniers had bested me, that I had made a fool of myself in front of my colleagues and let down the historical profession. Apparently, that was not the case. I have received hundreds of calls and letters from historians and the general public telling me that the deniers looked like cold-hearted buffoons and that I was the only one who kept his cool throughout the mayhem of the program.

I have also received letters and calls that focus on another issue. One Holocaust scholar was furious with me for accepting an invitation to "debate" the deniers (if you can call what happens on a talk show a debate). Had it not been for me, she argued mistakenly, there would have been no show. In a private correspondence, she told me that she was "amazed" that I "would be naive enough to allow yourself to be drawn into making them the other side." How one should respond to claims one finds repugnant is a personal matter. But we should consider the ramifications of not responding. For example, when I speak with Holocaust scholars, they occasionally will say something like "Off the record, I do not place much validity in survivors' testimony because their memories are faulty" or "Off the record, the deniers have identified some things that need further research." In my opinion, trying to keep these things off the record is going to backfire on historians. The deniers already know these things and are publicizing them. Do we want the public to think that we are covering up "problems" with the Holocaust story or that we have somehow missed these things? At every lecture I have ever given on Holocaust denial, when I state that the human soap story is generally a myth, audiences are shocked. No one but Holocaust historians and Holocaust deniers seems to know that the mass production of soap from Jews is a myth. (According to Berenbaum [1994] and Hilberg [1994], no bar of soap has ever tested positive for human fat.) Do we want the Bradley Smiths and the David Coles of the world explaining such things to the public? By keeping silent on such important issues, our inaction may come back to haunt us.

Of course, Holocaust historians are reluctant to speak out on such important issues because Holocaust deniers use such statements ruthlessly against the Holocaust. Consider the case of Elizabeth Loftus. In 1991, world-renowned memory expert and University of Washington psychology professor Elizabeth Loftus published her autobiographical work, *Witness for the Defense*. Loftus is well known for the stand she has taken against the abuse of "memory recovery" therapies. Through her research, she has shown that memory is not as reliable as we would like to think.

> As new bits and pieces of information are added into long-term memory, the old memories are removed, replaced, crumpled up, or shoved into corners. Memories don't just fade . . . they also grow. What fades is the initial perception, the actual experience of the events. But every time we recall an event, we must reconstruct the memory, and with each recollection the memory may be changed—colored by succeeding events, other people's recollections or suggestions. . . . Truth and reality, when seen through the filter of our memories, are not objective facts but subjective, interpretative realities. (Loftus and Ketcham 1991, p. 20)

In 1987, Loftus was asked to testify for the defense of John Demjanjuk, the Ukrainian-born Cleveland autoworker who was tried in Israel for allegedly helping to kill hundreds of thousands of Jews at Treblinka, where he was said to have been known as "Ivan the Terrible." The problem was in proving that Demjanjuk was Ivan. One witness, Abraham Goldfarb, first stated that Ivan was killed in a 1943 uprising but later identified Demjanjuk as Ivan. Another witness, Eugen Turowski, who initially had no recognition of Demjanjuk, announced after Goldfarb's testimony that Demjanjuk was Ivan. All five witnesses who positively identified Demjanjuk lived in Israel and had attended a commemoration of the Treblinka uprising in Tel Aviv. But twenty-three other Treblinka survivors did not make a positive identification.

Loftus was caught in a dilemma: "'If I take the case,' I explained, having talked this out with myself hundreds of times, 'I would turn my back on my Jewish heritage. If I don't take the case, I would turn my back on everything I've worked for in the last fifteen years. To be true to my work, I must judge the case as I have judged every case before it. If there are problems with the eyewitness identifications I must testify. It's the consistent thing to do'" (p. 232). Loftus then asked a close Jewish friend for advice. The answer was clear: "'Beth, please. Tell me you said no. Tell me you will not take this case.'" Loftus explained that there was a possibility of mistaken identity based on old and faulty memories. "'How could you?'" was the friend's reaction. "'Ilene, please try to understand. This is my work. I have to look beyond the emotions, to the issues here. I can't just automatically assume he's guilty.'" In the ultimate choice between loyalty to one's people and loyalty to the search for truth, Loftus's friend made it clear which she should choose. "I knew that in her heart she believed I had betrayed her. Worse than that, much worse, I had betrayed my people, my heritage, my race. I had betrayed them all for thinking that there might be a possibility that John Demjanjuk was innocent" (p. 229).

John Demjanjuk was indeed found innocent by the Israeli Supreme Court. Loftus went to Israel to watch the trial but chose not to testify. Her explanation reveals the human side of science: "As I looked around the audience filled with four generations of Jews . . . it was as if these were my relatives, and I, too, had lost someone I loved in the Treblinka death camp. With those kinds of feelings inside me, I couldn't suddenly switch roles and become a professional, an expert. . . . I couldn't do it. It was as simple and agonizing as that" (p. 237).

I have great respect for Loftus and her work, and considerable regard for the courage it took to make such an honest and soul-searching confession. But do you know how I heard about this story? From the deniers, who sent me a review of the book from their own journal, in which it was claimed that "Loftus is perhaps more culpable than the elderly persons who bore false witness against the defendant. For unlike the aging witnesses who were no longer able to distinguish truth from falsehood, and who had come to believe their own false testimony, Loftus knew better" (Cobden 1991, p. 249). I met Loftus at a conference and talked to her at length about how the deniers were using her work. She was shocked and had no idea this was happening. No wonder Holocaust historians are tempted to keep dilemmas under wraps.

Loftus is just one example among many of how personal and public censorship can backfire. Consider two more.

1. In the February 1995 issue (released in January) of *Marco Polo*, one of nine weekly and monthly magazines published by the highly respected Japanese publishing firm Bungei Shunju, appeared an article entitled "The Greatest Taboo of Postwar World History: There Were No Nazi 'Gas Chambers.'" The article was written by Dr. Masanori Nishioka, a thirty-eight-year-old physician, who called the Holocaust "a fabrication" and said "the story of 'gas chambers' was used as propaganda for the purposes of psychological warfare." Propaganda soon became history, Nishioka claims, and "The 'gas chambers' currently open to the public at the remains of the Auschwitz concentration camp in Poland are a postwar fabrication built either by the Polish Communist regime or by the Soviet Union, which controlled the country. Neither at Auschwitz nor anywhere else in the territory controlled by the Germans during the Second World War was there even one 'mass extermination' of Jews in 'gas chambers.'"

Reaction to the magazine article was swift. The Israeli government protested through its Tokyo embassy, while the Simon Wiesenthal Center suggested an economic boycott of the magazine by its major advertisers, including Mitsubishi Electric, Mitsubishi Motor, Cartier, Volkswagen, and Philip Morris. Within seventy-two hours these advertisers informed Bungei Shunju that if something was not done, they would pull their advertising not only from *Marco Polo* but from the publisher's other magazines as well. The editors first defended the article, then offered equal space for a rebuttal, an offer declined by the Wiesenthal Center. The Japanese government issued an official statement that called the article "extremely improper," and, under mounting economic strain, *Marco Polo*, circulation 250,000, folded on January 30. The company's president, Kengo Tanaka, explained, "We ran an article that was not fair to the Nazi massacre of Jewish people, and by running the article, we caused deep sorrow and hardship for Jewish society and related people." Some *Marco Polo* staff members were dismissed from their jobs, and remainders of the magazine were recalled from the newsstands. Two weeks later, on February 14, Tanaka resigned his presidency (although he remains chairman of Bungei Shunju).

Calling the publisher's decision "hara kiri," the March/April 1995 issue of the *Journal of Historical Review* claimed that "Jewish-Zionist groups responded to the article with characteristic speed and ruthlessness" and that "the publisher capitulated to an international Jewish-Zionist boycott and pressure campaign." Author Nishioka said, "*Marco Polo* was crushed by Jewish organizations using advertising [pressure], and Bungei obliged. They crushed room for debate." The *Journal of Historical Review* said the incident was "a great defeat for the cause of free speech and free inquiry" and concluded:

> American newspapers and magazines repeatedly assert that the Japanese hold "stereotyped" views about "the Jews," and frequently disparage them for thinking that Jews wield enormous power around the world, severely punishing anyone who defies their interests. The murder/suicide of *Marco Polo* magazine is unlikely to disabuse many Japanese of such "stereotyped" views. As in the United States, Japanese are expected to engage in a kind of Orwellian "doublethink," simultaneously taking to heart the harsh lesson of *Marco Polo's* demise, while regarding those who forced the execution as feeble victims. (pp. 2–6)

From the deniers' perspective, Jewish organizations did exactly what deniers have been accusing them of doing all along—wielding economic power and controlling the media. Simon Wiesenthal Center senior researcher Aaron Breitbart chose not to dignify their viewpoint with a serious rebuttal, responding only, "If it is not true, they have nothing to worry about. If it is true, they'd better be nice to us."

2. On May 7, 1995, fifty years to the day after the allies defeated Nazi Germany, the Toronto headquarters of Ernst Zündel, the noted neo-Nazi publisher and Holocaust denier, were set on fire, causing an estimated $400,000 in damage. Zündel was away on a speaking tour but swore that the attack, not the first, would not deter his efforts: "I have been beaten, bombed, spat at . . . but Ernst Zündel will not be run out of town. My work is legal and legitimate, and enjoys constitutional protection under the Canadian Charter of Rights and Freedoms." Zündel should know, as he defended these rights in two trials in 1985 and 1988, in which he was charged with "spreading false news" about the Holocaust. In 1992, Canada's Supreme Court acquitted Zündel on the grounds that the law under which Zündel had been charged was unconstitutional.

Claiming credit for the arson attack, according to the *Toronto Sun*, was "a shadowy offshoot of the Jewish Defense League" called the "Jewish Armed Resistance Movement." The group contacted the *Toronto Sun*, whose investigations revealed a connection "to yet another offshoot of the Jewish Defense League, Kahane Chai, an ultra-right Zionist group." Meir Halevi, leader of the Toronto Jewish Defense League, denied any connection with the attack, although a few days later, on May 12, Halevi and three companions, including Irv Rubin, leader of the Jewish Defense League in Los Angeles, tried to break into Zündel's home. Staff members photographed the would-be intruders and called the police, who, with Zündel in the car, chased them down and apprehended them. They were released, however, without being charged.

The point is this. Like the Loftus-Demjanjuk story, I heard about these events through the deniers themselves, who take such incidents and use them to prove their point about what "the Jews" are capable of doing. The Institute for Historical Review capitalized on the *Marco Polo* incident by citing it in a fund-raising letter asking for donations to support the fight against the so-called Jewish-Zionist conspiracy. Zündel plays to the hilt that it was "the Jews" who did this to him as he solicits funds to help him reconstruct his office.

My position regarding the freedom of speech of anyone on any subject is that while the government should never, under any conditions, limit the speech of anyone anytime, private organizations should also have the freedom to restrict the speech of anyone anytime within their own institution. Holocaust deniers should have the freedom to publish their own journals and books, and to try to have their views aired in other publications (e.g., college newspaper advertisements). But colleges, since they own their own newspapers, should have the freedom to restrict the deniers access to their readers.

Should they exercise this freedom? This is a question of strategy. Do you ignore what you know to be a false claim and hope it goes away, or do you stand it up and refute it for all to see? I believe that once a claim is in the public consciousness (as Holocaust denial undeniably is), it should be properly analyzed.

From a broader perspective there are, I believe, reasonable arguments for why we should not cover up, hide, suppress, or, worst of all, use the State to squelch someone else's belief system, no matter how wacky, unfounded, or venomous it may seem. Why?

- They might be completely right, and we would have just squashed the truth.

- They might be partially right, and we do not want to miss a part of the truth.

- They might be completely wrong, but by examining their wrong claims, we will discover and confirm the truth; we will also discover how thinking can go wrong, and thus improve our thinking skills.

- In science, it is not possible to know the absolute truth about anything, so we must always be on the alert for where we have gone wrong and where others have gone right.

- Being tolerant when you are in the majority means you have a greater chance of being tolerated when you are in the minority.

Once a mechanism for censorship of ideas is established, it can then work against you if and when the tables are turned. Let us pretend for a moment that the majority denies evolution and the Holocaust and that creationists and Holocaust deniers are in the positions of power. If a mechanism for censorship exists, then you, the believer in evolution and the Holocaust, may now be censored. The human mind, no matter what ideas it generates, must never be quashed. When evolutionists were in the minority in Tennessee in 1925, and politically powerful fundamentalists were successfully passing antievolution legislation making it a crime to teach evolution in public schools, Clarence Darrow made this brilliant observation in his closing remarks in the Scopes trial:

> If today you can take a thing like evolution and make it a crime to teach it in the public schools, tomorrow you can make it a crime to teach it in the private schools, and next year you can make it a crime to teach it in the church. At the

next session you can ban books and the newspapers. Ignorance and fanaticism are ever busy, indeed feeding, always feeding and gloating for more. Today it's the public school teachers, tomorrow the private. The next day the preachers and the lecturers, the magazines, the books, the newspapers. After awhile, your honor, it is the setting of man against man, creed against creed, until the flying banners and beating drums are marching backwards to the glorious ages of the sixteenth century when bigots lighted fagots to burn the man who dared to bring any intelligence, and enlightenment, and culture to the human mind. (in Gould 1983a, p. 278)

13

Who Says the Holocaust Never Happened, and Why Do They Say It?

An Overview of a Movement

The SS guards took pleasure in telling us that we had no chance of coming out alive, a point they emphasized with particular relish by insisting that after the war the rest of the world would not believe what happened; there would be rumors, speculation, but no clear evidence, and people would conclude that evil on such a scale was just not possible.

—Terrence des Pres, *The Survivor*, 1976

When historians ask, "How can anyone deny the Holocaust?" and deniers respond, "We are not denying the Holocaust," it becomes obvious that the two groups are defining the Holocaust in different ways. What deniers are explicitly denying are three points found in most definitions of the Holocaust:

1. There was intentionality of genocide based primarily on race.

2. A highly technical, well-organized extermination program using gas chambers and crematoria was implemented.

3. An estimated five to six million Jews were killed.

Deniers do not deny that antisemitism was rampant in Nazi Germany or that Hitler and many of the Nazi leaders hated Jews. Nor do they deny that Jews were deported, that the property of Jews was confiscated, or that Jews were rounded up and forced into concentration camps where, in general, they were very harshly treated and made the victims of over-crowding, disease, and forced labor. Specifically, as outlined in "The Holocaust Controversy: The Case for Open Debate" advertisements that Bradley Smith places in college newspapers, as well as in various other sources (Cole 1994; Irving 1994; Weber 1993a, 1994a, 1994b; Zündel 1994), the deniers are saying:

1. There was no Nazi policy to exterminate European Jewry. The Final Solution to the "Jewish question" was deportation out of the Reich. Because of early successes in the war, the Reich was con-fronted with more Jews than it could deport. Because of later failures in the war, the Nazis confined Jews in ghettos and, finally, camps.

2. The main causes of death were disease and starvation, caused pri-marily by Allied destruction of German supply lines and resources at the end of the war. There were shootings and hangings (and maybe even some experimental gassings), and the Germans did overwork Jews in forced labor for the war effort, but all this accounts for a very small percentage of the dead. Gas chambers were used only for delousing clothing and blankets, and the crema-toria were used only to dispose of the bodies of people who had died from disease, starvation, overwork, shooting, or hanging.

3. Between 300,000 and two million Jews died or were killed in ghet-tos and camps, rather than five to six million.

In the next chapter, I will address these claims in detail, but I wish to give brief answers here.

1. In any historical event, functional outcomes rarely match orig-inal intentions, which are always difficult to prove anyway, so historians should focus on contingent outcomes more than intentions. The functional process of carrying out the Final Solution evolved over time, driven by such contingencies as increasing political power, growing confidence in getting away with a variety of persecutions, the unfolding of the war (espe-cially against Russia), the inefficiency of transporting Jews out of the Reich, and the infeasibility of eliminating Jews by disease, exhaustion, overwork, random killings, and mass shootings. The outcome was millions of Jewish dead, whether extermination of European Jewry was explicitly and officially ordered or just tacitly approved.

2. Physical and documentary evidence corroborate that the gas chambers and crematoria were mechanisms of extermination. Regardless of the mechanism used for murder, however, murder is murder. Gas chambers and crematoria are not required for mass murder, as we have seen recently in Rwanda and Bosnia. In occupied Soviet territories, for example, the Nazis killed about 1.5 million Jews by means other than gassing.

3. Five to six million killed is a general but well-substantiated estimate. The figures are derived by collating the number of Jews reported living in Europe, transported to camps, liberated from camps, killed in Einsatzgruppen actions, and alive after the war. It is simply a matter of population demographics.

One of the things I commonly hear when I tell people about Holocaust deniers is that they must be raving racists or nutty fools on the lunatic fringe. Just who would say the Holocaust never happened? I wanted to find out, so I met with some of them to allow them to present their claims in their own words. In general, I found these deniers relatively pleasant. They were willing to talk about the movement and its members quite openly, and they generously provided a large sampling of their published literature.

After World War II, revisionism began in Germany with opposition to the Nuremberg trials, typically seen as "victor's trials" that were hardly fair and objective. Revisionism of the Holocaust itself took off in the 1960s and 1970s with Franz Scheidl's 1967 *Geschichte der Verfemung Deutschlands* (In Defense of the German Race), Emil Aretz's 1970 *Hexeneinmaleins einer Lüge* (The Six Million Lie), Thies Christophersen's 1973 *Die Auschwitz-Lüge* (The Auschwitz Lie), Richard Harwood's 1973 *Did Six Million Really Die?*, Austin App's 1973 *The Six Million Swindle*, Paul Rassinier's 1978 *Debunking the Genocide Myth*, and the bible of the movement, Arthur Butz's 1976 *The Hoax of the Twentieth Century*. It is in these volumes that the three pillars of Holocaust denial—no intentional genocide by race, gas chambers and crematoria not used for mass murder, many fewer than six million Jews killed—were crafted.

Except for Butz's book, which stays in circulation despite being disorganized beyond repair, these works have all given way to the *Journal of Historical Review* (*JHR*), the voice of the Institute for Historical Review (IHR). The institute's journal, along with its annual conference, has become the hub of the movement, which is populated by a handful of eccentric personalities including IHR director and *JHR* editor Mark Weber, author and biographer David Irving, gadfly Robert Faurisson, pro-Nazi publisher Ernst Zündel, and video producer David Cole. (See figure 17.)

Institute for Historical Review

In 1978, IHR was founded and organized primarily by Willis Carto, who also published *Right* and *American Mercury* (considered by some to have strong antisemitic themes) and now runs Noontide Press, publisher of controversial books including those denying the Holocaust. Carto also runs Liberty Lobby, which is classified by some as an ultra-right-wing organization. In 1980, IHR's promise to pay $50,000 for proof that Jews were gassed at Auschwitz made headlines. When Mel Mermelstein met this challenge, headlines and later a television movie detailed his collection of the award and an additional $40,000 for "personal suffering." IHR's first director, William McCalden (a.k.a. Lewis Brandon, Sandra Ross, David Berg, Julius Finkelstein, and David Stanford), was fired in 1981 due to conflicts with Carto and was succeeded by Tom Marcellus, a field staff member for the Church of Scientology who had been an editor for one of the church's publications. When Marcellus left IHR in 1995, *JHR*'s editor, Mark Weber, took over as its director.

Since the 1984 fire-bombing that destroyed its office, IHR is understandably cautious about revealing its location to outsiders. Situated in an industrial area of Irvine, California, the office has no sign and its glass door, entirely covered with one-way mirror coating, is dead-bolted at all times; one must be identified and admitted by the secretary working in a small office in front. Inside, there are several offices for the various staff members and a voluminous library. Not surprisingly, World War II and the Holocaust are the prime foci of its resources. In addition, IHR has a warehouse filled with back issues of *JHR*, pamphlets, and other promotional materials, as well as books and videotapes, all part of a catalogue business that, together with subscriptions, accounts for about 80 percent of revenues, according to Weber. The other 20 percent comes from tax-free donations (IHR is a registered nonprofit organization). Whatever funds the institute was receiving through Carto dried up after the 1993 falling out with (and subsequent filing of lawsuits against) the founder of IHR.

Before the break with Carto, IHR leaned heavily on the "Edison money," a total of about $15 million willed by Thomas Edison's granddaughter, Jean Farrel Edison. According to David Irving (1994), about $10 million of that money apparently was lost by Carto "in lawsuits by other members of the family in Switzerland" and the remaining $5 million was made available to Carto's Legion for the Survival of Freedom. "From that point on it vanishes into uncertainty. Certain sums of money have turned up. A lot of it is in a Swiss bank at present."

FIGURE 17:
Cover of the November/December 1994 issue of *JHR* featuring most of the key Holocaust deniers, including those discussed in this chapter: (*left to right*) Robert Faurisson, John Ball, Russ Granata, Carlo Mattogno, Ernst Zündel, Friedrich Berg, Greg Raven, David Cole, Robert Countess, Tom Marcellus, Mark Weber, David Irving, Jürgen Graf. [Reprinted from *The Journal of Historical Review*, Box 2739, Newport Beach, CA 92659 USA. Subscriptions: $40 per year (domestic).]

When the institute's board of directors voted to sever all ties with him, Carto apparently did not take it lying down. According to IHR, among many other things, Carto has "stormed IHR's offices with hired goons" and put out "the fantastic lie that the Zionist ADL [Anti-Defamation League] has been running IHR since last September" (Marcellus 1994). On December 31, 1993, IHR won a judgment against Carto. They are now suing him for damages incurred during his raid on the IHR office, which destroyed equipment and ended in fisticuffs, as well as for other moneys that, Weber claims, went "to Liberty Lobby and other Carto controlled enterprises. Probably the money has been frittered away by Carto but we are trying to track this down" (1994b).

In February 1994, Director Tom Marcellus sent a mass mailing to IHR members with "AN URGENT APPEAL FROM IHR" because it had "been forced to confront a threat to the editorial and financial integrity . . . that in the past several months has drained, and continues to drain, literally tens of thousands of dollars from our operations." Without help from its members, Marcellus wrote, "IHR may not survive." Carto was accused of becoming "increasingly erratic," both in personal matters and in business, and of involving "the corporation in three costly copyright violations." Most interesting, and in keeping with deniers' current attempts to disassociate themselves from earlier antisemitic connections and present themselves as objective historical scholars, the mailing condemned Carto for changing "the direction of IHR and its journal from serious, nonpartisan revisionist scholarship, reporting, and commentary to one of ranting, racialist-populist pamphleteering" (Marcellus 1994).

David Cole believes that the post-Carto "IHR is going to have to depend a lot more on journal and book sales" and thus on their right-wing, antisemitic backers:

> In order to keep the IHR in the black they have had to cater to the far right. I think if you were to look at their book sales you would see that some of the more complex, really solid historiographical works probably don't sell as well as Henry Ford's *International Jew* or the *Protocols of Zion*, or some of the other things they sell. If they had to rely on the sales of Holocaust revisionist works alone they'd be screwed. They have to cater to the money. There are a lot of elderly people with money saved or with social security checks, who want to spend the last years of their life fighting the Jews. Bradley [Smith] can get checks for $5,000, $7,000, $3,000. These people are very, very wealthy, and completely anonymous. There is a lot of money to be made by getting a really good ideological mailing list and the IHR has one that caters mainly to people of the far right. (1994)

As of 1996, IHR still holds conferences (attendance about 250), *JHR* continues to be published (circulation about 5,000 to 10,000), and promotional literature and book and videotape catalogues are regularly mailed out. Whether IHR survives the break with Carto or not, we must remember that the denier movement is not a homogeneous group held together by this organization alone.

Mark Weber

With the possible exception of David Irving, in the denier movement Mark Weber may know the most about history and historiography. Some people have claimed that Weber's master's degree in modern European history from Indiana University is fake, but I called the university and confirmed that his degree is real. Weber arrived on the denier scene when he appeared as a defense witness at Ernst Zündel's "free speech" trial in 1985. Weber denied any racist or antisemitic feelings and claimed, "I don't know anything more about the neo-Nazi movement in Germany than what I read in the papers" (1994b). Weber, however, was once the news editor of *National Vanguard*, the voice of the National Alliance, William Pierce's neo-Nazi, antisemitic organization. Weber also does not repudiate comments he made in a 1989 interview published by the *University of Nebraska Sower* about the United States becoming "a sort of Mexicanized, Puerto Ricanized country" due to the failure of "white Americans" to reproduce adequately. (Not that this sentiment is particularly unusual in our ever-increasingly segregationist society. Weber's wife told me at the 1995 IHR conference that these white guys should quit complaining about other races breeding too much and have more children themselves.) And on February 27, 1993, Weber was the object of a Simon Wiesenthal Center sting operation, secretly filmed by CBS, in which researcher Yaron Svoray, calling himself Ron Furey, met with Weber in a cafe to discuss *The Right Way*, a bogus magazine created to trick neo-Nazis into revealing their identities. Weber quickly figured out that Svoray "was an agent for someone" and "was obviously lying," and left (1994b). Subsequently, Weber was portrayed in an HBO movie about neo-Nazis in Europe and America, and he says that the Wiesenthal version of the event is greatly distorted.

Such clandestine operations by the Simon Wiesenthal Center raise many troubling questions. Nonetheless, one must wonder why, if he is trying to distance himself from the neo-Nazi fringe of denial (as he claims), Weber would agree to such a meeting. Even David Cole, who is his friend, admits that "Weber doesn't really see any problems with a society that is not only disciplined by fear and violence but also where a government feeds its people lies in order to keep them well-ordered." Says Cole, "Deniers criticize the Jews for lying to its people or the world, and yet a lot of these same revisionists will speak very complimentarily of what the Nazis did in feeding their people lies and falsehoods in order to keep morale up and to keep this notion of the master race" (1994).

Weber is extremely bright and very personable, and one could believe that he might be capable of good historical scholarship if he ended his fixation on Jews and the Holocaust. He knows history and current politics and is a formidable debater on any number of subjects. Unfortunately, one of these subjects is Jews, whom he continues to generalize into a unified whole and to fear as a unified threat to American and world culture. Weber cannot seem to discriminate between individual Jews, whose actions he may like or dislike, and "the Jews," whose supposed actions he generally dislikes, and he cannot seem to grasp the innate complexity of contemporary culture.

David Irving

David Irving has no professional training in history, but there is no disputing that he has mastered the primary documents of the major Nazi figures, and he is arguably the most historically sophisticated of the deniers. Although his attentions have spanned the Second World War—he is the author of histories such as *The Destruction of Dresden* (1963) and *The German Atomic Bomb* (1967), as well as biographies including *The Trail of the Fox* (1977, on Rommel), *Hitler's War* (1977), *Churchill's War* (1987), *Göring* (1989), and *Goebbels: Mastermind of the Third Reich* (1996)—his interest in the Holocaust is growing ever stronger. "I think that the Holocaust is going to be revised. I have to take my hat off to my adversaries and the strategies they have employed—the marketing of the very word Holocaust: I half expect to see the little 'TM' after it" (1994). For Irving, denial has become a war, which he has described in military language: "I'm presently in a fight for survival. My intention is to survive until five minutes past D-day rather than to go down heroically five minutes before the flag is finally raised. I'm convinced this is a battle we are winning" (1994). After completing his biography of Goebbels, Irving says, his publisher not only backed out of the contract because he had become a Holocaust denier but is trying to retrieve the "six-figure advance." The biography was published by Focal Point, Irving's own publishing house in London.

Irving's attitudes about the Holocaust have evolved, beginning with his 1977 offer to pay $1,000 to anyone who could provide proof that Hitler ordered the extermination of the Jews. After reading *The Leuchter Report* (1989), which argues that the gas chambers at Auschwitz were not used to commit homicide, Irving began to deny the Holocaust altogether, not just Hitler's involvement. Curiously, he sometimes wavers on the various points of Holocaust denial. He told me in 1994 that reading Eichmann's memoirs made him "glad I have not adopted the narrow-minded approach that there was no Holocaust" (1994). At the same time, he told me that only 500,000 to 600,000 Jews died as the unfortunate victims of war—the moral equivalent, he claimed, to the bombing of Dresden or Hiroshima. Yet on July 27, 1995, when asked by the host of an Australian radio show how many Jews died at the hands of the Nazis, Irving admitted that perhaps it was as many as four million: "I think like any scientist, I'd have to give you a range of figures and I'd have to say a minimum of one million, which is monstrous, depending on what you mean by killed. If putting people into a concentration camp where they die of barbarity and typhus and epidemics is killing, then I would say the four million figure because, undoubtedly, huge numbers did die in the camps in conditions that were very evident at the end of the war" (*Searchlight* editorial, 1995, p. 2).

Still, Irving testified for the defense in Ernst Zündel's "free speech" trial in 1985, after which various governments brought criminal charges against him. He has been deported from or denied entry into many countries, and his books have been removed from some stores and some stores that carry them have been vandalized. In May 1992, Irving told a German audience that the reconstructed gas chamber at Auschwitz I was "a fake built after the war." The following month, when he landed in Rome he was surrounded by police and put on the next plane to Munich where he was charged under German law for "defaming the memory of the dead." He was convicted and fined DM 3,000. When he appealed the conviction, it was upheld and the fine increased to DM 30,000 (about $20,000). In late 1992, while in California Irving received notice from the Canadian government that he would not be allowed into that country. He went anyway to accept the George Orwell award from a conservative free-speech organization, whereupon he was arrested by the Royal Canadian Mounted Police. He was led away in handcuffs and deported on the grounds that his German conviction made it likely that he would commit similar actions in Canada. He is presently barred from entering Australia, Canada, Germany, Italy, New Zealand, and South Africa.

Although Irving disclaims any official affiliation with IHR ("You will see that my name isn't on the masthead"), he is a regular speaker at IHR conventions and frequently lectures to denier groups around the world. At the 1995 IHR conference in Irvine, California, Irving was the featured speaker and was openly adored by many of the attendees. When not speaking, Irving staffed his own book table, selling and signing his many works. Purchasers of *Hitler's War* received a miniature swastika flag like the one mounted on Hitler's black Mercedes. During one conversation with a couple of fans, Irving explained that the worldwide Jewish cabal has been working against him to prevent his books from being published and him from giving talks. It is true that Irving has met with considerable resistance from Jewish groups when he has been asked to speak. For example, in 1995 Irving was brought to the University of California, Berkeley, by a free-speech group, but his lecture was picketed and he was not able to give the talk. But one must make a sharp distinction between local, spontaneous reactions to an event, and a worldwide, planned conspiracy. Irving seems unable to make this distinction.

In 1995, Irving attended a lecture against Holocaust denial by Deborah Lipstadt, after which, he claims, he stood up and announced his presence, whereupon he was swamped by audience members asking for his autograph. Irving says he brought a box of his biography, *Göring*, and gave them away so students could see "which of us is lying." Oh? If there was no plan to exterminate the Jews, then what will readers make of page 238 of *Göring*, where Irving writes: "Emigration was only one possibility that Göring foresaw. 'The second is as follows,' he said in November 1938, selecting his words with uncharacteristic care. 'If at any foreseeable time in the future the German Reich finds itself in a foreign political conflict, then it is self-evident that we in Germany will address ourselves first and foremost to effecting a grand settling of scores against the Jews.' " Since Irving claims that emigration is all the Nazis ever meant by *Ausrottung* (extermination) and the Final Solution, then just what did Göring mean here by "the second" plan? And what will readers think when they get to page 343 of *Göring*, where Irving writes:

> History now teaches that a significant proportion of those deported—particularly those too young or infirm to work—were being brutally disposed of on arrival. The surviving documents provide no proof that these killings were systematic; they yield no explicit orders from "above," and the massacres themselves were carried out by the local Nazis (by no means all of them German) upon whom the deported Jews had been dumped. That they were ad hoc extermination operations is suggested by such exasperated outbursts as that of Governor-General Hans Frank at a Krakau conference on December 16, 1941: "I have started negotiations with the aim of sweeping them [further] to the east. In January there is to be a big conference in Berlin on this problem . . . under SS Obergruppenführer Heydrich [the "Wannsee Conference" of January 20, 1942]. At any rate a big Jewish exodus will begin. . . . But what's to become of the Jews? Do you imagine they're going to be housed in neat estates in the Baltic provinces? In Berlin they tell us: What's bugging you—we've got no use for them either, liquidate them yourselves!"

"Berlin," says Irving, "more likely meant the party—or Himmler, Heydrich, and the SS." This passage, quoted verbatim from *Göring*, is Irving's own translation (Irving speaks fluent German) and interpretation. I fail to see how it can be taken to support an ad hoc interpretation of non-systematic killings with no order from above. From this passage, along with many others, it sounds like the killings were very systematic, the orders did come—directly or tacitly—from above, and the only thing ad hoc about the process was the contingent development of the final outcome. Finally, what can "liquidate" possibly mean other than exactly what Holocaust historians have always said that it means?

One factor that may be contributing to Irving's move into Holocaust denial is that he earns his living by lecturing and selling books, and the more he revises the Holocaust the more books he sells and the more invitations to lecture he receives from denier and right-wing groups. I believe that he has been slipping more and more into denial not so much because the historical evidence has taken him there but because he has found a profitable and welcoming home. The mainstream academy has rejected him, so he has created a niche on the margins. Irving is a first-rate documentarian and narrative historian, but he is not a good theoretician and does a lot of selective quoting to support his biases. First it was Hitler who was unaware of the Holocaust. Then it was Göring. Now it is Goebbels he is trying to exonerate.

Robert Faurisson

Once a legitimate professor of literature at the University of Lyon 2, Robert Faurisson has become the "Pope of Revisionism," a title bestowed by Holocaust deniers in Australia in response to his tireless efforts in holding up the major tenets of Holocaust denial. For his countless statements, letters, articles, and essays challenging Holocaust authorities to "show me or draw me a Nazi gas chamber," Faurisson lost his job, was physically beaten, and has been tried, convicted, fined $50,000, and barred from holding any government job. Faurisson's convictions came under the Fabius-Gayssot law passed in 1990 (inspired, in part, by Faurisson's activities), which made it a criminal offense "to contest by any means the existence of one or more of the crimes against humanity as defined by Article 6 of the Statutes of the International Military Tribunal, attached to the London Agreement of August 8, 1945, committed either by the members of an organization declared criminal in application of Article 9 of the same Statutes, or by a person held guilty of such a crime by a French or International jurisdiction."

Faurisson is the author of a number of works denying various aspects of the Holocaust, including *The Rumor of Auschwitz, Treatise in Defense Against Those Who Accuse Me of Falsifying History,* and *Is the Diary of Anne Frank Genuine?* After *The Rumor of Auschwitz* was published, famed MIT linguistics professor Noam Chomsky wrote an article in defense of Faurisson's freedom to deny whatever he wants, which triggered controversy over Chomsky's politics. Chomsky told the Australian magazine *Quadrant,* "I see no anti-Semitic implication in Faurisson's work." This was rather naive on Chomsky's part. During his 1991 trial in France, Faurisson summarized his feelings about Jews for the *Guardian Weekly*: "The alleged Hitlerian gas chambers and the alleged genocide of the Jews form one and the same historical lie, which permitted a gigantic financial swindle whose chief beneficiaries have been the State of Israel and international Zionism, and whose main victims have been the German people and the Palestinian people as a whole." (All quoted in Anti-Defamation League 1993.)

Faurisson likes to bait his opponents, whom he calls "exterminationists." On his way to the 1995 IHR conference in Irvine, California, for example, Faurisson visited the U.S. Holocaust Memorial Museum in Washington, D.C., and managed to arrange a meeting with one of its directors. By badgering him about the "lack of proof" that Nazi gas chambers were used for mass murder, Faurisson managed to trigger an emotional outburst from his host. At the conference, Faurisson invited me to his hotel room to discuss in private the gas chamber story. Faurisson harassed me incessantly for half an hour, getting in my face and wagging his finger, demanding "one proof, just one proof" that a Nazi gas chamber was used for mass murder. I simply asked over and over, "What would you consider 'proof'?" Faurisson was unwilling (or unable) to answer.

Ernst Zündel

Among the least subtle of all the Holocaust deniers is the pro-Nazi propagandist and publisher Ernst Zündel, whose self-proclaimed goal is "the rehabilitation of the German people." Zündel believes that "there are certain aspects of the Third Reich that are very admirable and I want to call people's attention to these," such as the eugenics and euthanasia programs (1994). To do so, Zündel publishes and distributes books, fliers, and video- and audiotapes through his Toronto-based Samisdat Publishers, Ltd. A small donation will net you an assortment of Zündelmania paraphernalia, including transcriptions of his trial court proceedings; copies of his publication *Power: Zündelists vs. Zionists*, with articles like "Is Spielberg's 'Schindler' a 'Schwindler'?"; video clips of his many media appearances; a video tour of Auschwitz with David Cole; and stickers that proclaim "GERMANS! STOP APOLOGIZING FOR THE THINGS YOU DID NOT DO!" and "TIRED OF THE HOLOCAUST? NOW YOU CAN STOP IT!" and so on (see figure 18).

I visited Zündel at his Toronto home/office just after the fire-bombing in September 1995 and found him to be at once jovial and friendly and at the same time deadly serious about his mission to free the German people "from the burden of the six million." In front of writer Alex Grobman and two other Jews, Zündel did not hesitate to speak his mind on all manners Semitic, including his belief that in the future the Jews are going to experience antisemitism the likes of which they have never seen before. Like other deniers, it bothers Zündel to no end that the Jews are the focus of so much attention, as he told me in a 1994 interview:

> Frankly, I don't think Jews should be so egotistical and think they are the navel of the universe. They're not. Only a people like them could think themselves so important that the whole world revolves around them. I tend to go with Hitler—the last thing that he was really worried about was what the Jews thought. To me Jews are just like any other person. That already will hurt them. They will be shrieking "Oy vey, that Ernst Zündel said Jews are just like normal people." Well, goddamn it, they are.

What the Holocaust has done to National Socialism, says Zündel, is to "bar so many thinkers from re-looking at the options that National

FIGURE 18:
Sampling of Ernst Zündel's stickers.

Socialism German style offers." Lift the Holocaust burden off the Germans' shoulders, and Nazism suddenly does not look so bad. Sound crazy? Even Zündel admits his ideas are a little extreme: "I know my ideas might be half-baked—I'm not exactly Einstein, and I know that. I'm not Kant. I'm not Goethe. I'm not Schiller. As a writer I'm not Hemingway. But goddamnit I'm Ernst Zündel. I walk on my hind legs and I have a right to express my viewpoints. I do the best I can in a kind way. My long term goal is to ring the bell of freedom and maybe in my lifetime I will achieve no more than I have achieved so far, which is not too bad." In 1994, Zündel said he was "presently negotiating a deal with an American satellite company who promised me that they can get a signal over Europe that can be picked up on satellite dishes." He wants to move denial into the mainstream in Europe and America, where, he thinks, "in another fifteen years revisionism will be discussed over pretzels and beer" (1994).

David Cole

The most paradoxical of the deniers is David Cole. His mother "was raised as a secular Jew" and his father "was raised Orthodox in London during the Blitz," and he proudly displays his Jewish heritage while simultaneously denying its most significant modern historical event. As he told me in a 1994 interview, "I am damned if I do and damned if I don't. That is, if I don't mention the Judaism I will be accused of being ashamed. If I mention it up front I will be accused of exploiting it." Cole's attentions center on the physical evidence, specifically on denying that gas chambers and crematoria were instruments of mass murder. For his views, he was physically beaten at the University of California, Los Angeles, during a debate on the Holocaust. He has received regular death threats from "a small group of people that genuinely hate me with a passion," and the Jewish Defense League, the Anti-Defamation League, and Jewish organizations in general "are a little harder on me because I am Jewish." He has been called a self-hating Jew, antisemitic, and a race traitor; and an editorial in *The Jewish News* compared him to Hitler, Hussein, and Arafat.

Although Cole's personality is affable and his attitude sanguine, he sees himself as a rebel in search of a cause. Where other deniers are political and racial ideologues, Cole's interests run deeper. He is a meta-ideologue—an atheist and an existentialist on a quest to understand how ideologues invent their realities. In the process, Cole has joined every conceivable fringe organization, including the Revolutionary Communist Party, Workers World Party, John Birchers, Lyndon LaRouchers, Libertarians, atheists, and humanists.

> I was everywhere. I ran a chapter of the Revolutionary Communist Party. I ran a John Birch Society chapter. I had about five different names, and there was, literally, not a part of the American political spectrum I wasn't involved in. I was a supporter of, and subscriber to, the ADL and the JDL. I have a World Jewish Congress card. I worked for the Heritage Foundation on the right, and the ACLU on the left. My point in doing this was that I felt superior to ideology and to the poor, brainwashed idiots who toil their lives away in pursuit of abstract concepts. (in Applebaum 1994, p. 33)

Holocaust denial, then, is just one in a long line of ideologies that have fascinated Cole since he was expelled from high school in southern California. With no college background but a parental stipend for self-education, Cole has a personal library that houses thousands of volumes, including a considerable Holocaust section. He knows his subject and can "debate the facts until the cows come home." Where other fringe claims only held his attention for a few months to a year, the Holocaust "is more about real physical things than some abstract concept that requires faith. We are talking about something for which much of the evidence still exists." And much of that physical evidence was filmed by Cole on a fact-finding mission over the summer of 1992, financed by denier Bradley Smith. "I figured I needed $15,000 to $20,000, and Bradley set to work—it took him about a month and a half to raise that amount." Cole's stated goal in his research is

> to try to move revisionism away from the fringe and into the mainstream. . . . I want to get people who are not right-wingers or neo-Nazis. Right now it is in a very dangerous position because there is a vacuum created by mainstream historians denouncing revisionism. The vacuum has been filled with the likes of Ernst Zündel. Zündel is a very likable human being, but he is a fascist and he is not the person I would like to see recognized as the world's leading Holocaust revisionist. (1994)

Cole states that he wants his video footage to be studied by professional scholars (he says he offered it to Yad Vashem in Jerusalem) but has edited it into a marketable product to be sold through IHR's catalogues, as he did his first video of Auschwitz, which he says has sold over 30,000 copies.

David Cole likes to stir things up, and not just for historians. Cole, for example, might take an African-American date to a denier social event where white supremacists will be present "just to watch them squirm and stare." Even though he disagrees mightily with many deniers' beliefs and most of their politics, he will introduce himself to the media as a "denier," knowing it will draw scorn and sometimes physical abuse. What is an outsider like Cole to do? He is angry that he has been locked out by historians who, he says, "are not gods, are not religious figures, and are not priests. We have a right to ask them for further explanations. I am not ashamed to ask the questions I am asking" (1994). One wonders, however, why such questions need to be asked, and why denial holds Cole's attention.

Interestingly, in 1995 Cole experienced something of a falling out with the deniers, triggered by a number of events, including an incident in Europe in October 1994, on another video tour of Nazi death camps. According to Bradley Smith, Cole was at the Natzweiler (Struthof) camp examining the gas chamber with Pierre Guillaume (Faurisson's French publisher), Henri Roques (author of The *"Confessions" of Kurt Gerstein*), Roques's wife, and denier Tristan Mordrel. While they were inside the building housing the gas chamber, one of the guards, according to Smith, "excused himself, went out, and locked the exit door from the outside." After about twenty minutes, the guard unlocked the door, and they returned to their cars, whereupon Cole discovered that "a front door window in his car had been smashed and his travel journals, papers, books, personal effects, videotapes and still camera film had all been stolen. In short, all his research. He was cleaned out" (Smith 1994). Smith claims the trip cost him $8,000 to fund, so he is now selling an eighty-minute video of Cole telling his story in order to dig himself out of the hole.

Ironically, Henri Roques denies Cole's story:

> The six of us were never locked from outside the gas chamber in order to be entrapped in it! Simply the guard locked the door from inside and he had to open it once because tourists were knocking at the door, and he told them that the visit was possible only for people with special permission (which was the case for our party). My wife and I remember only one guard. According to the guard and, later on, to the gendarmes in Schirmeck (near Struthof), this kind of theft is unfortunately common, especially in a car with a foreign license plate. Initially, I thought that it could have been a theft directed against revisionist people but I do not see anything which could substantiate this and, furthermore, the conversations I had with P. Guillaume and T. Mordrel tend to eliminate that possibility. Cole's version could make the readers believe in an anti-revisionist operation carried out with the complicity of the guards but I don't think it is fair to accuse the guards of having "entrapped" us or even perhaps participated in a theft. (1995, p. 2)

In another ironic twist, when Robert Faurisson claimed in the *Adelaide Institute Newsletter* that the Struthof gas chamber was never used for mass homicide, Cole, to his credit, rebuffed him:

> What evidence does Faurisson give us to "prove" that no homicidal gassings ever took place at Struthof? He tells us of an "expertise" that has "disappeared," but, "thanks to another piece of evidence," we know what it said. He refers us to a *Journal of Historical Review* article for more information. One would hope to find out in this article just *what* that other piece of evidence is that confirms the existence and conclusions of the 'expertise,' but sadly Faurisson refuses to enlighten us. So what do we have? A report that has disappeared and a revisionist who assures us that *he* knows what the report said, without feeling the need to provide us with any further evidence. How would a *revisionist* respond if an "exterminationist" acted this way? Revisionists routinely dismiss documents when the originals have vanished. We don't accept "hearsay," and we *certainly* don't take exterminationists on their word when it comes to the contents of documents. (1995, p. 3)

The Jewish Agenda of Holocaust Denial

Running throughout almost all denier literature—books, articles, editorials, reviews, monographs, guides, pamphlets, and promotional materials—

is fascination with Jews and everything Jewish. No issue of *JHR* fails to contain something on Jews. The January/February 1994 issue, for example, features a cover story on who killed the Romanovs and drove the Bolsheviks to power. Yes, it was the Jews, as Mark Weber explained: "Although officially Jews have never made up more than five percent of the country's total population, they played a highly disproportionate and probably decisive role in the infant Bolshevik regime, effectively dominating the Soviet government during its early years." But Lenin, who ordered the assassination of the Imperial family, wasn't Jewish. Weber gets around this fact by noting, "Lenin himself was of mostly Russian and Kalmuck ancestry, but he was also one-quarter Jewish" (1994c, p. 7). This is a typical denier line of reasoning. *Fact*: The Communists killed the Romanovs and instigated the Bolshevik Revolution. *Fact*: Some of the leading Communists were Jewish. *Conclusion*: The Jews killed the Romanovs and caused the Bolshevik Revolution. By the same logic: Ted Bundy was Catholic. Ted Bundy was a serial killer. Catholics are serial killers.

The Jewish focus is pervasive in *JHR*. Why? Mark Weber bluntly justified the IHR's attitude:

> We focus on the Jews because just about everyone else is afraid to. Part of the reason we exist, and part of the pleasure is to be able to deal with a subject that others are not dealing with in a way that we feel helps provide information on what is relevant. I wish that the same considerations were given in our society to talking about Germans, or Ukrainians, or Hungarians, that are given to talking about the Jews. At the Simon Wiesenthal so-called Museum of Tolerance there are constant references to what *the* Germans did to the Jews in the Second World War. We permit and encourage in our society what would be considered vicious stereotypes if applied to other groups, when they are applied to the Germans or the Hungarians. This is a double standard, of which the Holocaust campaign is the most spectacular manifestation. We have a museum in Washington, D.C., to the memorial of non-Americans victimized by other non-Americans. We don't have any comparable museum to the fate of American-Indians, the victims of blacks in slavery, the victims of communism, etc. The very existence of this museum points up this perverse sensitivity of Jewish concerns in our society. The IHR and those affiliated with us feel a sense of *liberation* in that we say, in effect, we don't give a damn if you criticize us or not. We're going to say it anyway. We don't have a job to lose because this is our job. (1994b)

There is not a lot of gray area in this statement. Sensitivity about Jews and the Holocaust "campaign" is "perverse," and taking them on provides "pleasure" and "liberation." Germans, however, are the victims who must be treated better.

The Conspiratorial Side of Holocaust Denial

Embedded in the Jewish agenda of Holocaust denial is a strong conspiratorial streak. The *"Holocaust" News*, published by the Centre for Historical Review (not to be confused with IHR), claims in its first issue that "the 'Holocaust' lie was perpetrated by Zionist-Jewry's stunning propaganda machine for the purpose of filling the minds of Gentile people the world over with such guilt feelings about the Jews that they would utter no protest when the Zionists robbed the Palestinians of their homeland with the utmost savagery" (n.d., p. 1). The more Holocaust deniers make their arguments, the more they believe them, and the more Jews and others argue against them, the more convinced Holocaust deniers are that there is some sort of Jewish conspiracy to "create" the Holocaust so that Jews can gain aid and sympathy for Israel, attention, power, and so on.

An early, classic example of conspiratorial thinking that influenced the modern denial movement is *Imperium: The Philosophy of History and Politics* ([1948] 1969), written by Francis Parker Yockey under the nom de plume Ulick Varange and dedicated to Adolf Hitler. The IHR catalogue describes the book as "a sweeping historico-philosophical treatise in the Spenglerian mold and a clarion call to arms in defense of Europe and the West." The book introduced Willis Carto, the founder of IHR, to Holocaust denial. *Imperium* details the "imperial" system modeled after Hitler's National Socialism in which democracy would whither away, elections would cease, power would be in the hands of the public, and businesses would be publicly owned. The problem, as Yockey saw it, was "the Jew," who "lives solely with the idea of revenge on the nations of the white European-American race." A conspiratorialist, Yockey described how the "Culture-Distorters" were undermining the West because of the covert operations of "the Church-State-Nation-People-Race of the Jew" (see Obert 1981, pp. 20–24) and how Hitler heroically defended the purity of the Aryan race against inferior racial-cultural aliens and "parasites" such as Jews, Asiatics, Negroes, and Communists (see McIver 1994).

Yockey's conspiratorial bent is not uncommon in America, an example of what Richard Hofstadter called the "paranoid style" in American politics. For instance, the German-American Anti-Defamation League of Washington, D.C., which "seeks to defend the rights of German-Americans, the forgotten minority," published a cartoon asking "How long can the Jews perpetrate the Holocaust myth?" over a vulgar caricature of

Jewish media moguls manipulating the press to perpetuate the hoax. The same organization produced an advertisement that asked, "Would Challenger have blown up if German scientists had still been in charge?" "We don't think so!" exclaims the ad, before explaining that Soviet "Fifth Columnists in the United States" have secretly worked to eliminate German scientists from NASA. For the conspiratorialist, all manner of demonic forces have been at work throughout history, including, of course, the Jews, but also the Illuminati, Knights Templar, Knights of Malta, Masons, Freemasons, Cosmopolitans, Abolitionists, Slaveholders, Catholics, Communists, Council on Foreign Relations, Trilateral Commission, Warren Commission, World Wildlife Fund, International Monetary Fund, League of Nations, United Nations, and many more (Vankin and Whalen 1995). In many of these, "the Jews" are seen to be at work behind the scenes.

John George and Laird Wilcox have outlined a set of characteristics of political extremists and fringe groups that is useful in considering the broader principles behind Holocaust denial (1992, p. 63):

1. Absolute certainty they have the truth.

2. America is controlled to a greater or lesser extent by a conspiratorial group. In fact, they believe this evil group is very powerful and controls most nations.

3. Open hatred of opponents. Because these opponents (actually "enemies" in the extremists' eyes) are seen as a part of or sympathizers with "The Conspiracy," they deserve hatred and contempt.

4. Little faith in the democratic process. Mainly because most believe "The Conspiracy" has such influence in the U.S. government, and therefore extremists usually spurn compromise.

5. Willingness to deny basic civil liberties to certain fellow citizens, because enemies deserve no liberties.

6. Consistent indulgence in irresponsible accusations and character assassination.

The Core and the Lunatic Fringe of Holocaust Denial

The development of the Holocaust denial movement has striking parallels with the development of other fringe movements. Since deniers are not consciously modeling themselves after, for example, the creationists, we may be tracking an ideological pattern common to fringe groups trying to move into the mainstream:

1. Early on, the movement includes a wide diversity of thought and members representing the extreme fringes of society, and it has little success in entering the mainstream (creationism in the 1950s; denial in the 1970s).

2. As the movement grows and evolves, some members attempt to disassociate themselves and their movement from the radical fringe and try to establish scientific or scholarly credentials (creationism in the 1970s when it became "creation-science"; denial in the 1970s with the founding of IHR).

3. During this drive toward acceptability, emphasis moves away from antiestablishment rhetoric and toward a more positive statement of beliefs (creationists abandoned the antievolution tactic and adopted "equal-time" arguments; IHR has broken with Carto and generally deniers are trying to shed their racist, antisemitic reputation).

4. To enter public institutions such as schools, the movement will use the First Amendment and claim that its "freedom of speech" is being violated when its views are not allowed to be heard (creationists legislated equal-time laws in several states in the 1970s and 1980s; Zündel's Canadian "free speech" trials [see figure 19]; and Bradley Smith's advertisements in college newspapers).

5. To get the public's attention, the movement tries to shift the burden of proof from itself to the establishment, demanding "just one proof" (creationists ask for "just one fossil" that proves transitional forms exist; deniers demand "just one proof" that Jews were killed in gas chambers).

The Holocaust denial movement has its extremes, and members of its lunatic fringe commonly hold neo-Nazi and white supremacist views. Holocaust denier and self-proclaimed white separatist Jack Wikoff, for example, publishes *Remarks* out of Aurora, New York. "Talmudic Jewry is at war with humanity," Wikoff explains. "Revolutionary communism and International Zionism are twin forces working toward the same goal: a despotic world government with the capital in Jerusalem" (1990). Wikoff also publishes statements such as this one, made in a letter from "R.T.K." from California: "Under Hitler and National Socialism, the German troops

FIGURE 19:
During his "free speech" trial in Canada, Ernst Zündel appeared in a concentration camp uniform among supporters holding placards proclaiming standard conspiratorial beliefs about Jews and the media, 1985. [Photograph courtesy Ernst Zündel.]

were taught White racism and never has this world seen such magnificent fighters. *Our job is re-education with the facts of genetics and history"* (1990). Interestingly, *Remarks* is endorsed by Bradley Smith, and Wikoff reviews books for *JHR.*

Another denier newsletter, *Instauration,* featured in its January 1994 issue an article titled "How to Cut Violent Crime in Half: An Immodest Proposal," with no byline. The author's solution is vintage Nazi:

There are 30 million blacks in the U.S., half of them male and about one-seventh of the males in the 16 to 26 age bracket, the violent sector of the black population. Half of 30 million is 15 million. One-seventh of 15 million is a little more than 2 million. This tells us that 2 million blacks, not 30 million, are committing the crimes. The Soviet Union had gulag populations that ran as high as 10 million at various times during the Stalin era. The U.S. with much more advanced technology should be able to contain and run camps that hold at least 20% of that number. Negroes not on drugs and with no criminal record would be released from the camps once psychological and genetic tests found no traces of violent behavior. As for most detainees, on their 27th birthday all but the most incorrigible "youths" would be let out, leaving room for the new contingent of 16-year-olds that would be replacing them. (p. 6)

The National Socialist German Workers Party, Foreign Organization (NSDAP/AO), hailing from Lincoln, Nebraska, publishes a bimonthly newspaper, *The New Order*. Here one can order swastika pins, flags, armbands, keychains, and medallions; SS songs and speeches; "White Power" T-shirts; and all manner of books and magazines promoting white power, neo-Nazis, Hitler, and antisemitism. The July/August 1996 issue, for instance, explains that "COMPLETE GLOBAL EXTINCTION of the NEGROID RACE (due to AIDS infection) will occur NO LATER than the year 2022 A.D." A happy face sits below this "good" news, with the slogan "Have a Nazi Day!" About Auschwitz, the reader is told, "With systematic German precision, each and every death was recorded and categorized. The small number of deaths over a three-year period is actually a testament to how humane, clean and healthy the conditions were at the SS labor camp in Poland!" The problem, of course, is that "the yids will use the truth to support THEIR evil lies and paranoid persecution complex" (p. 4).

Mark Weber, David Irving, and company have actively distanced themselves from this side of Holocaust denial. Weber, for instance, has protested, "Why is this relevant? [Lew] Rollins used to work for IHR. *Remarks* is on the cusp. They used to be more-or-less revisionist. But [publisher Jack Wikoff] is now getting engaged more and more into racialist matters. *Instauration* is racialist. I suppose they're affiliated so far as they agree with some of the things we might put out. But there is no relationship" (1994b). Yet these folks and others of their ilk also call themselves "Holocaust revisionists," and their literature is filled with references to standard denial arguments and to IHR Holocaust deniers. And, across the spectrum of Holocaust denial, Ernst Zündel is acknowledged as the spiritual leader of the movement.

For example, *Tales of the Holohoax* is dedicated to Robert Faurisson and Ernst Zündel and thanks Bradley Smith and Lew Rollins. After fourteen pages of gross cartoon depictions of Jews and the "Holohoax," the author states, "The wild fables about homicidal gas chambers loosely grouped under the Orwellian Newspeak heading of the 'Holocaust,' have become the informal state religion of the West. The government, the public schools and the corporate media promote the imposition of this morbid, funeral-home-of-the-mind on young people, to instill guilt as a form of group-libel/hate propaganda against the German people" (House 1989, p. 15).

Not all deniers are the same, but the fact remains that in all Holocaust denial there is a core of racist, paranoid, conspiratorial thinking that is clearly directed at Jews. It ranges from crass antisemitism to a more subtle and pervasive form of antisemitism that creeps into conversation as "Some of my best friends are Jews, but . . . " or "I'm not antisemitic, but . . . " followed by a litany of all the things "the Jews" are doing. This bias is what drives deniers to seek and find what they are looking for, and to confirm what they already believe. Why do they say the Holocaust never happened? Depending on whom you ask, interest in history, money, perversity, notoriety, ideology, politics, fear, paranoia, hate.

14

How We Know
the Holocaust Happened
Debunking the Deniers

The word *debunking* has negative connotations for most people, yet when you are presenting answers to claims of an extraordinary nature (and Holocaust denial surely qualifies), then debunking serves a useful purpose. There is, after all, a lot of bunk to be debunked. But I am attempting to do far more than this. In the process of debunking the deniers, I demonstrate how we know that the Holocaust happened, and that it happened in a particular way that most historians have agreed upon.

There is no immutable canon of truth about the Holocaust that can never be altered, as many deniers believe. When you get into the study of the Holocaust, and especially when you start attending conferences and lectures and tracking the debates among Holocaust historians, you discover that there is plenty of infighting about the major and minor points of the Holocaust. The brouhaha over Daniel Goldhagen's 1996 book, *Hitler's Willing Executioners*, in which he argued that "ordinary" Germans and not just Nazis participated in the Holocaust, is testimony to the fact that Holocaust historians are anything but settled on exactly what happened, when, why, and how. Nonetheless, an abyss lies between the points that Holocaust historians are debating and those that Holocaust deniers are promoting—their denial of intentional genocide based primarily on race, of programmatic use of gas chambers and crematoria for mass murder, and of the killing of five to six million Jews.

211

Methodology of Holocaust Denial

Before addressing the three main axes of Holocaust denial, let us look for a moment at the deniers' methodology, their modes of argument. Their fallacies of reasoning are eerily similar to those of other fringe groups, such as creationists.

1. They concentrate on their opponents' weak points, while rarely saying anything definitive about their own position. Deniers emphasize the inconsistencies between eyewitness accounts, for example.

2. They exploit errors made by scholars who are making opposing arguments, implying that because a few of their opponents' conclusions were wrong, *all* of their opponents' conclusions must be wrong. Deniers point to the human soap story, which has turned out to be a myth, and talk about "the incredible shrinking Holocaust" because historians have reduced the number killed at Auschwitz from four million to one million.

3. They use quotations, usually taken out of context, from prominent mainstream figures to buttress their own position. Deniers quote Yehuda Bauer, Raul Hilberg, Arno Mayer, and even leading Nazis.

4. They mistake genuine, honest debates between scholars about certain points within a field for a dispute about the existence of the entire field. Deniers take the intentionalist-functionalist debate about the development of the Holocaust as an argument about whether the Holocaust happened or not.

5. They focus on what is not known and ignore what is known, emphasize data that fit and discount data that do not fit. Deniers concentrate on what we do not know about the gas chambers and disregard all the eyewitness accounts and forensic tests that support the use of gas chambers for mass murder.

Because of the sheer quantity of evidence about the Holocaust—so many years and so much of the world involved, thousands of accounts and documents, millions of bits and pieces—there is enough evidence that some parts can be interpreted as supporting the deniers' views. The way that deniers treat testimony from the postwar Nuremberg trials of Nazis is typical of their handling of evidence. On the one hand, deniers dismiss the Nuremberg confessions as unreliable because it was a military tribunal run by the victors. The evidence, Mark Weber claims, "consists largely of extorted confessions, spurious testimonies, and fraudulent documents. The postwar Nuremberg trials were politically motivated proceedings meant more to discredit the leaders of a defeated regime than to establish truth" (1992, p. 201). Neither Weber nor anyone else has proven that most of the confessions were extorted, spurious, or fraudulent. But even if the deniers were able to prove that *some* of them were, this does not mean that they *all* were.

On the other hand, deniers cite Nuremberg trial testimony whenever it supports their arguments. For example, although deniers reject the testimony of Nazis who said there was a Holocaust and they participated in it, deniers accept the testimony of Nazis such as Albert Speer who said they knew nothing about it. But even here, deniers shy away from a deeper analysis. Speer indeed stated at the trials that he did not know about the extermination program. But his Spandau diary speaks volumes:

> *December 20, 1946.* Everything comes down to this: Hitler always hated the Jews; he made no secret of that at any time. He was capable of tossing off quite calmly, between the soup and the vegetable course, "I want to annihilate the Jews in Europe. This war is the decisive confrontation between National Socialism and world Jewry. One or the other will bite the dust, and it certainly won't be us." So what I testified in court is true, that I had no knowledge of the killings of Jews; but it is true only in a superficial way. The question and my answer were the most difficult moment of my many hours on the witness stand. What I felt was not fear but shame that I as good as knew and still had not reacted; shame for my spiritless silence at the table, shame for my moral apathy, for so many acts of repression. (1976, p. 27)

In addition, Matthias Schmidt, in *Albert Speer: The End of a Myth*, details Speer's activities in support of the Final Solution. Among other things, Speer organized the confiscation of 23,765 apartments from Jews in Berlin in 1941; he knew of the deportation of more than 75,000 Jews to the east; he personally inspected the Mauthausen concentration camp, where he ordered a reduction of construction materials and redirected supplies that were needed elsewhere; and in 1977 he told a newspaper reporter, "I still see my guilt as residing chiefly in the approval of the persecution of the Jews and the murder of millions of them" (1984, pp. 181–198). Deniers cite Speer's Nuremberg testimony and ignore all Speer's elaborations about that testimony.

Convergence of Evidence

No matter what we wish to argue, we must bring to bear additional evidence from other sources that corroborates our conclusions. Historians know that the Holocaust happened by the same general method that scientists in such historical fields as archeology or paleontology use— through what William Whewell called a "consilience of inductions," or a convergence of evidence. Deniers seem to think that if they can just find one tiny crack in the Holocaust structure, the entire edifice will come tumbling down. This is the fundamental flaw in their reasoning. The Holocaust was not a single event. The Holocaust was thousands of events in tens of thousands of places, and is proved by millions of bits of data that converge on one conclusion. The Holocaust cannot be disproved by minor errors or inconsistencies here and there, for the simple reason that it was never proved by these lone bits of data in the first place.

Evolution, for example, is proved by the convergence of evidence from geology, paleontology, botany, zoology, herpetology, entomology, biogeography, anatomy, physiology, and comparative anatomy. No one piece of evidence from these diverse fields says "evolution" on it. A fossil is a snapshot. But when a fossil in a geological bed is studied along with other fossils of the same and different species, compared to species in other strata, contrasted to modern organisms, juxtaposed with species in other parts of the world, past and present, and so on, it turns from a snapshot into a motion picture. Evidence from each field jumps together to a grand conclusion—evolution. The process is no different in proving the Holocaust. Here is the convergence of proof:

> *Written documents:* Hundreds of thousands of letters, memos, blueprints, orders, bills, speeches, articles, memoirs, and confessions.
>
> *Eyewitness testimony:* Accounts from survivors, Kapos, Sonderkommandos, SS guards, commandants, local townspeople, and even upper-echelon Nazis who did not deny the Holocaust.
>
> *Photographs:* Official military and press photographs and films, civilian photographs, secret photographs taken by prisoners, aerial photographs, and German and Allied film footage.
>
> *Physical evidence:* Artifacts found at the sites of concentration camps, work camps, and death camps, many of which are still extant in varying degrees of originality and reconstruction.
>
> *Demographics:* All those people who the deniers claim survived the Holocaust are missing.

Holocaust deniers ignore this convergence of evidence. They pick out what suits their theory and dismiss or avoid the rest. Historians and scientists do this too, but there is a difference. History and science have self-correcting mechanisms whereby one's errors are "revised" by one's colleagues in the true sense of the word. *Revision is the modification of a theory based on new evidence or a new interpretation of old evidence.* Revision should not be based on political ideology, religious conviction, or other human emotions. Historians are humans with emotions, of course, but they are the true revisionists because eventually the collective science of history separates the emotional chaff from the factual wheat.

Let us examine how the convergence of evidence works to prove the Holocaust, and how deniers select or twist the data to support their claims. We have an account by a survivor who says he heard about the gassing of Jews while he was at Auschwitz. The denier says that survivors exaggerate and that their memories are unsound. Another survivor tells another story different in details but with the core similarity that Jews were gassed at Auschwitz. The denier claims that rumors were floating throughout the camps and many survivors incorporated them into their memories. An SS guard confesses after the war that he actually saw people being gassed and cremated. The denier claims that these confessions were forced out of the Nazis by the Allies. But now a member of the Sonderkommando—a Jew who had helped the Nazis move dead bodies from the gas chambers and into the crematoria—says he not only heard about it and not only saw it happening, he had actually participated in the process. The denier explains this away by saying that the Sonderkommando accounts make no sense— their figures of numbers of bodies are exaggerated and their dates incorrect. What about the camp commandant, who confessed after the war that he not only heard, saw, and participated in the process but orchestrated it? He was tortured, says the denier. But what about his autobiography, written after his trial, conviction, and sentencing to death, when he had nothing to gain by lying? No one knows why people confess to ridiculous crimes, explains the denier, but they do.

No single testimony says "Holocaust" on it. But woven together they make a pattern, a story that holds together, while the deniers' story unravels. Instead of the historian having to present "just one proof," the denier must now disprove six pieces of historical data, with six different methods of disproof.

But there is more. We have blueprints of gas chambers and crematoria. Those were used strictly for delousing and body disposal, claims the denier; and thanks to the Allied war against Germany, the Germans were never given the opportunity to deport the Jews to their own homeland and instead had to put them into overcrowded camps where disease and lice were rampant. What about the huge orders for Zyklon-B gas? It was used strictly for delousing all those diseased inmates. What about those speeches by Adolf Hitler, Heinrich Himmler, Hans Frank, and Joseph Goebbels talking about the "extermination" of the Jews? Oh, they really meant "rooting out," as in deporting them out of the Reich. What about Adolf Eichmann's confession at his trial? He was coerced. Hasn't the German government confessed that the Nazis attempted to exterminate European Jewry? Yes, but they lied so they could rejoin the family of nations.

Now the denier must rationalize no less than fourteen different bits of evidence that converge to a specific conclusion. But the consilience continues. If six million Jews did not die, where did they go? They are in Siberia and Peoria, Israel and Los Angeles, says the denier. But why can't they find each other? They do—haven't you heard the stories of long-separated siblings making contact with one another after many decades? What about the photos and newsreels of the liberation of the camps with all those dead bodies and starving inmates? Those people were well taken care of until the end of the war when the Allies were mercilessly bombing German cities, factories, and supply lines, thus preventing food from reaching the camps; the Nazis tried valiantly to save their prisoners but the combined strength of the Allies was too much. But what about all the accounts by prisoners of the brutality of the Nazis—the random shootings and beatings, the deplorable conditions, the freezing temperatures, the death marches, and so on? That is the nature of war, replies the denier. The Americans interned Japanese-Americans and Japanese nationals in camps. The Japanese imprisoned Chinese. The Russians tortured Poles and Germans. War is hell. The Nazis were no different from anyone else.

We are now up to eighteen sets of evidence all converging toward one conclusion. The denier chips away at them all, determined not to give up his belief system. He is relying on what might be called *post hoc rationalization*—after-the-fact reasoning to justify contrary evidence—and then on demanding that the Holocaust historian disprove each of his rationalizations. But the convergence of positive evidence supporting the Holocaust means that the historian has already met the burden of proof, and when the denier demands that each piece of evidence independently prove the Holocaust he is ignoring the fact that no historian ever claimed that one piece of evidence proves the Holocaust or anything else. We must examine the evidence as part of a whole, and when we do so the Holocaust can be regarded as proven.

Intentionality

The first major axis of Holocaust denial is that genocide based primarily on race was not intended by Hitler and his followers.

Adolf Hitler

Deniers begin at the top, so I will too. In his 1977 *Hitler's War*, David Irving argued that Hitler did not know about the Holocaust. Shortly after, he put his money where his mouth is, promising to pay $1,000 to anyone who could produce documentary proof—specifically, a written document—that Hitler ordered the Holocaust. In a classic example of what I call the *snapshot fallacy*—taking a single frame out of a historical film—Irving reproduced, on page 505 of *Hitler's War*, Himmler's telephone notes of November 30, 1941, when the SS chief telephoned Reinhard Heydrich (deputy chief of the Reichssicherheitshaupamt [Head Office for Reich Security, or RSHA, of the SS]) "from Hitler's bunker at the Wolf's Lair, ordering that there was to be 'no liquidation' of Jews." From this, Irving concluded that "the Führer had ordered that the Jews were not to be liquidated" (1977, p. 504).

But we must see the snapshot in the context of the frames around it. As Raul Hilberg pointed out, in its entirety, the log entry says, "Jewish transport from Berlin. No liquidation." It was in reference to one particular transport, not all Jews. And, says Hilberg, "that transport *was* liquidated! That order was either ignored, or it was too late. The transport had already arrived in Riga [capital of Latvia] and they didn't know what to do with these thousand people so they shot them that very same evening" (1994). Moreover, for Hitler to veto an order for liquidation implies that liquidation was something that was ongoing. To that extent, David Irving's $1,000 challenge and Robert Faurisson's demand for "just one proof" are met. If Jews were not being exterminated, why would Hitler feel the need to halt the extermination of a particular transport? And this entry also proves that it was Hitler, and not Himmler or Goebbels, who ordered the Holocaust. As Speer observed regarding Hitler's role: "I don't suppose he had much to do with the technical aspects, but even the *decision* to proceed from shooting to gas chambers would have been his, for the simple reason, as I know only too well, that no major decisions could be made about *anything* without his approval" (in Sereny 1995, p. 362). As Yisrael Gutman noted, "Hitler interfered in all main decisions with regard to the Jews. All the people around Hitler came with their plans and initiatives because they knew that Hitler was interested [in solving the 'Jewish question'] and they wanted to please him and be the first to realize his intentions and his spirit" (1996).

Whether or not there was a specific order from Hitler for the extermination of the Jews does not matter, then, because it did not need to be spelled out. The Holocaust "was not so much a product of laws and commands as it was a matter of spirit, of shared comprehension, of consonance and synchronization" (Hilberg 1961, p. 55). This spirit was made plain in his speeches and writings. From his earliest political ramblings to the final Götterdämmerung of the end in his Berlin bunker, Hitler had it in for Jews. On April 12, 1922, in a speech given in Munich and later published in the newspaper *Völkischer Beobachter*, he told his audience, "The Jew is the ferment of the decomposition of people. This means that it is in the nature of the Jew to destroy, and he must destroy, because he lacks altogether any idea of working for the common good. He possesses certain characteristics given to him by nature and he never can rid himself of those characteristics. The Jew is harmful to us" (in Snyder 1981, p. 29). Twenty-three years later (1922–1945), with his world collapsing around him, Hitler said, "Against the Jews I fought open-eyed and in view of the whole world. . . . I made it plain that they, this parasitic vermin in Europe, will be finally exterminated" (February 13, 1945; in Jäckel 1993, p. 33), and "Above all I charge the leaders of the nation and those under them to scrupulous observance of the laws of race and to merciless opposition to the universal poisoner of all peoples, International Jewry" (April 29, 1945; in Snyder 1981, p. 521).

In between, Hitler made hundreds of similar statements. In a speech given January 30, 1939, for example, he said, "Today I want to be a prophet once more: If international finance Jewry inside and outside of Europe should succeed once more in plunging nations into another world war, the consequence will not be the Bolshevization of the earth and thereby the victory of Jewry, but the annihilation of the Jewish race in Europe" (in Jäckel 1989, p. 73). Hitler even told the Hungarian head of state, "In Poland this state of affairs has been . . . cleared up: if the Jews there did not *want* to work, they were shot. If they *could* not work, they were treated like tuberculosis bacilli with which a healthy body may become infected. This is not cruel if one remembers that even innocent creatures of nature, such as hares and deer when infected, have to be killed so that they cannot damage others. Why should the beasts who wanted to bring us Bolshevism be spared more than these innocents?" (in Sereny 1995, p. 420). How many more quotes do we need to prove that Hitler ordered the Holocaust—a hundred, a thousand, ten thousand?

Ausrotten Among the Nazi Elite

David Irving and other deniers make it sound like these speeches do not indicate a smoking gun, by playing a clever game of semantics with the word *ausrotten*, which according to modern dictionaries means "to exterminate, extirpate, or destroy." This word can be found in numerous Nazi speeches and documents referring to the Jews. But Irving insists that *ausrotten* really means "stamping or rooting out," arguing that "the word *ausrotten* means one thing now in 1994, but it meant something very different in the time Adolf Hitler uses it." Yet a check of historical dictionaries shows that *ausrotten* has always meant "to exterminate." Irving's rejoinder provides another example of *post hoc rationalization:*

Different words mean different things when uttered by different people. What matters is what that word meant when uttered by Hitler. I would first draw attention to the famous memorandum on the Four-Year Plan of August 1936. In that Adolf Hitler says, "We are going to have to get our armed forces in a fighting state within four years so that we can go to war with the Soviet Union. If the Soviet Union should ever succeed in overrunning Germany it will lead to the *ausrotten* of the German people." There's that word. There is no way that Hitler can mean the physical liquidation of 80 million Germans. What he means is that it will lead to the emasculation of the German people as a power factor. (1994)

I then pointed out that, at a December 1944 conference regarding the Ardennes attack against the Americans, Hitler ordered his generals "to *ausrotten* them division by division." Was Hitler giving the order to *transport* the Americans out of the Ardennes division by division? Irving countered:

> Compare that with a speech he made in August 1939, in which he says, with regard to Poland, "we are going to destroy the living forces of the Polish Army." This is the job of any commander—you have to destroy the forces facing you. How you destroy them, how you "take them out" is probably a better phrase, is immaterial. If you take those pawns off the chess board they are gone. If you put the American forces in captivity they are equally neutralized whether they are in captivity or dead. And that's what the word *ausrotten* means there. (1994)

But what about Rudolf Brandt's use of the word? To SS Gruppenführer Dr. Grawitz of the SS Reichsarzt in Berlin, SS Sturmbannführer Brandt wrote concerning "the *Ausrottung* of tuberculosis as a disease affecting the nation." A year later, now an SS Obersturmbannführer, he wrote to Ernst Kaltenbrunner, Heydrich's successor as chief of RSHA, "I am sending you the outline of a press announcement concerning the accelerated *Ausrottung* of the Jews in occupied Europe." The same man is using the same word to discuss the same process for tuberculosis and Jews (see figure 20). What else could *ausrotten* have meant in these contexts except "extermination"?

And what about Hans Frank's use of the word? In a speech to a Nazi assembly held on October 7, 1940, Frank summed up his first year of effort as head of the Generalgouvernement of occupied Poland: "I could not *ausrotten* all lice and Jews in only one year. But in the course of time, and if you help me, this end will be attained" (Nuremberg Doc. 3363-PS, p. 891). On December 16, 1941, Frank addressed a government session at the office of the governor of Krakau in conjunction with the upcoming Wannsee Conference:

> Currently there are in the Government Generalship approximately 2.5 million, and together with those who are kith and kin and connected in all kinds

Der Reichsführer-SS
Persönlicher Stab
Tgb.Nr. AR/236/7
Bra/H.

Führer-Hauptquartier
12. Febr. 42

1.) An den
Reichsarzt-SS
SS-Gruppenführer Dr. G r a w i t z
B e r l i n.

Lieber Gruppenführer!

Ich übersende Ihnen anliegend den Durchschlag einer Denkschrift, die ein Herr Dr. B l o m e an den Reichsleiter Bormann über die Ausrottung der Tuberkulose als Volkskrankheit eingereicht hat. Die Zusendung an den Reichsführer-SS ist von SS-Oberführer Prof. Dr. Gerlach erfolgt.

H e i l H i t l e r !

Ihr gez. R. B r a n d t

1 Anlage.

SS-Sturmbannführer

Der Reichsführer-SS
Persönlicher Stab
Tgb.-Nr. 39/13/43 g
Ma/O.

Feld-Kommandostelle, den 22. 2.43

Persönlicher Stab Reichsführer-SS
Schriftgutverwaltung
Akt. Nr. Geh. /

An den
Chef der Sicherheitspolizei und des SD
B e r l i n.

Im Auftrage des Reichsführer-SS übersende ich in der Anlage eine Pressemeldung über die beschleunigte Ausrottung der Juden im besetzten Europa.

I.A.

2 Anlagen. 14.2.43

SS-Obersturmbannführer.

The Reichsführer SS
Personal Staff Secret
(Diary Entry No.)

Field Command Post
Feb.22, 1943

To: Chief of Sicherheitspolizei [Security Police] and
SD [Security Service]
Berlin

As ordered by the SS Reichsführer, I am sending you the outline of a press announcement concerning the accelerated extermination of the Jews [Ausrottung der Juden...] in occupied Europe.

On behalf of

SS Obersturmbannführer

Two Enclosures

The "Ausrotten" Debate—the Meaning of "Extermination."

The February 12, 1942, memo from SS Sturmbannführer Rudolf Brandt to the SS Reichsdoctor Dr. Grawitz, proves that he means "to kill" TB when speaking of the "Ausrottung der Tuberkulose" in the first paragraph, as he does in the second document which translates ". . . concerning the accelerated extermination of the Jews in occupied Europe." The same man is using the same word to discuss the same process of extermination for both TB and Jews. Documents and translation courtesy of National Archives, Washington, DC.

FIGURE 20:
Rudolf Brandt writes about (*top*) "die Ausrottung die Tuberkulose" to SS Gruppenführer Dr. Grawitz of the SS Reichsarzt, February 12, 1942; and (*bottom*) "die beschleunigte Ausrottung der Juden" to Ernst Kaltenbrunner, chief of RSHA, February 22, 1943. *Ausrottung* means "extermination." [Documents and translation courtesy National Archives, Washington, D.C.]

of ways, we now have 3.5 million Jews. We cannot shoot these 3.5 million
Jews, nor can we poison them, yet we will have to take measures which will
somehow lead to the goal of annihilation, and that will be done in connection
with the great measures which are to be discussed together with the Reich.
The territory of the General Government must be made free of Jews, as is the
case in the Reich. Where and how this will happen is a matter of the means
which must be used and created, and about whose effectiveness I will inform
you in due time. (Original document and translation, National Archives,
Washington, D.C., T922, PS 2233)

If the Final Solution meant deportation out of the Reich, as Irving and
other deniers claim, does this mean that Frank was planning to send lice out
of Poland on trains? And why would Frank be making references to the
extermination of Jews through means other than shooting or poisoning?

And then there are entries from the diary of Joseph Goebbels,
Gauleiter (General) of Berlin, Reich Minister of Propaganda, and Reich
Plenipotentiary for total war effort, such as these:

> *August 8, 1941,* concerning the spread of spotted typhus in the Warsaw
> ghetto: "The Jews have always been the carriers of infectious diseases. They
> should either be concentrated in a ghetto and left to themselves or be liqui-
> dated, for otherwise they will infect the populations of the civilized nations."

> *August 19, 1941,* after a visit to Hitler's headquarters: "The Führer is con-
> vinced his prophecy in the Reichstag is becoming a fact: that should Jewry
> succeed in again provoking a new war, this would end with their annihilation.
> It is coming true in these weeks and months with a certainty that appears
> almost sinister. In the East the Jews are paying the price, in Germany they
> have already paid in part and they will have to pay more in the future."
> (Broszat 1989, p. 143)

Himmler also talks about the *ausrotten* of the Jews, and again there is
evidence that negates the deniers' definition of that word. For example, in
a lecture on the history of Christianity given in January 1937, Himmler
told his SS Gruppenführers, "I have the conviction that the Roman
emperors, who exterminated [*ausrotteten*] the first Christians, did precisely
what we are doing with the communists. These Christians were at that
time the vilest scum, which the city accommodated, the vilest Jewish people, the vilest Bolsheviks there were" (Padfield 1990, p. 188). In June
1941, Himmler informed Rudolf Hoess, the commandant of Auschwitz,
that Hitler had ordered the Final Solution (*Endlösung*) of the Jewish question, and that Hoess would play a major role at Auschwitz:

> It is a hard, tough task which demands the commitment of the whole person
> without regard to any difficulties that may arise. You will be given details by

Sturmbannführer Eichmann of the RSHA who will come to see you in the near future. The department taking part will be informed at the appropriate time. You have to maintain the strictest silence about this order, even to your superiors. The Jews are the eternal enemies of the German people and must be exterminated. All Jews we can reach now, during the war, are to be exterminated without exception. If we do not succeed in destroying the biological basis of Jewry, some day the Jews will annihilate the German Volk [people]. (Padfield 1990, p. 334)

Himmler made many similarly damning speeches. One of the most notorious is the October 4, 1943, speech to the SS Gruppenführer in Poznan (Posen), which was recorded on a red oxide tape. Himmler was lecturing from notes, and early in the talk he stopped the tape recorder to make sure it was working. He then continued, knowing he was being recorded, and spoke for over three hours on a range of subjects, including the military and political situation, the Slavic peoples and racial blends, how the racial superiority of Germans would help them win the war, and the like. Two hours into the speech, Himmler began to talk about the bloody 1934 purges of traitors in the Nazi Party and "the extermination of the Jewish people."

I also want to refer here very frankly to a very difficult matter. We can now very openly talk about this among ourselves, and yet we will never discuss this publicly. Just as we did not hesitate on June 30, 1934, to perform our duty as ordered and put comrades who had failed up against the wall and execute them, we also never spoke about it, nor will we ever speak about it. Let us thank God that we had within us enough self-evident fortitude never to discuss it among us, and we never talked about it. Every one of us was horrified, and yet every one clearly understood that we would do it next time, when the order is given and when it becomes necessary.

I am now referring to the evacuation of the Jews, to the extermination of the Jewish people. This is something that is easily said: "The Jewish people will be exterminated," says every Party member, "this is very obvious, it is in our program—elimination of the Jews, extermination, will do." And then they turn up, the brave 80 million Germans, and each one has his decent Jew. It is of course obvious that the others are pigs, but this particular one is a splendid Jew. But of all those who talk this way, none had observed it, none had endured it. Most of you here know what it means when 100 corpses lie next to each other, when 500 lie there or when 1,000 are lined up. To have endured this and at the same time to have remained a decent person—with exceptions due to human weaknesses—has made us tough. This is an honor roll in our history which has never been and never will be put in writing, because we know how difficult it would be for us if we still had Jews as secret saboteurs, agitators and rabble rousers in every city, what with the bombings, with the burden and with the hardships of the war. If the Jews were still part of the German nation, we would most likely arrive now at the state we were at in 1916/17. (Original document and translation, National Archives, Washington, D.C., PS Series 1919, pp. 64–67)

Irving's response to this quote was interesting:

Irving: I have a later speech he made on January 26, 1944, in which he is speaking to the same audience rather more bluntly about the *ausrotten* of Germany's Jews, when he announced that they had totally solved the Jewish problem. Most of the listeners sprang to their feet and applauded. "We were all there in Poznan," recalled a Rear Admiral, "when that man [Himmler] told us how he'd killed off the Jews. I can still recall precisely how he told us. 'If people ask me,' said Himmler, 'why did you have to kill the children too, then I can only say I am not such a coward that I leave for my children something I can do myself.'" Quite interesting—this is an Admiral afterwards recording this in British captivity without realizing he was being tape recorded, which is a very good summary of what Himmler actually said.

Shermer: That sounds to me like he means to kill Jews, not just transport them out of the Reich.

Irving: I agree, Himmler said that. He actually said, "We're wiping out the Jews. We're murdering them. We're killing them."

Shermer: What does that mean other than what it sounds like?

Irving: I agree, Himmler is admitting what I said happened to the 600,000. But, and this is the important point, nowhere does Himmler say, "We are killing millions." Nowhere does he even say we are killing hundreds of thousands. He is talking about solving the Jewish problem, about having to kill off women and children too. (1994)

Irving, once again, has fallen into the fallacy of *ad hoc rationalization*. Since Himmler never exactly said millions, therefore he really meant thousands. But, please note, Himmler never said thousands either. Irving is inferring what he wants to infer. The actual numbers come from other sources, which, in conjunction with Himmler's speeches and many other pieces of evidence, converge on the conclusion that he meant millions would be killed. And millions were killed.

The Einsatzgruppen

Finally, there is telling evidence about the extermination of Jews from lower down in the ranks. The Einsatzgruppen were mobile SS and police units for special missions in occupied territories. Their mandate included rounding up and killing Jews and other unwanted persons in towns and villages prior to occupation by Germans. For the winter of 1941–1942, for example, Einsatzgruppe A reported 2,000 Jews killed in Estonia, 70,000 in Latvia, 136,421 in Lithuania, and 41,000 in Belorussia. On November 14, 1941, Einsatzgruppe B reported 45,467 shootings, and on July 31, 1942, the governor of Belorussia reported that 65,000 Jews were killed during the previous two months. Einsatzgruppe C estimated they had killed 95,000 by December 1941, and Einsatzgruppe D reported on April 8, 1942, a total of 92,000 killed. The grand total is 546,888 dead in less than one year.

Numerous eyewitness accounts from members of the Einsatzgruppen can be found in *"The Good Old Days": The Holocaust as Seen by Its Perpetrators and Bystanders* (Klee, Dressen, and Riess 1991). For example, on Sunday, September 27, 1942, SS Obersturmführer Karl Kretschmer wrote to "My dear Soska," his wife. He apologizes for not writing more, is feeling ill and in "low spirits" because "what you see here makes you either brutal or sentimental." His "gloomy mood," he explains, is caused by "the sight of the dead (including women and children)." Which dead? Dead Jews, who deserve to die: "As the war is in our opinion a Jewish war, the Jews are the first to feel it. Here in Russia, wherever the German soldier is, no Jew remains. You can imagine that at first I needed some time to get to grips with this." In a subsequent letter, not dated, he explains to his wife that "there is no room for pity of any kind. You women and children back home could not expect any mercy or pity if the enemy got the upper hand. For that reason we are mopping up where necessary but otherwise the Russians are willing, simple and obedient. There are no Jews here any more." Finally, on October 19, 1942, in a letter signed "You deserve my best wishes and all my love, Your Papa," Kretschmer provides a paradigmatic example of what Hannah Arendt meant by the banality of evil:

> If it weren't for the stupid thoughts about what we are doing in this country, the Einsatz here would be wonderful, since it has put me in a position where I can support you all very well. Since, as I already wrote to you, I consider the last Einsatz to be justified and indeed approve of the consequences it had, the phrase: "stupid thoughts" is not strictly accurate. Rather it is a weakness not to be able to stand the sight of dead people; the best way of overcoming it is to do it more often. Then it becomes a habit. (pp. 163–171)

There may not have been a written order, but the Nazi's intentionality of genocide primarily by race was not only clear but also known rather widely.

The Intentionalist-Functionalist Controversy

For several decades following the war, historians debated the "intentionalism" versus the "functionalism" of the Holocaust. Intentionalists argued that Hitler intended the mass extermination of the Jews from the early 1920s, that Nazi policy in the 1930s was programmed toward this end, and that the invasion of Russia and the quest for *Lebensraum* were directly planned and linked to the Final Solution of the Jewish question. Functionalists, by contrast, argued that the original plan for the Jews was expulsion and that the Final Solution evolved as a result of the failed war against Russia. Holocaust historian Raul Hilberg, however, feels that these are artificial distinctions: "In reality it is more complicated than either of these interpretations. I believe Hitler gave a plenary order, but that order was itself the end product of a process. He said many things along the way which encouraged the bureaucracy to think along certain lines and to take initiatives. But on the whole I would say that any kind of systematic shooting, particularly of young children or very old people, and any kind of gassing, required Hitler's order" (1994).

Under the weight of historical evidence, intentionalism has not survived the test of time. The immediate reason, as outlined by Ronald Headland, was dawning recognition of "the competitive, almost anarchical and decentralized quality of the National Socialist system, with its rivalries, its ubiquitous personality politics, and the ever-present pursuit of power among its agencies. . . . Perhaps the greatest merit of the functionalist approach has been the extent to which it has delineated the chaotic character of the Third Reich and the often great complexity of factors involved in the decision-making process" (1992, p. 194). But the ultimate reason for acceptance of the functionalist view is that events, especially an event as complicated and contingent as the Holocaust, rarely unfold as historical actors plan. Even the famous Wannsee Conference of January 1942, at which the Nazis confirmed the implementation of the Final Solution, has been shown by Holocaust scholar Yehuda Bauer to be just one more contingent step down the road from original expulsion to final extermination. This is backed up by the existence of a realistic plan to deport the Jews to the island of Madagascar and attempts to trade Jews for cash after the Wannsee Conference. Bauer quotes Himmler's note to himself of December 10, 1942: "I have asked the Führer with regard to letting Jews go in return for ransom. He gave me full powers to approve cases like that, if they really bring in foreign currency in appreciable quantities from abroad" (1994, p. 103).

Does this discount the intentionality of the Nazis to exterminate the Jews? No, says Bauer, but it demonstrates the complexity of history and the expediency of the moment:

> In prewar Germany, emigration suited the circumstances best, and when that was neither speedy enough or complete enough, expulsion—preferably to some "primitive" place, uninhabited by true Nordic Aryans, the Soviet Union or Madagascar—was the answer. When expulsion did not work either, and the prospect of controlling Europe and, through Europe, the world arose in late

1940 and early 1941, the murder policy was decided on, quite logically, on the basis of Nazi ideology. All these policies had the same aim: removal. (Bauer 1994, pp. 252–253)

The functional sequence went from eviction of the Jews from German life (including confiscation of most of their property and homes), to concentration and isolation (often under overcrowded and filthy conditions, leading to disease and death), to economic exploitation (unpaid forced labor that often involved overwork, starvation, and death), to extermination. Gutman agrees with this contingent interpretation: "The Final Solution was an operation that started from the bottom, from a local basis, with a kind of escalation from place to place, until it was a comprehensive event. I don't know if I would call it a plan. I say it was a blueprint. Physical destruction was the outcome of a series of steps and attacks against the Jews" (1996).

The Holocaust can be modeled as a feedback loop fed by the flow of information, intentions, and actions (figure 21). From the time the

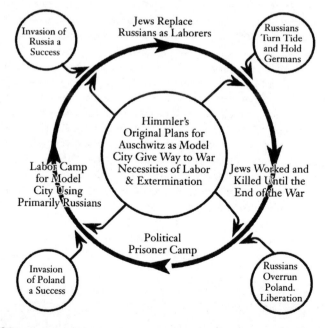

FIGURE 21:
Holocaust feedback loop. Interaction of internal psychological states and external social conditions may produce a genocidal feedback loop.

Nazis took power in 1933 and began passing legislation against Jews, to Kristallnacht and other acts of violence against Jews, to the deportation of Jews to ghettos and labor camps, to the extermination of Jews in labor and death camps, we can see at work such internal psychological components as xenophobia, racism, and violence, interacting with such external social components as a rigid hierarchical social structure, a strong central power, intolerance of diversity (religious, racial, ethnic, sexual, or political), built-in mechanisms of violence to handle dissenters, regular use of violence to enforce laws, and a low regard for civil liberties. Christopher Browning nicely summed up how this feedback loop worked in the Third Reich:

> In short, for Nazi bureaucrats already deeply involved in and committed to "solving the Jewish question," the final step to mass murder was incremental, not a quantum leap. They had already committed themselves to a political movement, to a career, and to a task. They lived in an environment already permeated by mass murder. This included not only programs with which they were not directly involved, like the liquidation of the Polish intelligentsia, the gassing of the mentally ill and handicapped in Germany, and then on a more monumental scale the war of destruction in Russia. It also included wholesale killing and dying before their very eyes, the starvation in the ghetto of Lodz and the punitive expeditions and reprisal shooting in Serbia. By the very nature of their past activities, these men had articulated positions and developed career interests that inseparably and inexorably led to a similar murderous solution to the Jewish question. (1991, p. 143)

History addresses the complexities of human acts, but within these complexities are simplicities of essences. Hitler, Himmler, Goebbels, Frank, and other Nazis were quite serious in their intention to solve the Jewish question, mainly because they were virulently antisemitic. They may have begun with resettlement, but they ended up at genocide because history's final pathways are determined by the functions of any given moment interacting with the intentions that came before. Hitler and his followers built out of their functions and intentions a road that led to camps, gas chambers and crematoria, and the extermination of millions.

Gas Chambers and Crematoria

The second major axis of Holocaust denial is that gas chambers and crematoria were not used for mass killings. How can anyone deny that the Nazis used gas chambers and crematoria? After all, these facilities still exist in many camps. To debunk the deniers can't you just go there and see for yourself? What about the evidence? In 1990, Arno Mayer noted in *Why Did the Heavens Not Darken?* that "sources for the study of the gas chambers are at once rare and unreliable." Deniers cite this sentence as vindication of their position. Mayer is a highly respected diplomatic historian at Princeton University, so one can see why deniers might be delighted by having him seemingly reinforce what they have always believed. But the entire paragraph reads:

> Sources for the study of the gas chambers are at once rare and unreliable. Even though Hitler and the Nazis made no secret of their war on the Jews, the SS operatives dutifully eliminated all traces of their murderous activities and instrument. No written orders for gassing have turned up thus far. The SS not only destroyed most camp records, which were in any case incomplete, but also razed nearly all killing and cremating installations well before the arrival of Soviet troops. Likewise, care was taken to dispose of the bones and ashes of the victims. (1990, p. 362)

Clearly, Mayer is not arguing that gas chambers were not used for mass extermination. Mayer's paragraph also neatly summarizes why the physical evidence for mass murder is not quite as overwhelmingly obvious as one might expect.

Deniers do not deny the use of gas chambers and crematoria, but they claim that gas chambers were used strictly for delousing clothing and blankets, and crematoria were used solely to dispose of bodies of people who died of "natural" causes in the camps. Before examining the evidence that the Nazis used gas chambers for mass murder in detail, consider in general the convergence of evidence from various sources:

Official Nazi documents: Orders for large quantities of Zyklon-B (the trade name of hydrocyanic acid gas), blueprints for gas chambers and crematoria, and orders for building materials for gas chambers and crematoria.

Eyewitness testimony: Survivor accounts, Jewish Sonderkommando diaries, and confessions of guards and commandants all tell of gas chambers and crematoria being used for mass murder.

Photographs: Photographs not only of the camps but also secret photographs of the burning of bodies at Auschwitz and Allied aerial reconnaissance photographs of prisoners being marched to the gas chambers at Auschwitz-Birkenau.

The camps themselves: Buildings and artifacts at the camps and the results of modern forensic tests that point to the use of both gas chambers and crematoria for killing large numbers of people.

No one source by itself proves that gas chambers and crematoria were used for genocide. It is the convergence of these sources that leads inexorably to this conclusion. For example, delivery of Zyklon-B to the camps in accordance with the written orders is corroborated by the remains of Zyklon-B canisters at the camps and by eyewitness accounts of the use of Zyklon-B in the gas chambers.

About the gassings themselves, deniers ask why no extermination victim has given an eyewitness account of an actual gassing (Butz 1976). This is like asking why no one from the killing fields of Cambodia or Stalin's purges came back to tell tales on their executioners. What we do have are hundreds of eyewitness accounts not only from SS men and Nazi doctors but from Sonderkommandos who dragged the bodies from the gas chambers and into the crematoria. In his *Eyewitness Auschwitz: Three Years in the Gas Chambers*, Filip Müller describes the deception and gassing process as follows:

> Two of the SS men took up positions on either side of the entrance door. Shouting and wielding their truncheons, like beaters at a hunt, the remaining SS men chased the naked men, women and children into the large room inside the crematorium. A few SS men were leaving the building and the last one locked the entrance door from the outside. Before long the increasing sound of coughing, screaming and shouting for help could be heard from behind the door. I was unable to make out individual words, for the shouts were drowned by knocking and banging against the door, intermingled with sobbing and crying. After some time the noise grew weaker, the screams stopped. Only now and then there was a moan, a rattle, or the sound of muffled knocking against the door. But soon even that ceased and in the sudden silence each one of us felt the horror of this terrible mass death. (1979, pp. 33–34)
>
> Once everything was quiet inside the crematorium, Unterscharführer Teuer, followed by Stark, appeared on the flat roof. Both had gas-masks dangling round their necks. They put down oblong boxes which looked like food tins; each tin was labeled with a death's head and marked Poison! What had been just a terrible notion, a suspicion, was now a certainty: the people inside the crematorium had been killed with poison gas. (p. 61)

We also have the confessions of guards. SS Unterscharführer Pery Broad was captured on May 6, 1945, by the British in their zone of occupation in Germany. Broad began work at Auschwitz in 1942 in the "Political Section" and stayed there until the liberation of the camp in January 1945. After his capture, while working as an interpreter for the British, he wrote a memoir that was passed on to the British Intelligence Service in July 1945. In December 1945, he declared under oath that what he wrote was true. On September 29, 1947, the document was translated into English and used at the Nuremberg trials regarding the gas chambers as mechanisms of mass murder. Later in 1947, he was released. When called to testify at a trial of Auschwitz SS men in April 1959, Broad acknowledged his authorship of the memoir, confirmed its validity, and retracted nothing.

I give this context for Broad's memoir because deniers dismiss damning Nazi confessions as either coerced or made up for bizarre psychological reasons (while accepting without hesitation confessions that support deniers' views). Broad was never tortured, and he had little to gain and everything to lose by confessing. When given the opportunity to recant, which he certainly could have in the later trial, he did not. Instead, he described in detail the gassing procedure, including the use of Zyklon-B, the early gassing experiments in Block 11 of Auschwitz, and the temporary chambers set up in the two abandoned farms at Birkenau (Auschwitz II), which he correctly called by their jargon name, "Bunkers I and II." He also recalled the construction of Kremas II, III, IV, and V at Birkenau, and accurately depicted (by comparison with blueprints) the design of the undressing room, gas chamber, and crematorium. Then Broad described the process of gassing in gruesome detail:

> The disinfectors are at work . . . with an iron rod and hammer they open a couple of harmless looking tin boxes, the directions read Cyclon [*sic*] Vermin Destroyer, Warning, Poisonous. The boxes are filled with small pellets which look like blue peas. As soon as the box is opened the contents are shaken out through an aperture in the roof. Then another box is emptied in the next aperture, and so on. After about two minutes the shrieks die down and change to a low moaning. Most of the men have already lost consciousness. After a further two minutes . . . it is all over. Deadly quiet reigns. . . . The corpses are piled together, their mouths stretched open. . . . It is difficult to heave the interlaced corpses out of the chamber as the gas is stiffening all their limbs. (in Shapiro 1990, p. 76)

Deniers point out that Broad's total of four minutes for the process is at odds with the statements of others, such as Commandant Hoess, who claim it was more like twenty minutes. Because of such discrepancies, deniers dismiss the account entirely. A dozen different accounts give a dozen different figures for time of death by gassing, so deniers believe no one was gassed at all. Does this make sense? Of course not. Obviously, the gassing process would take different amounts of time due to variations in conditions, including the temperature (the rate of hydrocyanic acid gas evaporation from the pellets depends on air temperature), the number of people in the room, the size of the room, and the amount of Zyklon-B poured into the room—not to mention that each observer would perceive time differently. If the time estimates were exactly the same, in fact, we would have to be suspicious that they were all taking their stories from a single account. In this case, discrepancy tends to support the veracity of the evidence.

Compare Broad's testimony with that of the camp physician, Dr. Johann Paul Kremer:

> September 2, 1942. Was present for first time at a special action at 3 A.M. By comparison Dante's Inferno seems almost a comedy. Auschwitz is justly called an extermination camp!
>
> September 5, 1942. At noon was present at a special action in the women's camp—the most horrible of all horrors. Hschf. Thilo, military surgeon, was right when he said to me today that we are located here in the *anus mundi* [anus of the world]. (1994, p. 162)

Deniers seize upon the fact that Kremer says "special action," not "gassing," but at the trial of the Auschwitz camp garrison in Krakau in December 1947, Kremer specified what he meant by "special action":

> By September 2, 1942, at 3 A.M. I had already been assigned to take part in the action of gassing people. These mass murders took place in small cottages situated outside the Birkenau camp in a wood. The cottages were called "bunkers" in the SS-men's slang. All SS physicians on duty in the camp took turns to participate in the gassings, which were called *Sonderaktion* [special action]. My part as physician at the gassing consisted in remaining in readiness near the bunker. I was brought there by car. I sat in front with the driver and an SS hospital orderly sat in the back of the car with oxygen apparatus to revive SS-men, employed in the gassing, in case any of them should succumb to the poisonous fumes. When the transport with people who were destined to be gassed arrived at the railway ramp, the SS officers selected from among the new arrivals persons fit to work, while the rest—old people, all children, women with children in their arms and other persons not deemed fit to work—were loaded onto lorries and driven to the gas chambers. There people were driven into the barrack huts where the victims undressed and then went naked to the gas chambers. Very often no incidents occurred, as the SS-men kept people quiet, maintaining that they were to bathe and be deloused. After driving all of them into the gas chamber the door was closed and an SS-man in a gas mask threw the contents of a Cyclon [*sic*] tin through an opening in the side wall. The shouting and screaming of the victims could be heard through that opening and it was clear that they were fighting for their lives. These shouts were heard for a very short while. (1994, p. 162n)

The convergence of Broad's and Kremer's accounts—and there are plenty more—provides evidence that the Nazis used gas chambers and crematoria for mass extermination.

We have hundreds of accounts of survivors describing the unloading and separation process of Jews at Auschwitz, and we have photographs of the process. We also have eyewitness accounts of the Nazis burning bodies

FIGURE 22:
Open pit burning of bodies at Auschwitz. Sonderkommandos took this picture secretly and
smuggled it out of the camp. [Photograph © Yad Vashem. All rights reserved.]

in open pits after gassing (the crematoria often broke down), and we have a
photograph of such a burning, taken secretly by a Greek Jew named Alex
(figure 22). Alter Fajnzylberg, a French Sonderkommando at Auschwitz,
recalled how this photograph was obtained:

> On the day on which the pictures were taken we allocated tasks. Some of us
> were to guard the person taking the pictures. At last the moment came. We
> all gathered at the western entrance leading from the outside to the gas cham-
> ber of Crematorium V: we could not see any SS men in the watch-tower over-
> looking the door from above the barbed wire, nor near the place where the

pictures were to be taken. Alex, the Greek Jew, quickly took out his camera, pointed it toward a heap of burning bodies, and pressed the shutter. This is why the photograph shows prisoners from the Sonderkommando working at the heap. (Swiebocka 1993, pp. 42–43)

Deniers also focus on the lack of photographic proof of gas chamber and crematoria activity in aerial reconnaissance photographs taken of the camps by the Allies. In 1992, denier John Ball actually published an entire book documenting this lack of evidence. The book is a high-quality, slick publication printed on glossy paper in order to hold the detail of the aerial photographs. Ball spent tens of thousands of dollars on the book, did all the layout and typesetting, and even printed the book himself. The project cost him more than just his savings. His wife gave him an ultimatum: her or the Holocaust. He chose the latter. Ball's book is a response to a 1979 CIA report on the aerial photographs—*The Holocaust Revisited: A Retrospective Analysis of the Auschwitz-Birkenau Extermination Complex*—in which the two authors, Dino A. Brugioni and Robert G. Poirier, present aerial photographs taken by the Allies that they claim prove extermination activities. According to Ball, the photographs were tampered with, marked, altered, faked. By whom? By the CIA itself, in order to match the story as depicted in the television mini-series *Holocaust*.

Thanks to Dr. Nevin Bryant, supervisor of cartographic applications and image processing applications at Caltech/NASA's Jet Propulsion Laboratory in Pasadena, California, I was able to get the CIA photographs properly analyzed by people who know what they are looking at from the air. Nevin and I analyzed the photographs using digital enhancement techniques not available to the CIA in 1979. We were able to prove that the photographs had not been tampered with, and we indeed found evidence of extermination activity. The aerial photographs were shot in sequence as the plane flew over the camp (on a bombing run toward its ultimate target—the IG Farben Industrial works). Since the photographs of the camp were taken a few seconds apart, stereoscopic viewing of two consecutive photographs shows movement of people and vehicles and provides better depth perception. The aerial photograph in figure 23 shows the distinctive features of Krema II. Note the long shadow from the crematorium chimney and, on the roof of the adjacent gas chamber at right angles to the crematorium building, note the four staggered shadows. Ball claims these shadows were drawn in, but four small structures that match the shadows are visible on the roof of the gas chamber in figure 24, a picture taken by an SS photographer of the back of Krema II (if you look directly below the chimney of Krema II, you will see two sides of the rectangular underground gas chamber structure protruding a few feet above the ground).

FIGURE 23:
Aerial photograph of Krema II, August 25, 1944. Note the four staggered shadows on the gas chamber roof in this photograph and compare them to the four small structures visible on the roof of the gas chamber in figure 24. These photographs support eyewitness accounts of Nazis pouring Zyklon-B pellets through the roof of the gas chamber—an example of how separate lines of evidence converge to a single conclusion. [Negative courtesy National Archives, Washington, D.C. (Film 3185); enhancement courtesy Nevin Bryant.]

This photographic evidence converges nicely with eyewitness accounts describing SS men pouring Zyklon-B pellets through openings in the roof of the gas chamber. The aerial photograph in figure 25 shows a group of prisoners being marched into Krema V for gassing. The gas chamber is at the end of the building, and the crematorium has double chimneys. From the camp's daily logs, it is clear that these are Hungarian Jews from an RSHA transport, some of whom were selected for work and the rest sent for extermination. (Additional photographs and detailed discussion appear in Shermer and Grobman 1997.)

For obvious reasons, there are no photographs recording an actual gassing, and the difficulty with photographic evidence is that any photograph of activity at a camp cannot by itself prove anything, even if it has not been tampered with. One photograph shows Nazis burning bodies at Auschwitz. So what, say deniers. Those are bodies of prisoners who died of natural causes, not of prisoners who were gassed. Several aerial photographs show the details of the Kremas at Birkenau and record prisoners being marched into them. So what, say deniers. The prisoners are going to work to clean up after bodies of people who died of natural causes were burned; or they are going for delousing. Again, it is context and conver-

FIGURE 24:
Back view of Krema II taken by an SS photographer, 1942. [Photograph © Yad Vashem. All rights reserved.]

gence with other evidence that make such photographs telling—and the fact that none of the photographs records activities at variance with the accounts of life in the camps supports the Holocaust and the use of gas chambers and crematoria for mass murder.

How Many Jews Died?

The final major axis of Holocaust denial is the number of Jewish victims. Paul Rassinier concluded his *Debunking the Genocide Myth: A Study of the Nazi Concentration Camps and the Alleged Extermination of European Jewry* by claiming "a minimum of 4,419,908 Jews succeeded in leaving Europe between 1931 and 1945" (1978, p. x) and therefore far fewer than six million Jews died at the hands of the Nazis. Most Holocaust scholars, however, place the total number of Jewish victims between 5.1 and 6.3 million.

While estimates do vary, historians using different methods and different source materials independently arrive at five to six million Jewish victims of the Holocaust. The fact that the estimates vary actually adds

FIGURE 25:
Aerial photograph of prisoners being marched into Krema V, May 31, 1944. [Negative courtesy National Archives, Washington, D.C. (Film 3055); enhancement courtesy Nevin Bryant.]

credibility; that is, it would be more likely that the numbers were "cooked" if the estimates all came out the same. The fact that the estimates do not come out the same yet all are within a reasonable range of error variance means somewhere between five and six million Jews died in the Holocaust. Whether it is five or six million is irrelevant. It is a large number of people. And it was not just several hundred thousand or "only" one or two million, as some deniers suggest. More accurate estimates will be made in the future as new information arrives from Russia and former Soviet territories. The overall figure, however, is not likely to change by more than a few tens of thousands, and certainly not by hundreds of thousands or millions.

The table below presents estimated Jewish losses in the Holocaust by country. The figures were compiled by a number of scholars, each working in his or her own geographic area of specialty, and then combined by Yisrael Gutman and Robert Rozett for the *Encyclopedia of the Holocaust*. The figures were derived from population demographics, taking the number of Jews registered living in every village, town, and city in Europe, the number

<table>
<tr><td colspan="4" align="center">ESTIMATED LOSS OF JEWS IN THE HOLOCAUST</td></tr>
<tr><th>Country</th><th>Initial
Jewish Population</th><th>Minimum
Loss</th><th>Maximum
Loss</th></tr>
<tr><td>Austria</td><td>185,000</td><td>50,000</td><td>50,000</td></tr>
<tr><td>Belgium</td><td>65,700</td><td>28,900</td><td>28,900</td></tr>
<tr><td>Bohemia and Moravia</td><td>118,310</td><td>78,150</td><td>78,150</td></tr>
<tr><td>Bulgaria</td><td>50,000</td><td>0</td><td>0</td></tr>
<tr><td>Denmark</td><td>7,800</td><td>60</td><td>60</td></tr>
<tr><td>Estonia</td><td>4,500</td><td>1,500</td><td>2,000</td></tr>
<tr><td>Finland</td><td>2,000</td><td>7</td><td>7</td></tr>
<tr><td>France</td><td>350,000</td><td>77,320</td><td>77,320</td></tr>
<tr><td>Germany</td><td>566,000</td><td>134,500</td><td>141,500</td></tr>
<tr><td>Greece</td><td>77,380</td><td>60,000</td><td>67,000</td></tr>
<tr><td>Hungary</td><td>825,000</td><td>550,000</td><td>569,000</td></tr>
<tr><td>Italy</td><td>44,500</td><td>7,680</td><td>7,680</td></tr>
<tr><td>Latvia</td><td>91,500</td><td>70,000</td><td>71,500</td></tr>
<tr><td>Lithuania</td><td>168,000</td><td>140,000</td><td>143,000</td></tr>
<tr><td>Luxembourg</td><td>3,500</td><td>1,950</td><td>1,950</td></tr>
<tr><td>Netherlands</td><td>140,000</td><td>100,000</td><td>100,000</td></tr>
<tr><td>Norway</td><td>1,700</td><td>762</td><td>762</td></tr>
<tr><td>Poland</td><td>3,300,000</td><td>2,900,000</td><td>3,000,000</td></tr>
<tr><td>Romania</td><td>609,000</td><td>271,000</td><td>287,000</td></tr>
<tr><td>Slovakia</td><td>88,950</td><td>68,000</td><td>71,000</td></tr>
<tr><td>Soviet Union</td><td>3,020,000</td><td>1,000,000</td><td>1,100,000</td></tr>
<tr><td>**Total**</td><td>**9,796,840**</td><td>**5,596,029**</td><td>**5,860,129**</td></tr>
</table>

SOURCE: *Encyclopedia of the Holocaust*, editor in chief Yisrael Gutman (New York: Macmillan, 1990), p. 1799.

reported transported to camps, the number liberated from camps, the number killed in "special actions" by the Einsatzgruppen, and the number remaining alive after the war. The minimum and maximum loss figures represent the range of error variation.

Finally, one might ask the denier one simple question: If six million Jews did not die in the Holocaust, where did they all go? The denier will say they are living in Siberia and Kalamazoo, but for millions of Jews to suddenly appear out of the hinterlands of Russia or America or anywhere else is so unlikely as to be nonsensical. The Holocaust survivor who does turn up is a rare find indeed.

Conspiracies

There were many millions more killed by the Nazis, including Gypsies, homosexuals, mentally and physically handicapped persons, political prisoners, and especially Russians and Poles, but Holocaust deniers do not worry about the numbers of these dead. This fact has something to do with the widespread lack of attention to non-Jewish victims of the Holocaust, yet it also has something to do with the antisemitic core of Holocaust denial.

Coupled with deniers' obsession with "the Jews" is an obsession with conspiracies. On the one hand, they deny that the Nazis had a plan (i.e., a conspiracy) to exterminate the Jews. They reinforce this argument by pointing out how extreme conspiratorial thinking can become (à la JFK conspiracy theories). They demand powerful evidence before historians can conclude that Hitler and his followers conspired to exterminate European Jewry (Weber 1994b). Fine. But they cannot then claim, on the other hand, that the idea of the Holocaust was a Zionist conspiracy to obtain reparations from Germany in order to fund the new State of Israel, without meeting their own demands for proof.

As a part this latter argument, deniers claim that if the Holocaust really happened as Holocaust historians say it did, then it would have been widely known during the war (Weber 1994b). It would be as obvious as, say, the D-day landing was. Plus, the Nazis would have discussed their murderous plans among themselves. Well, for obvious reasons, D-day was kept a secret and the D-day landing was not widely known until after it began. Likewise for the Holocaust. It was not something that was casually discussed even between fellow Nazis. Albert Speer, in fact, wrote about this in his Spandau diary:

December 9, 1946. It would be wrong to imagine that the top men of the regime would have boasted of their crimes on the rare occasions when they met. At the trial we were compared to the heads of a Mafia. I recalled movies in which the bosses of legendary gangs sat around in evening dress chatting about murder and power, weaving intrigues, concocting coups. But this atmosphere of back room conspiracy was not at all the style of our leadership. In our personal dealings, nothing would ever be said about any sinister activities we might be up to. (1976, p. 27)

Speer's observation is corroborated by SS guard Theodor Malzmueller's description of his introduction to mass murder upon his arrival at the Kulmhof (Chelmno) extermination camp:

When we arrived we had to report to the camp commandant, SS-Hauptsturmführer Bothmann. The SS-Hauptsturmführer addressed us in his living quarters, in the presence of SS-Untersturmführer Albert Plate. He explained that we had been dedicated to the Kulmhof extermination camp as guards and added that in this camp the plague boils of humanity, the Jews, were exterminated. We were to keep quiet about everything we saw or heard, otherwise we would have to reckon with our families' imprisonment and the death penalty. (Klee, Dressen, and Riess 1991, p. 217)

The answer to the deniers' overall contention that there was a conspiracy by Jews to concoct a Holocaust in order to finance the State of Israel (Rassinier 1978) is straightforward. The basic facts about the Holocaust were established before there was a State of Israel and before the United States or any other country gave it one cent. Moreover, when reparations were established, the amount Israel received from Germany was not based on numbers killed but on Israel's cost of absorbing and resettling the Jews who fled Germany and German-controlled countries before the war and the survivors of the Holocaust who came to Israel after the war. In March 1951, Israel requested from the Four Powers reparations, to be calculated on this basis.

The government of Israel is not in a position to obtain and present a complete statement of all Jewish property taken or looted by the Germans, and said to total more than $6 thousand million. It can only compute its claim on the basis of total expenditures already made and the expenditure still needed for the integration of Jewish immigrants from Nazi-dominated countries. The number of these immigrants is estimated at some 500,000, which means a total expenditure of $1.5 thousand million. (Sagi 1980, p. 55)

Needless to say, if reparations were based on the total number of survivors, then any Zionist conspirators should have exaggerated not the number of Jews killed by the Nazis but the number of survivors. In fact, given the

provisions of the reparation settlement, if the deniers are right and only a
few hundred thousand Jews died, then Germany owes Israel far more in
reparations, for where else could those five to six million survivors have
gone? Deniers might argue that the Zionist conspirators traded reparation money from Germany for a greater prize: money and long-term
sympathy from all over the world. But here we really go off the deep end.
Why should the supposed conspirators have risked sure money for some
uncertain future payoff? In reality, the State of Israel as the recipient of
German money is a myth. Most of it went to individual survivors, not to
the Israeli government.

Moral Equivalency

When all else fails, deniers shift from wrangling about intentionality,
gassings and crematoria, and the number of Jews killed to arguing that the
Nazi's treatment of the Jews is really no different from what other nations
do to their perceived enemies. Deniers point out, for example, that the
U.S. government obliterated with atomic weapons two entire Japanese
cities filled with civilians (Irving 1994) and forced Japanese-Americans into
camps, which is just what the Germans did to their perceived internal
enemy—the Jews (Cole 1994).

The response to this is twofold. First, just because another country
does evil does not make your own evil right. Second, there is a difference
between war and the systematic state-organized killing of unarmed people
within your own country, not in self-defense, not to gain more territory,
raw materials, or wealth, but simply because they are perceived as a type of
Satanic force and inferior race. At his trial in Jerusalem, Adolf Eichmann,
SS Obersturmbannführer of the RSHA and one of the chief implementers
of the Final Solution, tried to make the moral equivalence argument. But
the judge didn't buy it, as this sequence from the trial transcript shows
(Russell 1963, pp. 278–279):

> **Judge Benjamin Halevi to Eichmann:** You have often compared the
> extermination of the Jews with the bombing raids on German cities and
> you compared the murder of Jewish women and children with the death of
> German women in aerial bombardments. Surely it must be clear to you that
> there is a basic distinction between these two things. On the one hand the
> bombing is used as an instrument of forcing the enemy to surrender. Just as
> the Germans tried to force the British to surrender by their bombing. In
> that case it is a war objective to bring an armed enemy to his knees.

On the other hand, when you take unarmed Jewish men, women, and children from their homes, hand them over to the Gestapo, and then send them to Auschwitz for extermination it is an entirely different thing, is it not?

Eichmann: The difference is enormous. But at that time these crimes had been legalized by the state and the responsibility, therefore, belongs to those who issued the orders.

Halevi: But you must know surely that there are internationally recognized Laws and Customs of War whereby the civilian population is protected from actions which are not essential for the prosecution of the war itself.

Eichmann: Yes, I'm aware of that.

Halevi: Did you never feel a conflict of loyalties between your duty and your conscience?

Eichmann: I suppose one could call it an internal split. It was a personal dilemma when one swayed from one extreme to the other.

Halevi: One had to overlook and forget one's conscience.

Eichmann: Yes, one could put it that way.

During his trial, Eichmann never denied the Holocaust. His argument was that "these crimes had been legalized by the state" and therefore the people that "issued the orders" are responsible. This was the classic defense used at the Nuremberg trials by most of the Nazis. Since the higher-ups all committed suicide—Hitler, Himmler, Goebbels, and Hermann Göring—they were off the hook, or so they thought.

We are not off the hook either. Like evolution denial, Holocaust denial is not simply going to go away and it is not benign or trivial. It has had and will have ugly and dire consequences, not only for Jews but for all of us and for future generations. We must provide answers to the claims of Holocaust deniers. We have the evidence and we must stand up and be heard.

15

Pigeonholes and Continuums

An African-Greek-German-American Looks at Race

Science books rarely make the best-seller lists, but when they do they usually have something to do either with our cosmological origins and destiny—Stephen Hawking's *A Brief History of Time*—or with the metaphysical side of our existence—Fritjof Capra's *The Tao of Physics*. How, then, did Free Press sell over 500,000 copies of a $30 book (yes, that's $15 million) filled with graphs, charts, curves, and three hundred pages of appendices, notes, and references, all on the obscure topic of psychometrics? Because one of those curves illustrates a fifteen-point difference in IQ scores between white and black Americans. In America, nothing sells like racial controversy. *The Bell Curve* (1994), by Richard Herrnstein and Charles Murray, generated a furor among scientists, intellectuals, and activists throughout the country that continues to this day—the Bell Curve Wars, as one debunking book is titled.

The arguments in *The Bell Curve* are not novel, in our time or any other. In fact, earlier that same year, the prestigious journal *Intelligence* published an article by another controversial scientist, Philippe Rushton, in which he claimed that not only do blacks and whites differ in intelligence but also in maturation rate (age of first intercourse, age of first pregnancy), personality (aggressiveness, cautiousness, impulsivity, sociability), social organization (marital stability, law abidingness, mental health), and reproductive effort (permissiveness, frequency of sexual intercourse, size of male genitalia). In addition to lower IQs, Rushton believes that blacks have earlier maturation rates, higher impulsivity and aggressiveness, less mental health and law abidingness, more permissive attitudes and greater frequency of intercourse, and larger male genitalia (inversely proportional to IQ, the data for which he collected through condom distributors).

In both *The Bell Curve* and Rushton's article, the Pioneer Fund is acknowledged. This caught my attention because of its connections to Holocaust denial. The Pioneer Fund was established in 1937 by textile millionaire Wycliffe Preston Draper to fund research that promotes "race betterment" and that proves blacks are inferior to whites, the repatriation to Africa of blacks, and educational programs for children "descended predominantly from white persons who settled in the original thirteen states . . . and/or from related stocks" (in Tucker 1994, p. 173; the Pioneer Fund denies that these are its current goals). William Shockley, a Nobel laureate in physics, for example, received $179,000 over ten years for his research on the heritability of IQ. Shockley believed that white Europeans are "the most competent population in terms of social management and general capacity for organization" and that "the most brutal selective mechanisms" of colonial life made the white race superior (in Tucker 1994, p. 184). Rushton's work was financed by the Pioneer Fund to the tune of several hundred thousand dollars.

The Pioneer Fund also supports the journal *Mankind Quarterly*. One of the early editors of the journal, Roger Pearson, when he immigrated to the United States in the 1960s worked with Willis Carto, organizer of the Liberty Lobby and founder of the *Journal of Historical Review*, the leading publication of Holocaust denial. Over the past twenty-three years, Pearson and his organization have received no less than $787,400 from the Pioneer Fund. According to William Tucker, Pearson and Carto "regularly blamed the 'New York money changers' for causing the 'Second Fratricidal War' and the subsequent 'Allied War Crimes' against the Reich out of a desire to impose financial slavery on Germany and the world" (1994, p. 256). Carto's Noontide Press, publisher of racist and eugenics tracts as well as books denying the Holocaust, also featured Pearson's *Race and Civilization*, which describes "how the aristocratic Nordic, the 'symbol . . . of human dignity,' had been forced by 'taxes against landholders . . . to intermarry with Jewish and other non-Nordic elements,' thus securing the wealth necessary to retain their family estates but sacrificing their 'biological heritage' and 'thereby renouncing their real claim to nobility'" (in Tucker 1994, p 256). *Race and Civilization*, Pearson acknowledges, was based on the work of Hans Gunther, who was a leading German racial theoretician before, during, and after the Third Reich, although Pearson claims he was de-Nazified after the war. Pearson has also been on the advisory committee for Nouvelle Ecole, called by some "a French highbrow neo-Nazi group" but by Pearson merely "right wing" (1995).

I telephoned Roger Pearson. When I interviewed him, Pearson confirmed that he did work with Willis Carto for three months when he first came to America, editing Carto's journal *Western Destiny*, but he explicitly denied having used phrases such as "New York money changers." He also repudiated other charges, including the one that he "once reportedly boasted of helping to hide Josef Mengele" (see Tucker 1994, p. 256). This rumor seems to have spread far and wide, and Pearson is especially perturbed by it since at the time of Mengele's escape in March 1945, Pearson was seventeen-and-a-half and undergoing basic infantry training in the British Army. He has never had any contact whatsoever with Mengele and believes that the charge is like an urban legend, recycling itself through books and articles without anyone being able to cite a primary source for it.

I found Pearson a kind, soft-spoken man who has given considerable thought to the major issues of our time. He presently holds an honorary position as president of the Institute for the Study of Man (he is sixty-eight and semi-retired), and he is the publisher of *Mankind Quarterly*, which the institute took over in 1979. At that time, Pearson broadened the journal to include sociology, psychology, and mythology, adding appropriate new board members such as psychometrician Raymond Cattell and mythologist Joseph Campbell. During his reign, Pearson claims, neither the institute nor the journal has endorsed the repatriation of blacks or white supremacy.

Then where did the idea come from that they do endorse such racialist beliefs? Pearson admits that before his time the journal did endorse such ideas, and that he himself believes that societies ideally should be as homogeneous as possible (i.e., WASP), with the elite running the show. The problem, as he explained, is that this "natural" process is being interfered with by modern war and politics, a belief he developed from personal experiences:

> I served in the British Army in World War II. On May 29, 1942, my only sibling, a 21-year-old Battle of Britain fighter pilot, was killed in combat in North Africa against Rommel. This had a great impact on me and until I was about 32—when I got married and started my own family—I had dreams of my brother returning. In that war I also lost four cousins and three close school friends, all young and without children. And lots of people I knew were killed before they had children. What I was seeing was that the more talented individuals were being selected against in modern warfare and it left me with an acute feeling that there is something deeply wrong with the world where you have wholesale over-breeding by people who are not as competent as others, while the more competent are killed off. Today I am very much against war because it disproportionally selects and destroys the more intelli-

gent people. Plus it destroys culture. Look what we did to the great cities of Europe in World War II. A good example of this can be seen in the book *War and the Breed*, written in 1915 by the chancellor of Stanford University, David Starr Jordan. It is a story of young, childless Englishmen who were killed in World War I, and how warfare was destroying the West. I republished this book to show that the Europeans were a warlike bunch of people who didn't know what was good for them. Through centuries they destroyed themselves by fighting each other and consequently, from an evolutionary perspective, they did not deserve to survive.

I was a great nationalist who believed, in those days, in the purity of the gene pool. Nations used to be seen as breeding pools. Not any longer. The nation as a kinship unit is a thing of the past. We are moving into multi-cultural, multi-racial units. I question how desirable this is from an evolutionary point of view. I think it is a reversal of the evolutionary process. (1995)

To help me better understand his views, Pearson sent me copies of some of his books and a selection of back issues of *Mankind Quarterly*. He was convinced I would see that the racialist tone of decades past has subsided in recent years. There are many interesting articles in this journal that have nothing to do with race, but there are also plenty that do, and these exhibit the same old slant now tricked out in more technical and less provocative jargon. Here are a few of the many instances I could cite. The Fall/Winter 1991 issue contains an article by Richard Lynn, titled "The Evolution of Racial Differences in Intelligence," in which he concludes that Caucasoids and Mongoloids living in temperate and cold climates "encountered the cognitively demanding problems of survival" and thus "a selection pressure favoring enhanced intelligence explains why the Caucasoids and the Mongoloids are the races which have evolved the highest intelligence" (p. 99). I guess Egyptians, Greeks, Phoenicians, Jews, Romans, Aztecs, Mayans, and Incans—a rather mixed group of races all living in "unchallenging" warm environments—were not particularly smart; and the Neanderthals who inhabited cold northern Europe long ago must have been very intelligent, even though modern humans allegedly outsmarted them. To be fair, the journal did publish critiques of this argument in the same issue.

The Summer 1995 issue features Glayde Whitney's Presidential Address to the Behavior Genetics Association, delivered on June 2, 1995, complete with graphs and charts demonstrating a dramatic ninefold black-white difference in murder rates, about which Whitney concludes, "Like it or not, it is a reasonable scientific hypothesis that some, perhaps much, of the race difference in murder rate is caused by genetic differences in contributory variables such as low intelligence, lack of empathy, aggressive acting out, and impulsive lack of foresight" (p. 336). What is his evidence for this hypothesis? Nothing whatsoever. Not even a single citation. And this in an address given to a room full of behavior geneticists and printed in a scientific journal read by anthropologists, psychologists, and geneticists. In this same issue, Pearson concludes a twenty-eight-page history titled "The Concept of Heredity in Western Thought" by bewailing the dysgenics of the modern world in which the elite are being selected against and outbred by the hoi polloi: "Heavily dysgenic trends have dominated this century as a result of the selective elimination of air crews and other talented personnel involved in modern warfare in Europe; the genocidal slaughter of the elite in Europe, the Soviet Union and Maoist China; and the general tendency for the more creative members of modernized societies around the world to have fewer children than the less creative" (p. 368).

I am not quoting selectively here. Pearson's latest book, *Heredity and Humanity: Race, Eugenics and Modern Science*, elaborates the same theme, ending with this dramatic prediction about what will happen if we do not do something about this so-called problem: "Any species that adopts patterns of behavior that run counter to the forces that govern the universe is doomed to decline until it either undergoes a painful, harshly enforced and totally involuntary eugenic process of evolutionary reselection and readaptation, or is subjected to an even more severe penalty—extinction" (1996, p. 143). Just what does "total involuntary eugenic reselection" mean? State-enforced segregation, repatriation, sterilization, or perhaps even extermination? I asked him. "No! I simply mean that nature selects and eliminates and that if we continue on our present course the species will go extinct. Evolution itself is an exercise in eugenics. Natural selection in the long run tends to be eugenic" (1995). But following on the heels of lengthy discussions about racial differences in intelligence, criminality, creativity, aggression, and impulsiveness, the implication seems to be that it is nonwhites who are the potential cause of the extinction of the species, and therefore something needs to be done about *them*.

The End of Race

Is it possible to prevent interbreeding and preserve genetic integrity? Has any nation ever been or could any nation ever be a "breeding unit," in Pearson's terminology? Perhaps a worldwide Nazi state might be able to legislate such biological walls, but nature certainly has not, as Luca Cavalli-Sforza and his colleagues, Paolo Menozzi and Alberto Piazza, demonstrate in *The History and Geography of Human Genes*, lauded by *Time* magazine as the study that "flattens *The Bell Curve*" (appropriate, since it weighs in at eight pounds and runs 1,032 pages). In this book, the authors present evidence from fifty years of research in population genetics, geography, ecology, archeology, physical anthropology, and linguistics that, "from a scientific point of view, the concept of race has failed to obtain any consensus; none is likely, given the gradual variation in existence" (1994, p. 19). In other words, the concept of race is biologically meaningless.

But don't we all know a black person or a white person when we see one? Sure, agree the authors: "It may be objected that the racial stereotypes have a consistency that allows even the layman to classify individuals." But, they continue, "the major stereotypes, all based on skin color, hair color and form, and facial traits, reflect superficial differences that are not confirmed by deeper analysis with more reliable genetic traits and whose origin dates from recent evolution mostly under the effect of climate and perhaps sexual selection" (p. 19). Traditional popular racial categories are literally skin deep.

But aren't races *supposed* to blend into one another as fuzzy sets, while retaining their uniqueness and separateness (see Sarich 1995)? Yes, but how these groups are classified depends on whether the classifier is a "lumper" or "splitter"—seeing similarities or differences. Darwin noted that naturalists in his time cited anywhere from two to sixty-three different races of *Homo sapiens*. Today there are anywhere from three to sixty races, depending on the taxonomist. Cavalli-Sforza and his colleagues conclude, "Although there is no doubt that there is only one human species, there are clearly no objective reasons for stopping at any particular level of taxonomic splitting" (1994, p. 19). One might think that Australian Aborigines, for example, would be more closely related to African blacks than southeast Asians, since they certainly *look* more alike (and facial features, hair type, and skin color are what everyone focuses on in identifying race). Genetically, however, Australians are most *distant* from Africans and *closest* to Asians. This makes sense from an evolutionary perspective, even if it goes against our perceptual intuitions, since humans first migrated out of Africa, then moved through the Middle and Far East, down Southeast Asia, and into Australia, taking tens of thousands of years to do so. Regardless of what they look like, Australians and Asians should be more closely related evolutionarily, and they are. And who would intuit, for example, that Europeans are an intermediate hybrid population of 65 percent Asian genes and 35 percent African genes? But this is not surprising from an evolutionary perspective.

Part of the problem of race classification is that within-group variability is greater than between-group variability, as Cavalli-Sforza and his colleagues argue: "Statistically, genetic variation within clusters is large compared with that between clusters." In other words, individuals within a group vary more than individuals between groups. Why? The answer is an evolutionary one:

> There is great genetic variation in all populations, even in small ones. This individual variation has accumulated over very long periods, because most polymorphisms observed in humans antedate the separation into continents, and perhaps even the origin of the species, less than half a million years ago. The same polymorphisms are found in most populations, but at different frequencies in each, because the geographic differentiation of humans is recent, having taken perhaps one-third or less of the time the species has been in existence. There has therefore been too little time for the accumulation of a substantial divergence. (1944, p. 19)

And, the authors repeat (it cannot be overstated), "The difference between groups is therefore small when compared with that within the major groups, or even within a single population" (1994, p. 19). Recent research shows, in fact, that if a nuclear war exterminated all humans but a small band of Australian Aborigines, a full 85 percent of the variability of *Homo sapiens* would be preserved (Cavalli-Sforza and Cavalli-Sforza 1995).

The End of Racism

It is always the individual that matters, not the group; and it is always how individuals differ that matters, not how groups differ. This is not liberal hope or conservative hype. It is a fact of evolution, as one entomologist noted in 1948: "Modern taxonomy is the product of an increasing awareness among biologists of the uniqueness of individuals, and of the wide range of variation which may occur in any population of individuals." This entomologist believed that taxonomists' generalizations of species, genera, and even higher categories "are too often descriptions of unique individuals and structures of particular individuals that are not quite like anything that any other investigator will ever find." Psychologists are equally guilty of such hasty generalizations, he adds: "A mouse in a maze, today, is taken as a sample of all individuals, of all species of mice under all sorts of conditions, yesterday, today, and tomorrow." Worse still, these collective conclusions are extrapolated to humans: "A half dozen dogs, pedigrees unknown and breeds unnamed, are reported upon as 'dogs'—meaning all kinds of dogs—if, indeed, the conclusions are not explicitly or at least implicitly applied to you, to your cousins, and to all other kinds and descriptions of humans" (p. 17).

If he had only talked about bugs, this entomologist would be relatively unknown. But midway through his career, he switched from studying an obscure species of wasp to a very well-known species of WASP—the human variety. In fact, he concluded, if wasps showed so much variation, how much more might humans? Accordingly, in the 1940s, he began the most thorough study ever conducted on human sexuality, and in 1948 Alfred Kinsey, entomologist turned sexologist, published *Sexual Behavior in the Human Male*. In this book, Kinsey observed that "the histories which have been available in the present study make it apparent that the heterosexuality or homosexuality of many individuals is not an all-or-none proposition" (Kinsey, Pomeroy, and Martin 1948, p. 638). One can be both simultaneously. Or neither temporarily. One can start as heterosexual and become homosexual, or vice versa. And the percentage of time spent in either state varies considerably amongst individuals in the population. "For instance," Kinsey wrote, "there are some who engage in both heterosexual and homosexual activities in the same year, or in the same month or week, or even in the same day" (p. 639). One might add, "at the same time." Therefore, Kinsey concluded, "One is not warranted in recognizing merely two types of individuals, heterosexual and homosexual, and that the characterization of the homosexual as a third sex fails to describe any actuality" (p. 647). Extrapolating this to taxonomy in general, Kinsey deduced the uniqueness of individuals (in a powerful statement tucked away in the midst countless tables):

> Males do not represent two discrete populations, heterosexual and homosexual. The world is not to be divided into sheep and goats. Not all things are black nor all things white. It is a fundamental of taxonomy that nature rarely deals with discrete categories. Only the human mind invents categories and tries to force facts into separate pigeonholes. The living world is a continuum in each and every one of its aspects. The sooner we learn this concerning human sexual behavior the sooner we shall reach a sound understanding of the realities of sex. (p. 639)

Kinsey saw the implications of this variation for moral and ethical systems. If variation and uniqueness are the norm, then what form of morality can possibly envelope *all* human actions? For human sexuality alone, Kinsey measured 250 different items for each of over ten thousand people. That is 2.5 million data points. Regarding the variety of human behavior, Kinsey concluded, "Endless recombinations of these characters in different individuals swell the possibilities to something which is, for all essential purposes, infinity" (in Christenson 1971, p. 5). Since all moral systems are absolute, yet the variation of these systems is staggeringly broad, then all absolute moral systems are actually relative to the group conferring (usually imposing) it upon others. At the end of the volume on males, Kinsey concluded that there is virtually no evidence for "the existence of such a thing as innate perversity, even among those individuals whose sexual activities society has been least inclined to accept." On the contrary, as he demonstrated with his vast statistical tables and in-depth analyses, the evidence leads to the conclusion "that most human sexual activities would become comprehensible to most individuals, if they could know the background of each other individual's behavior" (Kinsey, Pomeroy, and Martin 1948, p. 678).

Variation is what Kinsey called "the most nearly universal of all biologic principles," but it is one that most seem to forget when they "expect their fellows to think and behave according to patterns which may fit the lawmaker, or the imaginary ideals for which the legislation was fashioned, but which are ill-shaped for all real individuals who try to live under them." Kinsey demonstrated that while "social forms, legal restrictions, and moral codes may be, as the social scientist would contend, the codification of human experience," they are, like all statistical and population generalizations, "of little significance when applied to particular individuals" (in Christenson 1971, p. 6). These laws tell us more about the lawmakers than they do about the laws of human nature:

> Prescriptions are merely public confessions of prescriptionists. What is right for one individual may be wrong for the next; and what is sin and abomination to one may be a worthwhile part of the next individual's life. The range of individual variation, in any particular case, is usually much greater than is generally understood. Some of the structural characters in my insects vary as much as twelve hundred percent. In some of the morphologic and physiologic characteristics which are basic to the human behavior which I am studying, the variation is a good twelve thousand percent. And yet social forms and moral codes are prescribed as though all individuals were identical; and we pass judgments, make awards, and heap penalties without regard to the diverse difficulties involved when such different people face uniform demands. (in Christenson 1971, p. 7)

Kinsey's conclusions may be applied to race. How can we pigeonhole "blacks" as "permissive" or "whites" as "intelligent" when such categories as black and white, permissive and intelligent, are actually best described as a continuum, not a pigeonhole? "Dichotomous variation is the exception and continuous variation is the rule, among men as well as among insects," Kinsey concluded. Likewise, for behavior we identify right and wrong "without allowance for the endlessly varied types of behavior that are possible between the extreme right and the extreme wrong." That being the case, the hope for cultural evolution, like that of biological evolution, depends on the recognition of variation and individualism: "These individual differences are the materials out of which nature achieves progress, evolution in the organic world. In the differences between men lie the hopes of a changing society" (in Christenson 1971, pp. 8–9).

In America, we tend to confound race and culture. For instance, "white or Caucasian" is not parallel to "Korean-American" but to "Swedish-American." The former roughly indicates a supposed racial or genetic make-up, while the latter roughly acknowledges cultural heritage. In 1995, the Occidental College school newspaper announced that almost half (48.6 percent) of the Frosh class were "people of color." For the life of me, however, I have a difficult time identifying most students by the traditional external signs of race because there has been so much blending over the years and centuries. I suspect most of them would be hyphenated races, a concept even more absurd than "pure" races. Checking a box on a form for race—"Caucasian," "Hispanic," "African-American," "Native American," or "Asian-American"—is untenable and ridiculous. For one thing, "American" is not a race, so labels such as "Asian-American" and "African-American" are still exhibits of our confusion of culture and race. For another thing, how far back does one go in history? Native Americans are really Asians, if you go back more than twenty or thirty thousand years to before they crossed the Bering land bridge between Asia and America. And Asians, several hundred thousand years ago probably came out of Africa, so we should really replace "Native American" with "African-Asian-Native American." Finally, if the Out of Africa (single racial origin) theory holds true, then *all* modern humans are from Africa. (Cavalli-Sforza now thinks this may have been as recently as seventy thousand years ago.) Even if that theory gives way to the Candelabra (multiple racial origins) theory, ultimately all hominids came from Africa, and therefore everyone in America should simply check the box next to "African-American." My maternal grandmother was German and my maternal grandfather was Greek. The next time I fill out one of those forms I am going to check "Other" and write in the truth about my racial and cultural heritage: "African-Greek-German-American."

And proud of it.

PART 5

HOPE

SPRINGS

ETERNAL

Hope springs eternal in the human breast;
Man never Is, but always To be blest.
The soul, uneasy, and confin'd from home,
Rests and expatiates in a life to come.
Lo, the poor Indian! whose untutor'd mind
Sees God in clouds, or hears him in the wind;
His soul proud Science never taught to stray
Far as the solar walk or milky way;
Yet simple Nature to his hope has giv'n,
Behind the cloud-topp'd hill, an humbler heav'n.

—Alexander Pope, *An Essay on Man*, 1733

16

Dr. Tipler Meets Dr. Pangloss

Can Science Find the Best of All Possible Worlds?

lfred Russel Wallace, the nineteenth-century British naturalist whose name is permanently tethered to Charles Darwin's for his co-discovery of natural selection, got himself into trouble in his quest to find a purpose for every structure and every behavior he observed. For Wallace, natural selection shaped every organism to be well adapted to its environment. His overemphasis on natural selection led to his hyper-adaptationism. He argued in the April 1869 issue of the *Quarterly Review*, much to Darwin's dismay, that the human brain could not entirely have been the product of evolution because in nature there is no reason to have a human-size brain, capable of such unnatural abilities as higher math and aesthetic appreciation. No purpose, no evolution. His answer? "An Overruling Intelligence has watched over the action of those laws, so directing variations and so determining their accumulation, as finally to produce an organization sufficiently perfect to admit of, and even to aid in, the indefinite advancement of our mental and moral nature" (p. 394). The theory of evolution proves the existence of God.

Wallace fell into hyper-adaptationism because he believed evolution should have created the best possible organisms in this best of all possible worlds. Since it had not, there had to be another active agent—a higher intelligence. Ironically, the natural theologians whose beliefs Wallace's evolutionary theories helped to overturn made a similar argument, the

255

most famous of which is William Paley's 1802 *Natural Theology,* which opens with this passage:

> In crossing a heath, suppose I pitched my foot against a stone, and were asked how the stone came to be there; I might possibly answer, that, for any thing I knew to the contrary, it had lain there for ever. . . . But suppose I had found a watch upon the ground, and it should be inquired how the watch happened to be in that place; I should hardly think of the answer which I had before given—that, for any thing I knew, the watch might have always been there. Yet why should not this answer serve for the watch as well as for the stone? For this reason, and for no other, viz. that, when we come to inspect the watch, we perceive that its several parts are framed and put together for a purpose.

For Paley, a watch is purposeful and thus must have been created by a being with a purpose. A watch needs a watchmaker, just as a world needs a world-maker—God. Yet both Wallace and Paley might have heeded the lesson from Voltaire's *Candide* (1759), in which Dr. Pangloss, a professor of "metaphysico-theology-cosmolonigology," through reason, logic, and analogy "proved" that this is the best of all possible worlds: "'Tis demonstrated that things cannot be otherwise; for, since everything is made for an end, everything is necessarily for the best end. Observe that noses were made to wear spectacles; and so we have spectacles. Legs were visibly instituted to be breeched, and we have breeches" (1985, p. 238). The absurdity of this argument was intended on the part of the author, for Voltaire firmly rejected the Panglossian paradigm that all is best in the best of all possible worlds. Nature is not perfectly designed, nor is this the best of all possible worlds. It is simply the world we have, quirky, contingent, and flawed as it may be.

For most people, hope springs eternal that if this is not the best of all possible worlds, it soon will be. That hope is the wellspring of religions, myths, superstitions, and New Age beliefs. We are not surprised to find such hopes at large in the world, of course, but we expect science to rise above wish fulfillment. But should we? After all, science is done by human scientists, complete with their own hopes, beliefs, and wishes. As much as I admire Alfred Russel Wallace, with hindsight it is easy to see where his hopes for a better world biased his science. But surely science has progressed since then? Nope. A plethora of books, mostly by physicists and cosmologists, testifies to the fact that hope continues to spring eternal in science as well as religion. Fritjof Capra's *The Tao of Physics* (1975) and especially *The Turning Point* (1982) unabashedly root for the blending of science and spirituality and hope for a better world. *The Faith of a Physicist* (1994) by the Cambridge University theoretical physicist turned Anglican priest, John Polkinghorne, argues that physics proves the Nicene Creed, which is based on a fourth-century formula of Christian faith. In 1995, physicist Paul Davies won the $1 million Templeton Prize for the advancement of religion, in part for his 1991 book, *The Mind of God*. The nod for the most serious attempts, however, has to go to John Barrow and Frank Tipler's 1986 *Anthropic Cosmological Principle* and Frank Tipler's 1994 *The Physics of Immortality: Modern Cosmology, God and the Resurrection of the Dead*. In the first book, the authors claim to prove that the universe was intelligently designed and thus there is an intelligent designer (God); in the second, Tipler hopes to convince readers that they and everyone else will be resurrected in the future by a supercomputer. These attempts provide a case study in how hope shapes belief, even in the most sophisticated science.

As I read *The Physics of Immortality* and talked with its author, I was struck by the parallels between Tipler, Wallace, and Paley. Tipler, I came to realize, is Dr. Pangloss in disguise. He is a modern hyper-adaptationist, a twentieth-century natural theologian. (Upon hearing this analogy, Tipler admitted to being a "progressive" Panglossian.) Tipler's highly tutored mind has brought him full circle to Alexander Pope's Indian in his *Essay on Man* (see the epigraph on the opening page of Part 5), although Tipler finds God not only in the clouds and wind but also on his own solar walk through the cosmos in pursuit of not a humbler heaven but a vainglorious one.

What in Tipler's background might explain his Panglossian tendencies— his need to make this the best of all possible worlds? From his youth, Tipler was sold on the DuPont motto, "Better living through chemistry," and all that it stood for—unalloyed progress through science. Fascinated by the Redstone rocket program and the possibility of sending a man to the moon, for instance, at age eight Tipler wrote a letter to the great German rocket scientist, Wernher von Braun. "The attitude of unlimited technological progress is what drove Wernher von Braun and it is what has motivated me all my life" (1995).

Raised in the small rural town of Andalusia, Alabama, where he graduated from high school in 1965 as class valedictorian, Tipler intended to speak out in his graduation speech against segregation—not a popular position to take in the Deep South of the mid-1960s, especially for a youth of seventeen. Tipler's father, an attorney who routinely represented individuals against large corporations and who also opposed segregation, insisted that Frank not go public with such a controversial position since the family had to continue living in the town after Frank went away to college. Despite (or perhaps because of) the fact that he was raised a Southern Baptist with a strong fundamentalist influence, Tipler says he was an agnostic by the age of sixteen. Brought up in an upper-middle-class environment by a politically liberal father and apolitical mother, Tipler is a firstborn with one brother four years his junior.

What difference does birth order make? Frank Sulloway (1996) has conducted a multivariate correlational study, examining the tendency toward rejection of or receptivity to heretical theories based on such variables as "date of conversion to the new theory, age, sex, nationality, socioeconomic class, sibship size, degree of previous contact with the leaders of the new theory, religious and political attitudes, fields of scientific specialization, previous awards and honors, three independent measures of eminence, religious denomination, conflict with parents, travel, education attainment, physical handicaps, and parents' ages at birth." Using multiple regression models, Sulloway discovered, in analyzing over one million data points, that birth order was the strongest factor in intellectual receptivity to innovation in science.

Consulting over a hundred historians of science, Sulloway had them evaluate the stances taken by 3,892 participants in twenty-eight disparate scientific controversies dating from 1543 to 1967. Sulloway, himself a laterborn, found that the likelihood of accepting a revolutionary idea is 3.1 times greater for laterborns than firstborns; for radical revolutions, the likelihood is 4.7 times higher. Sulloway noted that "the likelihood of this happening by chance is virtually nil." Historically, this indicates that "laterborns have indeed generally introduced and supported other major conceptual transformations over the protests of their firstborn colleagues. Even when the principal leaders of the new theory occasionally turn out to be firstborns—as was the case with Newton, Einstein, and Lavoisier—the opponents as a whole are still predominantly firstborns, and the converts continue to be mostly laterborns" (p. 6). As a "control group" of sorts, Sulloway examined data from only children and found only children wedged between firstborns and laterborns in their support for radical theories.

Why are firstborns more conservative and influenced by authority? Why are laterborns more liberal and receptive to ideological change? What is the connection between birth order and personality? Firstborns, being first, receive considerably more attention from their parents than laterborns, who tend to receive greater freedom and less indoctrination into the ideologies of and obedience to authorities. Firstborns generally have greater responsibilities, including the care of younger siblings, and thus become surrogate parents. Laterborns are frequently a step removed from parental authority, and thus less inclined to obey and adopt the beliefs of the higher authority. Sulloway has taken this a step further by applying a Darwinian sibling-competition model in which children must compete for limited parental resources and recognition. Firstborns are larger, faster, and older, and so receive the lion's share of the goodies. Laterborns, in order to maximize parental benefits, diversify into new areas. This explains why firstborns tend to go into more traditional careers, whereas laterborns seek out less traditional ones.

Developmental psychologists J. S. Turner and D. B. Helms noted that "usually, firstborns become their parents' center of attraction and monopolize their time. The parents of firstborns are usually not only young and eager to romp with their children but also spend considerable time talking to them and sharing their activities. This tends to strengthen bonds of attachment between the two" (1987, p. 175). Quite obviously, this attention would include more rewards and punishment, thus reinforcing obedience to authority and controlled acceptance of the "right way" to think. R. Adams and B. Phillips (1972) and J. S. Kidwell (1981) report that this distribution of attention causes firstborns to strive harder for approval than laterborns, and H. Markus (1981) concluded that firstborns tend to be more anxious, dependent, and conforming than laterborns. I. Hilton (1967), in a mother-child interactive experiment with twenty firstborn, twenty laterborn, and twenty only children, found that at four years of age firstborns were significantly more dependent on and asked more frequently for help or reassurance from their mothers than the laterborn or only children. In addition, mothers were most likely to interfere with a firstborn child's task (constructing a puzzle). Finally, R. Nisbett (1968) showed that laterborns are far more likely to participate in relatively dangerous sports than firstborns, which is linked to risk taking and thus to "heretical" thinking.

Sulloway is not suggesting that birth order alone determines receptivity to radical ideas. Far from it, in fact, as he notes that "birth order is hypothesized to be the occasion for psychologically formative influences operating within the family" (p. 12). In other words, birth order is a predisposing variable that sets the stage for numerous other variables, such as age, sex, and social class, to influence receptivity. Not all scientific theories are equally radical, of course, and in taking this into consideration, Sulloway discovered a correlation between laterborns and the degree of "liberal or radical leanings" of the controversy. He noted that laterborns tended "to prefer statistical or probabilistic views of the world (Darwinian natural selection and quantum mechanics, for example) to a worldview premised on predictability and order." By contrast, he found that when firstborns did accept new theories, they were typically theories of the most conservative type, "theories that typically reaffirm the social, religious, and political status quo and that also emphasize hierarchy, order, and the possibility of complete scientific certainty" (p. 10).

Frank Tipler's theory, far from being the radical idea he thinks it is, is actually ultra-conservative, reaffirming a hierarchical, ordered worldview and the ultimate religious status quo of God and immortality. Tipler may have rejected God at sixteen, but as he approaches fifty, he is arguing with all his scientific acumen for the existence of Paley's Divine Watchmaker and Wallace's Over-ruling Intelligence. "It's a return to the great chain of being," Tipler asserted. "The difference is that it is a temporal chain." Even his physics is conservative:

> My theory is very conservative from the physics point of view. What I say is take the standard equations—the old traditional equations of quantum mechanics and general relativity—and all we have to do is change the boundary conditions from the past to the future to understand the universe. It is counter-intuitive because we human beings always move from past to present to future, so we tacitly assume that the universe has to work the same way. What I'm saying is that there is no reason the universe should work in our way. Once you take the point of view of the future, the universe becomes much more comprehensible to physicists, just as the solar system did when looked at from the sun's point of view. (1995)

The firstborn son is using his advanced science to conserve his parents' religion. "My father always vaguely believed in God, and since he has always been a rationalist himself and he likes a rational foundation for religious belief, he naturally liked the book. And my mother was happy because it defends, in many ways, the traditional view of Christianity" (1995). Indeed, Tipler's fundamentalist background shines through in his continued literal use of "God," "heaven," "hell," and "resurrection," despite the fact that many of his fellow physicists advised him to avoid using such terms (1994, p. xiv). But what are the chances that modern physics *really* describes Judeo-Christian doctrines? Pretty good, says Tipler: "If you look back and think about all the possible explanations there are for things like a soul, for instance, there aren't very many. A soul is either a pattern in matter or a mysterious soul substance. That's about it. Plato took the position that the soul consists of this soul substance, whereas Thomas Aquinas took the attitude that resurrection was going to be reproducing the pattern, which is what I argue in my book. With only two possibilities someone is bound to get it right" (1995). There is, of course, a third possibility, that there is *no* soul, if by soul one means something that survives the physical body. If this is the case, then no one "got it right" because there is nothing to get right. (Tipler says if "soul" is defined like this, then he agrees that there is no soul. But he claims the ancients defined "soul" operationally as that which makes a living being different from a corpse, and then argues only two choices exist. But this is not what most contemporary theologians mean by soul.)

Whereas most scientists do not dare publish such controversial notions until late in their careers, by the time he began studying physics at MIT Tipler was already entertaining ideas in the borderlands between science and science fiction:

> I became aware of time travel in the dorm when a bunch of us physics students discussed it. We would talk about the real far-out ideas in physics, such as the many-histories interpretation of physics. I read Gödel's paper on closed time–like curves. I was fascinated by that and went and got a copy of the second volume of *Albert Einstein, Philosopher/Scientist*. I read that Einstein became aware of this possibility when he was generating the general theory of relativity, and he even discussed the Gödel paper. That gave me confidence because the majority of the community of physicists may not believe in the possibility of time travel, but Kurt Gödel and Albert Einstein did, and those were not lightweight scientists. (1995)

Tipler's first published paper appeared in the prestigious *Physical Review*. Written while he was a graduate student, it proposed that a time machine might actually be possible. "Rotating Cylinders and the Possibility of Global Causality Violation" was revolutionary for its time; it was even adapted for a short story by science fiction author Larry Niven.

While earning his Ph.D. in physics, working with the general relativity group at the University of Maryland, Tipler was laying the groundwork for his later books. In 1976, Tipler began postdoctoral work at the University of California, Berkeley, where he met British cosmologist John Barrow, also a postdoc. Tipler and Barrow discussed a manuscript by Brandon Carter which described the Anthropic Principle. "We thought it would be a good idea to take the idea and expand it out. And that became the *Anthropic Cosmological Principle*. In our last chapter we combined the idea from Freeman Dyson [1979] of life going on forever, with physical reductionism and global general relativity; the Omega Point Theory then follows." Tipler's steps in reasoning sound logical, but his conclusions push the limits of science:

> I wanted our book to be completely general, so I said to myself, well, what about the flat universe and the closed universe [instead of an open universe]? One of the problems in the closed universe is communication because we have event horizons everywhere. So I said to myself, that wouldn't be a problem if there were no event horizons. If there were no event horizons, what would the *c*-boundary be like? *Aha*, it would be a single point, and a single-point end of time reminded me of Teilhard's Omega Point, which he identified with God. So I thought maybe there is a religious connection here. (1995)

Barrow and Tipler's work is an attack on the Copernican Principle, which states that man has no special place or purpose in the cosmos. According to the Copernican Principle, our sun is merely one of a hundred billion stars on the outskirts of an average galaxy, itself one of a hundred billion (or more) galaxies in the known universe that cares not one iota for humanity. By contrast, Carter, Barrow, and Tipler's Anthropic Principle insists that humans do have a significant role in the cosmos, both in its observation and its existence. Carter (1974) takes the part of Heisenberg's Uncertainty Principle that says that the observation of an object changes it and extrapolates this part from the atomic level (where Heisenberg was operating) to the cosmological level: "What we can expect to observe must be restricted by the conditions necessary for our presence as observers." In its weak form—the Weak Anthropic Principle—Barrow and Tipler contend quite reasonably that for the cosmos to be observed, it must be structured in such a way as to give rise to observers: "The basic features of the Universe, including such properties as its shape, size, age and laws of change, must be observed to be of a type that allows the evolution of observers, for if intelligent life did not evolve in an otherwise possible universe, it is obvious that no one would be asking the reason for the observed shape, size, age and so forth of the Universe" (1986, p. 2). The principle is tautological: in order for the universe to be observed, there must be observers. Obviously. Who would disagree? The controversy generated by Carter, Barrow, and Tipler lies not with the Weak Anthropic Principle but with the Strong Anthropic Principle, the Final Anthropic Principle, and the Participatory Anthropic Principle. Barrow and Tipler define the Strong Anthropic Principle as "The Universe must have those properties which allow life to develop within it at some stage in its history" and the Final Anthropic Principle as "Intelligent information-processing must come into existence in the Universe, and, once it comes into existence, it will never die out" (pp. 21–23).

That is, the universe must be exactly like it is or there would be no life; therefore, if there were no life, there could be no universe. Further, the Participatory Anthropic Principle states that once life is created (which is inevitable), it will change the universe in such a way that it assures its, and all life's, immortality: "The instant the Omega Point is reached life will have gained control of all matter and forces not only in a single universe, but in all universes whose existence is logically possible; life will have spread into all spatial regions in all universes which could logically exist, and will have stored an infinite amount of information, including all bits of knowledge which it is logically possible to know. And this is the end" (p. 677). This Omega Point, or what Tipler calls a "singularity" of space and time, corresponds to "eternity" in traditional religion. Singularity is also the term used by cosmologists to describe the theoretical starting point of the Big Bang, the center point of a black hole, and the possible ending point of the Big Crunch. Everything and everyone in the universe will converge at this final end point.

Like Dr. Pangloss, Barrow and Tipler relate their incredible claims to a number of seemingly coincidental conditions, events, and physical constants that *must* be a certain way or else there could be no life. For example, they find great meaning in the fact that

$$\frac{\text{electrical force between a proton and an electron}}{\text{gravitational force between a proton and an electron}} \quad or \quad \frac{e^2}{Gm_p m_e} \approx 2.3 \times 10^{39}$$

approximately equals the

$$\frac{\text{age of the universe}}{\text{time for light to cross an atom}} \quad or \quad \frac{t_u}{e^2/m_e c^3} \approx 6 \times 10^{39}$$

They also think it significant that

$$\frac{\text{Planck's constant} \times \text{speed of light}}{\text{Newton's gravitational constant} \times \text{squared mass of a proton}} \quad or \quad \frac{hc}{Gm_p^2} \approx 10^{39}$$

approximately equals the

square root of the number of protons in the observable universe *or*

$$\sqrt{p_u} = 10^{39}$$

Change these relationships significantly and our universe and life as we know it could not exist; thus, they conclude, this is not just the best of all possible worlds, it is the *only* possible world. Barrow and Tipler assume that this relationship, known as Dirac's Large Numbers Hypothesis, is no coincidence. Change any of the constants and the universe would be different enough that life as we know it could not exist, and neither could the universe. There are two problems with this argument.

1. *The Lottery Problem.* Our universe may only be one bubble among many bubble universes (with the whole thing being a multiverse), each one of which has slightly different laws of physics. According to this controversial theory recently pioneered by Lee Smolin (1992) and Andrei Linde (1991), each time a black hole collapses, it collapses into a singularity like the entity out of which our universe was created. But as each collapsing black hole creates a new baby universe, it alters the laws of physics slightly within that baby universe. Since there have probably been billions of collapsed black holes, there are billions of bubbles with slightly different laws

of physics. Only those bubbles with laws of physics like ours can give rise to our types of life. Whoever happens to be in one of these bubbles will think that theirs is the only bubble and thus that they are unique and specially designed. It's like the lottery—it is extremely unlikely that any one person will win, but someone *will* win! Astrophysicist and science writer John Gribbin even suggests an analogy with evolution, where each new bubble is mutated to be slightly different from its parent, and the bubbles are competing with one another, "jostling for spacetime elbow room within superspace" (1993, p. 252). Caltech scientist Tom McDonough and science writer David Brin (1992) wrote melodramatically, "Perhaps we owe our existence, and the convenient perfection of our physical laws, to the trial-and-error evolution of untold generations of prior universes, a chain of mother-and-child cosmoses, each of them spawned in the nurturing depths of black holes."

Much is explained by this model. Our particular bubble universe is unique, but it is not the only bubble nor is it in itself unique in any designed sense. The set of conditions that came together to create life is merely contingent—a conjuncture of events without design. There is no need to posit a higher intelligence. In the long term, this model makes historical sense. From the time of Copernicus, our perspective on the cosmos has been expanding: solar system, galaxy, universe, multiverse. The bubble universe is the next logical step, and it is the best explanation yet for the apparent design of the laws of physics.

2. *The Design Problem.* As David Hume argued in his brilliant analysis of causality in *An Enquiry Concerning Human Understanding* (1758), an orderly world with everything in its rightful place only seems that way because of our experience of it as such. We have perceived nature as it is, so for us this is how the world *must* be designed. Alter the universe and the world, and you alter life in such a way that *its* universe and world would appear as it must be for that observer, and no other. The Weak Anthropic Principle says the universe must be as it is to be observed, but it should include the modifier "by its *particular* observers." As Richard Hardison noted, "Aquinas considered two eyes to be the ideal number and this was evidence of God's existence and benevolence. However, is it not likely that two seems the proper number of eyes simply because that is the pattern to which we have become accustomed?" (1988, p. 123). The so-called coincidental relationships between the physical constants and large numbers of the universe can be found just about anywhere by someone with patience and a turn for numbers. For example, John Taylor, in his book *The Great Pyramid* (1859), observed that if you divide the height of the pyramid into

twice the side of its base, you get a number close to π; he also thought he had discovered the length of the ancient cubit as the division of the Earth's axis by 400,000—both of which Taylor found to be too incredible to be coincidental. Others discovered that the base of the Great Pyramid divided by the width of a casing stone equals the number of days in the year and that the height of the Great Pyramid multiplied by 10^9 approximately equals the distance from the Earth to the Sun. And so on. Mathematician Martin Gardner analyzed the Washington Monument, "just for fun," and "discovered" the property of fiveness to it: "Its height is 555 feet and 5 inches. The base is 55 feet square, and the windows are set at 500 feet from the base. If the base is multiplied by sixty (or five times the number of months in a year) it gives 3,300, which is the exact weight of the capstone in pounds. Also, the word 'Washington' has exactly ten letters (two times five). And if the weight of the capstone is multiplied by the base, the result is 181,500—a fairly close approximation of the speed of light in miles per second" (1952, p. 179). After musing that "it should take an average mathematician about 55 minutes to discover the above 'truths,'" Gardner notes "how easy it is to work over an undigested mass of data and emerge with a pattern, which at first glance, is so intricately put together that it is difficult to believe it is nothing more than the product of a man's brain" (p. 184). The skeptics' skeptic, Gardner leaves "it to readers to decide whether they should opt for OPT [the Omega Point Theory] as a new scientific religion superior to Scientology . . . or opt for the view that OPT is a wild fantasy generated by too much reading of science fiction" (1991b, p. 132).

None of this deterred Tipler, who continued without John Barrow in *The Physics of Immortality*. He submitted a rough draft to his publisher, Oxford University Press, who sent it out for review. The book was rejected. Tipler received the "anonymous" reviews, but by accident their names were not blocked out on the photocopy. One of them, a physicist who is one of the world's leading proponents of integrating science and religion, "said he could recommend this book be published only if I would write it as if I didn't really believe this stuff" (1995).

A longer, more detailed manuscript was submitted to and accepted by Doubleday for publication. While sales were better in Europe (especially Germany) than in America, the reviews for the most part were devastating. Well-known German theologian Wolfhart Pannenberg, who believes in God as a future being, offered his support in *Zygon* (Summer 1995), but most scientists and theologians echoed astronomer Joseph Silk's review in *Scientific American*: "Tipler, however, takes the search for a science of God to a ridiculous extreme. Humility in the face of the persistent, great unknowns is the true philosophy that modern physics has to offer" (July 1995, p. 94).

Frank Tipler faces the great unknowns not with humility but with eternal optimism. When asked to summarize his book in a single sentence, Tipler offered, "Rationality increases without limit; progress goes on forever; life never dies out." How? Tipler's complex arguments may be summarized as three points. (1) In the far future of the universe, humans— the only life in the universe, says Tipler—will have left Earth, populating the rest of the Milky Way galaxy and eventually all other galaxies. If we don't, we are doomed when the Sun expands to envelope the Earth and burn it to a cinder. Therefore, if we must we will. (2) If science and technology continues progressing at its current rate (consider how far we have come from room-size computers in the 1940s to today's laptops), in a thousand or a hundred thousand years, not only will populating the galaxy and universe be possible, but supercomputers with supermemories and super-virtual realities will essentially replace biological life (life and culture are just information systems—genes and memes—to be reproduced in these supercomputers). (3) When the universe eventually collapses, humans and their supercomputers will utilize the energy of the collapsing process to re-create every human who ever lived (since this is a finite number, the supercomputer will have enough memory to accomplish this feat). Since this supercomputer is, for all intents and purposes, omniscient and omnipotent, it is like God; and since "God" will re-create us all in its virtual reality, we are, for all intents and purposes, immortal.

Like Wallace and Paley, Tipler attempts to ground his arguments in pure rationality—no appeals to mysticism, no leaps of religious faith. But can it be pure coincidence that their conclusions create a cosmology in which humankind has had and will continue to have a place . . . forever? "Wouldn't it be better if it were true that you actually made a difference to universal history rather than if whatever you do is ultimately pointless?" Tipler insisted. "The universe would be a happier place if that were true, and I think it is irrational not to at least entertain the possibility that the universe is this way" (1995).

This may sound like hope springing eternal, but Tipler claims that it "is a logical consequence of my own area of research in global general relativity." And though he thinks that part of the problem is that his colleagues "are trained to detest religion so ferociously that even the suggestion that there might be some truth to the statements of religion is an outrage," Tipler says "the only reason the bigger names in the field of global general relativity, like Roger Penrose and Stephen Hawking, have not come to the same conclusion is that they draw back when they realize the outlandish consequences of the equations." Although Penrose and Hawking may retreat in deep understanding, in a revealing comment Tipler explained that most simply will not get it because "the essence of the Omega Point Theory is global general relativity. You have to be trained to think of the universe in the largest possible scale and to automatically view the cosmos in its temporal entirety—you envision the mathematical structure of the future as well as the past. That means you have got to be a global relativist. And there are only three out there better than I am, and only two that are my peers" (1995).

A prominent astronomer I spoke with said that Tipler must have needed money to have written such a ridiculous book. But anyone who talks with Tipler about his book for any length of time quickly realizes that he is not in it for the money or fame. He is deadly serious about his arguments and was fully prepared to take the heat he knew he would get. Frank Tipler is a man who, in my opinion, cares deeply for humanity and its future. His book is dedicated to the grandparents of his wife, "the great-grandparents of my children," who were killed in the Holocaust but "who died in the hope of the Universal Resurrection, and whose hope, as I shall show in this book, will be fulfilled near the End of Time." Here is a deeper motivation. Perhaps Tipler never really abandoned his Baptist, fundamentalist upbringing after all. Through hard work, honest living, and, now, good science, immortality is ours. But we will have to wait. In the meantime, how can we restructure the social, political, economic, and moral systems of society to ensure that we survive long enough to resurrect ourselves? The Dr. Pangloss of his time, Frank Tipler, will venture an answer in his next book, tentatively titled *The Physics of Morality*.

I enjoyed reading Tipler's book. On any number of subjects—space exploration, nanotechnology, artificial intelligence, quantum mechanics, relativity—he writes with clarity and confidence. But I found six problems, the first four of which could be applied to any number of controversial claims. These problems do not prove that Tipler's theory, or any other theory, is wrong. They just alert us to exercise skepticism. Although Tipler may very well be right, the burden of proof is on him to provide empirical data rather than relying almost exclusively on clever, logical reasoning.

1. *The Hope Springs Eternal Problem.* On the first page of *The Physics of Immortality*, Tipler claims that his Omega Point Theory is a "testable physical theory for an omnipotent, omniscient, omnipresent God who

will one day in the far future resurrect every single one of us to live in an abode which is in all essentials the Judeo-Christian Heaven" and that "if any reader has lost a loved one, or is afraid of death, modern physics says: 'Be comforted, you and they shall live again.'" So, everything we always believed to be true based on faith turns out to be true based on physics. What are the chances? Not good, I am afraid. And, after 305 pages of concise and cogent argumentation, Tipler finally admits, "The Omega Point Theory is a viable scientific theory of the future of the physical universe, but the only evidence in its favor at the moment is theoretical beauty." Beauty by itself does not make a theory right or wrong, but when a theory fulfills our deepest wishes we should be especially cautious about rushing to embrace it. When a theory seems to match our eternal hopes, chances are that it is wrong.

2. *The Faith in Science Problem.* When confronting a limitation in one's scientific theory, it is not enough to argue that someday science will solve it just because science has solved so many other problems in the past. Tipler states that to colonize our galaxy and eventually all galaxies, we will have to be able to accelerate spacecraft to near the speed of light. How are we going to do this? No problem. Science will find a way. Tipler spends twenty pages chronicling all the amazing advances in computers, spacecraft, and spacecraft speeds, and in his "Appendix for Scientists" he explains precisely how a relativistic antimatter rocket could be built. All of this is relevant and fascinating but in no way proves that because it *could* happen it *will* happen. Science does have its limitations, and the history of science is replete with failures, wrong turns, and blind alleys. Just because science has been enormously successful in the past does not mean that it can or will solve all problems in the future. And can we really predict what beings in the far future are going to do based on what we think (and hope) they will do?

3. *The If-Then Argument Problem.* Tipler's theory runs something like this: *If* the density parameter is greater than 1 and thus the universe is closed and will collapse; *if* the Bekenstein bound is correct; *if* the Higgs boson is 220 ± 20 GeV; *if* humans do not cause their own extinction before developing the technology to permanently leave the planet; *if* humans leave the planet; *if* humans develop the technology to travel interstellar distances at the required speeds; *if* humans find other habitable planets; *if* humans develop the technology to slow down the collapse of the universe; *if* humans do not encounter forms of life hostile to their goals; *if* humans build a computer that approaches omniscience and omnipotence at the end of time; *if* Omega/God wants to resurrect all previous lives; *if* . . . ; *then* his theory is right. The problem is obvious: if any one of these steps fails, the

entire argument collapses. What if the density parameter is less than 1 and the universe expands forever (as some evidence indicates it will)? What if we nuke or pollute ourselves into oblivion? What if we allocate resources to problems on Earth instead of to space exploration? What if we encounter advanced aliens who intend to colonize the galaxy and Earth, thus dooming us to slavery or extinction?

No matter how rational, an if-then argument without empirical data to support each step in the argument is more philosophy (or protoscience or science fiction) than it is science. Tipler has created an extremely rational argument for God and immortality. Each step follows from the previous step. But so many of the steps might be wrong that the theory is essentially speculative. In addition, his clever switch of the temporal frame of reference to the far future contains a logical flaw. He *first* assumes the existence of God and immortality toward the end of time (his Omega Point boundary conditions—what he previously called the Final Anthropic Principle) and *then* works backward to derive what he has already assumed to be true. Tipler claims this is how all general relativists work (i.e., when they analyze black holes). Even if true, I suspect that most general relativists withhold confidence in their assumptions until there is empirical data to support them, and I have seen no other theories by general relativists which attempt to encompass God, immortality, heaven, and hell. Tipler has made a few testable predictions, but he is a long, *long* way from proving our immortality, and the end of the universe is, well, a long, *long* time away.

4. *The Problem of Analogies.* In *The Tao of Physics: An Exploration of the Parallels Between Modern Physics and Eastern Mysticism* (1975), physicist Fritjof Capra claims that these "parallels" are not accidental. Instead, he argues, there is a single underlying reality that both ancient Eastern philosophers and modern Western physicists have discovered. Although the language of description is different, Capra can see that both groups are really talking about the same thing. (See Gary Zukav's *The Dancing Wu Li Masters* for a similar analysis.) Really? Or is it more likely that the human mind orders the universe in only so many ways and that there are bound to be vague similarities between ancient myths and modern theories, especially if one wants to find them.

Tipler has one-upped Capra. He is not just finding similarities between ancient Judeo-Christian doctrines and modern physics and cosmology, he is redefining both to *make* them fit together: "Every single term in the theory—for example, 'omnipresent,' 'omniscient,' 'omnipotent,' 'resurrection (spiritual) body,' 'Heaven'—will be introduced as pure physics concepts" (1994, p. 1). With each, the reader finds Tipler straining to make the term fit his physics, or vice versa. In starting with God and immortality and reasoning backward, Tipler is not so much discovering these connections between physics and religion as he is creating them. He claims this is both good physics and good theology. I claim that without empirical evidence it is good philosophy and good speculative science fiction. Just because two ideas from separate realms seem to resemble each other does not mean that a meaningful connection between the two exists.

5. *The Problem of Memory and Identity.* Tipler argues that Omega/God, toward the end of the universe, will reconstruct everyone who ever lived or ever could have lived in a super-virtual reality that will include their memories. The first problem is that if memory is a product of neuronal connections and our flawed and ever-changing reconstruction of these neuronal connections, how will Omega/God reconstruct something that does not really exist? There is a vast difference between every memory that *could* be reconstructed and an individual's actual set of memory patterns, the vast majority of which are lost to time. The controversy over false memory syndrome is a case in point. We have very little understanding of how memory works, much less how to reconstruct it. Memories cannot be reconstructed in the sense of playing back a videotape. The event occurs. A selective impression of the event is made on the brain through the senses. Then the individual rehearses the memory and in the process changes it a bit, depending on emotions, previous memories, subsequent events and memories, and so on. This process recurs thousands of times over the years, to the point where we must ask whether we have memories or just memories of memories of memories.

We have another problem, too. If Omega/God resurrects me with all of my memories, which memories will they be? The memories I had at a particular point in my lifetime? Then, that won't be all of me. All the memories I had at every point in my life? That won't be me either. Thus, whatever would be resurrected by Omega/God, it cannot possibly be me, with my very own memories. And if a Michael Shermer is resurrected, and he does not have my memories, who will he be? For that matter, who am I? These problems of memory and identity must be worked through before we can even begin to speculate well about resurrecting an actual person.

6. *The Problem of History and the Lost Past.* A human being may be only a computer consisting of DNA and neuronal memories, but a human *life*, that is, the *history* of a human, is much more than DNA and neuronal memories. It is a product of all a person's interactions with other lives and life histories, plus the environment, itself a product of countless interactions as a function of countless conjunctures of events in a complex matrix with so many variables that it is inconceivable that even Tipler's computer, which can store 10 to the power of 10 to the power of 123 bits (a 1 fol-

lowed by 10^{123} zeros), could represent it. (This figure depends on the Bekenstein bound being real, which cosmologist Kip Thorne says is highly questionable.) Even if it had the computational power to reconstruct all the innumerable historical necessities—climate, geography, population immigrations and emigrations, wars, political revolutions, economic cycles, recessions and depressions, social trends, religious revolutions, paradigm shifts, ideological revolutions, and the like—how does Omega/God recapture all the individual conjunctures, all the interactions between the contingencies and necessities of history?

Tipler's answer is that quantum mechanics tells us there can be only a finite number of these memories, events, and historical conjunctures, and because the computers of the far future will have unlimited computing power, they will be able to resurrect every possible variation of you at all given times in your life. But, on page 158, Tipler confesses to a significant problem with an aspect of this answer: "I should warn the reader that I have ignored the problem of opacity and the problem of loss of coherence of the light. Until these are taken into account, I cannot say exactly how much information can in fact be extracted from the past." The problem of the irrecoverable past is serious, since history is a conjuncture of events compelling a certain course of action by constraining prior events. History often turns on tiny contingencies, very few of which we know about. Given the sensitive dependence on initial conditions—the butterfly effect—how does Omega/God resurrect all the butterflies?

This perception of history derails Drs. Tipler and Pangloss, as Voltaire noted at the end of *Candide*:

> Pangloss sometimes said to Candide. "All events are linked up in this best of all possible worlds; for, if you had not been expelled from the noble castle by hard kicks in your backside for love of Mademoiselle Cunegonde, if you had not been clapped into the Inquisition, if you had not wandered about America on foot, if you had not stuck your sword in the Baron, if you had not lost all your sheep from the land of Eldorado, you would not be eating candied citrons and pistachios here." "'Tis well said," replied Candide, "but we must cultivate our gardens." (1985, p. 328)

Namely, whatever the sequence of contingencies and necessities in our lives and in history, the outcome would have seemed equally inevitable. But in Candide's response is another kernel of truth. We can never know all of the contingencies and necessities guiding history at any given point in time, let alone the initial conditions of any historical sequence, and from this methodological weakness comes philosophical strength. Human freedom—cultivating our gardens—may be found not only in our inability to process all the data of the past and present but also in our ignorance of the initial conditions and conjunctures of events that shape our actions. We are free in our ignorance, free in the knowledge that most of the causes that determine us are lost to the past . . . forever. It is in this knowledge, rather than in the physics of immortality and resurrection by supercomputers, that hope springs eternal.

17

Why Do *People Believe Weird Things?*

O n the evening of Thursday, May 16, 1996, I walked across burning coals barefoot for an episode of the PBS show, *Bill Nye "The Science Guy."* The producers of this splendid science education series geared toward children wanted to do a segment on pseudoscience and the paranormal, and they thought a scientific explanation for firewalking would make for dramatic television. Since Bill Nye is my daughter's hero, I agreed to host the firewalk. Bernard Leikind, a plasma physicist and one of the world's leading experts on firewalking, got the fire going, spread out the coals, and strolled across, sans shoes, socks, or blisters. As I made my way to the edge of the coals, Leikind reminded me that the temperature in the middle of the raked-out path was about 800°F. I tried to focus on his assurance that this was not a matter of the power of positive thinking but of physics. When you bake a cake in an oven, by way of analogy, the air, the cake, and the metal pan are all at 400°F, but only the pan will burn your skin. Hot coals, even at 800°F, are like cake—they do not conduct heat very quickly—so as long as I strode across the bed without delay I should be safe. My naked toes, inches away from the glowing red coals, were skeptical. This was no cakewalk, they told my brain. It wasn't, but six feet and three seconds later, they were none the worse for wear. My confidence in science was restored, right down to my toes.

Firewalking. What a weird thing to do. I have filing cabinets and bookshelves filled with the records of such weird things. But what constitutes a weird thing? I have no formal definition. Weird things are like pornography—difficult to define but obvious when you see them. Each claim, case, or person must be examined individually. One person's weird thing might be another's cherished belief. Who's to say?

Well, one criteria—the criteria of choice for me and millions of others—is science. What, we ask, is the scientific evidence for a claim? Infomercial megastar Tony Robbins, the self-help guru who got his start in the early 1980s by holding weekend seminars climaxing in a firewalk, queries his audience: "What would happen if you were to discover a way to achieve any goal you desire now?" If you can walk on hot coals, says Robbins, you can accomplish anything. Can Tony Robbins really walk barefoot over hot coals without burning his feet? Sure he can. So can I. So can you. But you and I can do it without meditating, chanting, or paying hundreds of dollars for a seminar because firewalking has nothing to do with mental power. Belief that it does is what I would call a weird thing.

Firewalkers, psychics, UFOlogists, alien abductees, cryonicists, immortalists, Objectivists, creationists, Holocaust deniers, extreme Afrocentrists, racial theorists, and cosmologists who believe science proves God—we have met a lot of people who believe a lot of weird things. And I can assure you after two decades of tracking such people and beliefs that I have only scratched the surface in this book. What are we to make of these?

- Whole Life Expo workshops on such topics as "Electromagnetic Ghostbusting," "Megabrain: New Tools for Mind Expansion," "The Revolutionary Energy Machine," and "Lazaris," the 35,000-year-old guru channeled by Jach Pursel.

- The Brain/Mind Expansion Intensive Dome "designed by John-David for a broad range of brain/mind expansion applications, including brain damage re-education." The dome comes complete with a "comprehensive sound training and Certification Training, stereo decks, amplifiers, switchers, cables and the Brain/Mind Matrix Mixer (pat. pending). Soundproofing materials and consulting also included." The price? Only $65,000.

- A bulk-mailing card instructing you to rub a purple spot on the card with your index finger and then to "press your finger firmly in the ball below and roll it from left to right. You are now ready to call THE COSMIC CONNECTION!" The connection is a 900 number, of course, costing only $3.95 per minute. "An experienced psychic will enlighten you on all matters PAST, PRESENT AND FUTURE!"

Can Jach Pursel actually speak to someone who has been dead for tens of thousands of years? This seems rather unlikely. More likely is that we are listening to Jach Pursel's active imagination. Can the Brain/Mind Expansion Intensive Dome really cure brain damage? Let's see the evidence for this remarkable claim. None is offered. Can a psychic really give me deep and meaningful insights over the phone (or even in person)? I doubt it.

What is going on in our culture and thinking that leads to such beliefs? Theories proffered by skeptics and scientists abound: no education, miseducation, lack of critical thinking, rise of religion, decline of religion, displacement of traditional religion by cults, fear of science, the New Age, the Dark Ages revisited, too much television, not enough reading, reading the wrong books, poor parenting, lousy teachers, and plain old ignorance and stupidity. A correspondent from Ontario, Canada, sent me what he called "the vilest embodiment of what you are up against." It was a Day-Glo orange cardboard sign from his local bookstore on which was scrawled: NEW AGE SECTION MOVED TO SCIENCE SECTION. "I am truly frightened by the ease with which society is substituting voodoo and superstition for inquiry and critical examination," he wrote. "If there was ever to be an icon showing how far this phenomenon has ingrained itself into our culture, then this sign would surely be it." As a culture we seem to have trouble distinguishing science from pseudoscience, history from pseudohistory, and sense from nonsense. But I think the problem lies deeper than this. To get to it we must dig through the layers of culture and society into the individual human mind and heart. There is not a single answer to the question of why people believe weird things, but we can glean some underlying motivations, all linked to one another, from the diverse examples I have discussed in this book:

Credo Consolans. More than any other, the reason people believe weird things is because they want to. It feels good. It is comforting. It is consoling. According to a 1996 Gallup poll, 96 percent of American adults believe in God, 90 percent in heaven, 79 percent in miracles, and 72 percent in angels (*Wall Street Journal*, January 30, p. A8). Skeptics, atheists, and militant antireligionists, in their attempts to undermine belief in a higher power, life after death, and divine providence, are butting up against ten thousand years of history and possibly one hundred thousand years of evolution (if religion and belief in God have a biological basis, which some anthropologists believe they do). Throughout all of recorded history, everywhere on the globe, such beliefs and similar percentages are common. Until a suitable secular substitute surfaces, these figures are unlikely to change significantly.

Skeptics and scientists are not immune. Martin Gardner—one of the founders of the modern skeptical movement and slayer of all manner of weird beliefs—classifies himself as a philosophical theist or, a broader term, a fideist. Gardner explains,

> Fideism refers to believing something on the basis of faith, or emotional reasons rather than intellectual reasons. As a fideist I don't think there are any arguments that prove the existence of God or the immortality of the soul. More than that I think the better arguments are on the side of the atheists. So it is a case of quixotic emotional belief that really is against the evidence. If you have strong emotional reasons for metaphysical belief and it's not sharply contradicted by science or logical reasoning, you have a right to make a leap of faith if it provides sufficient satisfaction. (1996)

Similarly, to the frequently asked question, "What is your position on life after death?" my standard response is "I'm for it, of course." The fact that I am *for* life after death does not mean I'm going to get it. But who wouldn't want it? And that's the point. It is a very human response to believe in things that make us feel better.

Immediate Gratification. Many weird things offer immediate gratification. The 900 number psychic hotline is a classic example. A magician/ mentalist friend of mine works one such hotline, so I have been privileged to hear how the system operates from the inside. Most companies charge $3.95 per minute, with the psychic receiving 60¢ per minute; that's $36.00 an hour for the psychic, if the psychic works continuously, and $201 an hour for the company. The goal is to keep callers on the line long enough to turn a good profit but not so long that they refuse to pay the phone bill. Currently, my friend's record for a single call is 201 minutes, for a total of $793.95! People call for one or more of four reasons: love, health, money, career. Using cold-reading techniques, the psychic begins broad and works toward specifics. "I sense there is some tension in your relationship—one of you is more committed than the other." "I'm getting the feeling that financial pressures are causing problems for you." "You have been thinking about changing careers." Such trite statements are true for almost everyone. If your psychic chooses the wrong one, the psychic only has to say it *will* happen—in the future. And the psychic only has to be right occasionally. Callers forget the misses and remember the hits, and, most important, they *want* the psychic to be right. Skeptics don't spend $3.95 a minute on psychic hotlines, believers do. Calling mostly at night and on weekends, most need someone to talk to. Traditional psychotherapy is formal, expensive, and time-consuming. Deep insight and improvement may take months or years. Delay of gratification is the norm, instant satisfaction the exception. By contrast, the psychic is only a telephone call away. (Many 900 number psychics, including my friend, justify it as "poor man's counseling." At $3.95 a minute, I beg to differ. Interestingly, the two major psychic associations are in conflict, with the so-called "real" psychics feeling that the psychic "entertainers" are making them look phony.)

Simplicity. Immediate gratification of one's beliefs is made all the easier by simple explanations for an often complex and contingent world. Good and bad things happen to both good and bad people, seemingly at random. Scientific explanations are often complicated and require training and effort to work through. Superstition and belief in fate and the supernatural provide a simpler path through life's complex maze. Consider the following example from Harry Edwards, head of the Australian Skeptics Society.

As an experiment, on March 8, 1994, Edwards published a letter in his local newspaper in St. James, New South Wales, about his pet chicken, which perches on his shoulder, occasionally leaving its calling card there. Keeping track of the time and location of the chicken's "deposits," and correlating them with subsequent events, Edwards told readers that he was the recipient of good luck. "Over the past few weeks, I have won the lotto, had money returned to me that I had completely forgotten about and received a large order for my recently published books." Edwards's son, who also dons the chicken and its markings, on one wearing "found wallets containing sums of money which he has returned to owners and received rewards, on another a wrist watch, an unused phone card, a pensioner's card and a clock." Edwards then explained that he took the chicken's feathers to a palmist, "checked its horoscope and consulted a past lives reader who confirmed that it was a reincarnated philanthropist and that I should spread the good luck around by selling the product." He ended his letter by offering to sell his "lucky chicken crap" and providing an address where readers should send their money. Edwards wrote to me exuberantly, "As a firm believer that one can sell anything as long as it is associated with 'good luck,' believe it or not I received two orders and $20 for my 'lucky chicken crap'!" I believe it.

Morality and Meaning. At present, scientific and secular systems of morality and meaning have proved relatively unsatisfying to most people. Without belief in some higher power, people ask, why be moral? What is the basis for ethics? What is the ultimate meaning of life? What's the point of it all? Scientists and secular humanists have good answers to these good questions, but for many reasons these answers have not reached the population at large. To most people, science seems to offer only cold and brutal logic in its presentation of an infinite, uncaring, and purposeless universe. Pseudoscience, superstition, myth, magic, and religion offer simple, immediate, and consoling canons of morality and meaning. Because I used to be a born-again Christian, I empathize with those who feel threatened by science. Who feels threatened?

Like other magazines, every so often *Skeptic* sends a mass mailing to tens of thousands of people in order to increase circulation. Our mailings include a "Business Reply Mail" envelope, along with literature about the Skeptics Society and *Skeptic*. Never in these mailings do we discuss religion, God, theism, atheism, or anything whatsoever to do with such subjects. Yet every mailing we receive dozens of our postage-paid envelopes back from people obviously offended by our existence. Some of the envelopes are stuffed with trash or shredded newspaper; one was glued to a box filled with rocks. Some contain our own literature scrawled with messages of doom and gloom. "No thank you—there is none so blind as he who will not see," reads one. "No thanks, I will pass on your anti-Christian bigotry," says another. "Including you skeptics every knee'll bow, every tongue confess that Jesus Christ is Lord," warns a third. Many are filled with religious pamphlets and literature. One person sent me "FREE TICKET NO. 777 ETERNAL ADMITTANCE TO SPEND ETERNITY IN HEAVEN WITH JESUS CHRIST THE SON OF GOD." The "price of admission" is simple. I merely have to acknowledge "Jesus Christ as YOUR Savior and Lord. THAT VERY MOMENT you are saved FOREVER!" And if I don't? The flip side is another ticket, this one a "FREE TICKET TO SPEND ETERNITY IN THE LAKE OF FIRE WITH THE DEVIL AND HIS ANGELS." Can you guess the number of this ticket? That's correct: 666.

If there were only one thing skeptics, scientists, philosophers, and humanists could do to address the overall problem of belief in weird things, constructing a meaningful and satisfying system of morality and meaning would be a good place to start.

Hope Springs Eternal. Linking all these reasons together is the title of the final part of this book. It expresses my conviction that humans are, by nature, a forward-looking species always seeking greater levels of happiness and satisfaction. Unfortunately, the corollary is that humans are all too often willing to grasp at unrealistic promises of a better life or to believe that a better life can only be attained by clinging to intolerance and ignorance, by lessening the lives of others. And sometimes, by focusing on a life to come, we miss what we have in this life. It is a different source of hope, but it is hope nonetheless: hope that human intelligence, combined with compassion, can solve our myriad problems and enhance the quality of each life; hope that historical progress continues on its march toward greater freedoms and acceptance for all humans; and hope that reason and science as well as love and empathy can help us understand our universe, our world, and ourselves.

18

Why Smart *People Believe Weird Things*

"When men wish to construct or support a theory, how they torture facts into their service!"
—John Mackay, *Extraordinary Popular Delusions and the Madness of Crowds*, 1852

Contingency: "A conjuncture of events occurring without design." (Oxford English Dictionary)

Consider the following conjuncture of events that led me to an answer to the question suggested in the title of this chapter. During the month of April, 1998, when I was on a lecture tour for the first edition of this book, the psychologist Robert Sternberg (best known for his pioneering work in multiple intelligences) attended my presentation at the Yale Law School. His response to the lecture was both enlightening and troubling. It is certainly entertaining to hear about other people's weird beliefs, Sternberg reflected, because we are confident that we would never be so foolish as to believe in such nonsense as alien abductions, ghosts, ESP, Big Foot, and all manner of paranormal ephemera. But, he retorted, the interesting question is not why other people believe weird things, but why you and I believe weird things; and, as a subset of Us (versus Them), why smart people believe weird things. Sternberg then proceeded to rattle off a number of beliefs held by his colleagues in psychology—by all accounts a reasonably smart cohort—that might reasonably be considered weird. And, he wondered with wry irony, which of his own beliefs . . . and mine . . . would one day be considered weird?

My contingency came the following day when I was in Boston for a lecture at MIT. Speaking at the same time in the same building just a few doors down from me was Dr. William Dembski, a mathematician and philosopher lecturing on the inference of design signals within the noise of a system. By the criteria that counts in the academy Dembski is smart. He has a Ph.D. in mathematics from the University of Chicago, a second Ph.D. in philosophy from the University of Illinois at Chicago, and a master's degree in theology from Princeton Theological Seminary. His 1998 book, *The Design Inference*, is published by Cambridge University Press. Yet the subject of his lecture and book—in fact, the subject of his full-time occupation as a research fellow for the Center for the Renewal of Science and Culture at the Discovery Institute in Seattle—is to show that science proves God's existence (design inferred in nature implies a grand designer). In my pantheon of "weird things" to believe this one is toward the top of the list (Darwin debunked Paley's design argument nearly a century and a half ago), yet as we chatted for several hours at a quaint Boston pub following our joint lectures I was struck by just how thoughtful, rational, and intelligent Dembski is. Why would someone with such talent and credentials bypass a promising career in favor of chasing the chimera of proving what is inherently unprovable—God? (For a full defense of this position see my 1999 book *How We Believe*.)

To be fair to William Dembski, he is not alone among highly intelligent and educated scholars and scientists who share his beliefs. Although old-guard creationists like Henry Morris and Duane T. Gish sport Ph.D.s after their names, they are in fields outside the biological sciences and they have no mainstream academic affiliations. But the new breed of creationists are coming from more traditional venues, such as Philip Johnson, a law professor at the flagship campus of the University of California at Berkeley, whose 1991 book, *Darwin on Trial*, helped launch the latest wave of evolution deniers. Hugh Ross earned his Ph.D. in astronomy from the University of Toronto and had a position as a research fellow at the California Institute of Technology (Caltech) before founding Reasons to Believe, an organization whose stated purpose (implied in the name) is to provide Christians with scientific reasons for their faith (see Ross 1993, 1994, and 1996). Even more impressive is Michael Behe, a Lehigh University biochemistry professor and the author of the 1996 book *Darwin's Black Box* that has become something of a bible of the "Intelligent Design" movement. And both received the ultimate endorsement of the conservative intelligentsia when they were invited by William F. Buckley to join his team in a television PBS debate on evolution and creation. (Buckley's PBS *Firing Line* show aired in December 1997, where it was resolved that "Evolutionists should acknowledge creation." The debate was emblematic of the new creationism, employing new euphemisms such as "intelligent-design theory," "abrupt appearance theory," and "initial complexity theory," where it was argued that the "irreducible complexity" of life proves it was created by an intelligent designer, or God.)

For my money, however, the quintessential example of a smart person believing a weird thing is Frank Tipler, a professor of theoretical mathematics at Tulane University and one of the world's leading cosmologists and global general relativists. Tipler enjoys close friendships with such luminaries as Stephen Hawking, Roger Penrose, and Kip Thorne. He has published hundreds of technical papers in leading physics journals, and when he is doing traditional physics he is held in high regard among his colleagues. Yet Tipler also authored the 1996 book, *The Physics of Immortality: Modern Cosmology, God and the Resurrection of the Dead*, in which he claims to prove (through no fewer than 122 pages of mathematical equations and physics formulas in an "Appendix for Scientists") that God exists, the afterlife is real, and we will all be resurrected in the far future of the universe through a super computer with a memory large enough to re-create a reality virtually indistinguishable from our own. This is Star Trek's holodeck writ large.

How can we reconcile this belief with the fact of Tipler's towering intellect? I posed this question to a number of his colleagues. Caltech's Kip Thorne shook his head in utter befuddlement, noting in an exchange with Tipler at Caltech that while each step in Tipler's argument was scientifically sound, the leaps between the steps were wholly unfounded. A UCLA cosmologist said she thought Tipler must have needed the money, for why else would anyone write such nonsense? Others offered less printable assessments. I even asked Stephen Hawking's opinion, who said (through his now-infamous voice synthesizer): "My opinion would be libelous."

Of course, to be sure, both Tipler and Dembski would see me as the one with the weird belief—a dogmatic skepticism in the face of their overwhelming empirical evidence and logical reasoning. "You can't libel the laws of physics," Tipler responded when I told him of Hawking's assessment. "If I didn't think there was something to these design arguments I wouldn't be making them," Dembski told me. So it is reasonable to be skeptical even of the skeptics, although we would do well to remember that the burden of proof is on those making the original claims, not on the skeptics who challenge them. My aim here, however, is not to assess the validity of these claims (I know Dembski and Tipler and consider them friends, yet I critique Dembski's ideas in my book *How We Believe*, and I made Tipler's theory the penultimate chapter of this book). Rather, my purpose is to explore the relationship between intelligence (and other psychological variables) and beliefs—particularly beliefs that, by almost any standard (and regardless if they turn out to be right or wrong) are considered to be on the fringe.

Weird Things, Smart People

Through my work as the editor-in-chief of *Skeptic* magazine, the executive director of the Skeptics Society, and as the "Skeptic" columnist for *Scientific American*, the analysis and explanation of what we loosely refer to as "weird things" are a daily routine. Unfortunately, there is no formal definition of a weird thing that most people can agree upon, because it depends so much on the particular claim being made in the context of the knowledge base that surrounds it and the individual or community proclaiming it. One person's weird belief might be another's normal theory, and a weird belief at one time might subsequently become normal. Stones falling from the sky were once the belief of a few daffy Englishmen; today we have an accepted theory of meteorites. In the jargon of science philosopher Thomas Kuhn (1962, 1977), revolutionary ideas that are initially anathema to the accepted paradigm, in time may become normal science as the field undergoes a paradigm shift.

Still, we can formulate a general outline of what might constitute a weird thing as we consider specific examples. For the most part, what I mean by a "weird thing" is: (1) a claim unaccepted by most people in that particular field of study, (2) a claim that is either logically impossible or highly unlikely, and/or (3) a claim for which the evidence is largely anecdotal and uncorroborated. In my introductory example, most theologians recognize that God's existence cannot be proven in any scientific sense, and thus Dembski's and Tipler's goal of using science to prove God is not only unacceptable to most members of his knowledge community, it is uncorroborated because it is logically impossible. Cold fusion, to pick another example, is unaccepted by almost all physicists and chemists, is highly unlikely, and positive results have not been corroborated. Yet there is a handful of smart people (Arthur C. Clarke is the most notable) who hold out hope for cold fusion's future.

"Smart people" suffers from a similar problem in operational definition, but at least here our task is aided by achievement criteria that most would agree, and the research shows, require a minimum level of intelligence. Graduate degrees (especially the Ph.D.), university positions (especially at recognized and reputable institutions), peer-reviewed publications, and the like, allow us to concur that, while we might quibble over how smart some of these people are, the problem of smart people believing weird things is a genuine one that is quantifiable through measurable data. Additionally, there is a subjective evaluation that comes from the experiences I have had in dealing directly with so many people whose claims I have evaluated. While I have not had the opportunity to administer intelligence tests to my various subjects, through numerous television and radio appearances and personal interviews I have conducted with such claimants, and especially through the lecture series that I organize and host at Caltech, I have had the good fortune to meet a lot of really smart people, some out-and-out brilliant scholars and scientists, and even a handful of geniuses so far off the scale that they strike me as wholly Other. All of these factors combined affords me a reasonable assessment of my subjects' intelligence.

An Easy Answer to a Hard Question

"The gentleman has eaten no small quantity of flapdoodle in his lifetime."
"What's that, O'Brien?" replied I . . .
"Why, Peter," rejoined he, "it's the stuff they feed fools on."
—P. Simple, *Marryat*, 1833

It is a given assumption in the skeptical movement—elevated to a maxim really—that intelligence and education serve as an impenetrable prophylactic against the flimflam that we assume the unintelligent and uneducated masses swallow with credulity. Indeed, at the Skeptics Society we invest considerable resources in educational materials distributed to schools and the media under the assumption that this will make a difference in our struggle against pseudoscience and superstition. These efforts do make a difference, particularly for those who are aware of the phenomena we study but have not heard a scientific explanation for them, but are the cognitive elite protected against the nonsense that passes for sense in our culture? Is flapdoodle the fodder only for fools? The answer is no. The question is why?

For those of us in the business of debunking bunk and explaining the unexplained, this is what I call the Hard Question: Why do *smart* people believe weird things? My Easy Answer will seem somewhat paradoxical at first: *Smart people believe weird things because they are skilled at defending beliefs they arrived at for non-smart reasons.*

That is to say, most of us most of the time come to our beliefs for a variety of reasons having little to do with empirical evidence and logical reasoning (that, presumably, smart people are better at employing). Rather, such variables as genetic predispositions, parental predilections, sibling influences, peer pressures, educational experiences, and life impressions all shape the personality preferences and emotional inclinations that, in conjunction with numerous social and cultural influences, lead us to make certain belief choices. Rarely do any of us sit down before a table of facts, weigh them pro and con, and choose the most logical and rational belief, regardless of what we previously believed. Instead, the facts of the world come to us through the colored filters of the theories, hypotheses, hunches, biases, and prejudices we have accumulated through our lifetime. We then sort through the body of data and select those most confirming what we already believe, and ignore or rationalize away those that are disconfirming.

All of us do this, of course, but smart people are better at it through both talent and training. Some beliefs really are more logical, rational, and supported by the evidence than others, of course, but it is not my purpose here to judge the validity of beliefs; rather, I am interested in the question of how we came to them in the first place, and how we hold on to them in the face of either no evidence or contradictory evidence.

The Psychology of Belief

There are a number of principles of the psychology of belief that go to the heart of fleshing out my Easy Answer to the Hard Question.

1. Intelligence and Belief

Although there is some evidence that intelligent people are slightly less likely to believe in some superstitions and paranormal beliefs, overall conclusions are equivocal and limited. A study conducted in 1974 with Georgia high school seniors, for example, found that those who scored higher on an IQ test were significantly less superstitious than students with lower IQ scores (Killeen et al. 1974). A 1980 study by psychologists James Alcock and L. P. Otis found that belief in various paranormal phenomena was correlated with lower critical thinking skills. In 1989, W. S. Messer and R. A. Griggs found that belief in such psychic (psi) phenomena as out-of-body experiences, ESP, and precognition was negatively correlated with classroom performance as measured by grades (as belief goes up, grades go down).

But it should be noted that these three studies are using three different measures: IQ, critical thinking skills, and educational performance. These may not always be indicative of someone being "smart." And what we mean by "weird things" here is not strictly limited to superstition and the paranormal. For example, cold fusion, creationism, and Holocaust revisionism could not reasonably be classified as superstitions or paranormal phenomena. In his review of the literature in one of the best books on this subject (*Believing in Magic*), psychologist Stuart Vyse (1997) concludes that while the relationship between intelligence and belief holds for some populations, it can be just the opposite in others. He notes that the New Age movement in particular "has led to the increased popularity of these ideas among groups previously thought to be immune to superstition: those with higher intelligence, higher socioeconomic status, and higher educational levels. As a result, the time-honored view of believers as less intelligent than nonbelievers may only hold for certain ideas or particular social groups."

For the most part intelligence is orthogonal to and independent of belief. In geometry, orthogonal means "at right angles to something else"; in psychology orthogonal means "statistically independent. Of an experimental design: such that the variates under investigation can be treated as statistically independent," for example, "the concept that creativity and intelligence are relatively orthogonal (i.e., unrelated statistically) at high levels of intelligence" (OED). Intuitively it seems like the more intelligent people are the more creative they will be. In fact, in almost any profession significantly affected by intelligence (e.g., science, medicine, the creative arts), once you are at a certain level among the population of practitioners (and that level appears to be an IQ score of about 125), there is no difference in intelligence between the most successful and the average in that profession. At that point other variables, independent of intelligence, take over, such as creativity, or achievement motivation and the drive to succeed (see Hudson 1966; Getzels and Jackson 1962).

Cognitive psychologist Dean Keith Simonton's research on genius, creativity, and leadership (1999), for example, has revealed that the raw intelligence of creative geniuses and leaders is not as important as their ability to generate a lot of ideas and select from them those that are most likely to succeed. Simonton argues that creative genius is best understood as a Darwinian process of variation and selection. Creative geniuses generate a massive variety of ideas from which they select only those most likely to survive and reproduce. As the two-time Nobel laureate and scientific genius Linus Pauling observed, one must "have lots of ideas and throw away the bad ones. . . . You aren't going to have good ideas unless you have lots of ideas and some sort of principle of selection." Like Forest Gump, genius is as genius does, says Simonton: "these are individuals credited with creative ideas or products that have left a large impression on a particular domain of intellectual or aesthetic activity. In other words, the creative genius attains eminence by leaving for posterity an impressive body of contributions that are both original and adaptive. In fact, empirical studies have repeatedly shown that the single most powerful predictor of eminence within any creative domain is the sheer number of influential products an individual has given the world." In science, for example, the number one predictor of receiving the Nobel Prize is the rate of journal citation, a measure, in part, of one's productivity. As well, Simonton notes, Shakespeare is a literary genius not just because he was good, but because "probably only the Bible is more likely to be found in English-speaking homes than is a volume containing the complete works of Shakespeare." In music, Simonton notes that "Mozart is considered a greater musical genius than Tartini in part because the former accounts for 30 times as much music in the classical repertoire as does the latter. Indeed, almost a fifth of all classical music performed in modern times was written by just three composers: Bach, Mozart, and Beethoven." In other words, it is not so much that these creative geniuses were smart, but that they were productive and selective. (See also Sulloway, 1996.)

So intelligence is also orthogonal to the variables that go into shaping someone's beliefs. Think of this relationship visually as follows:

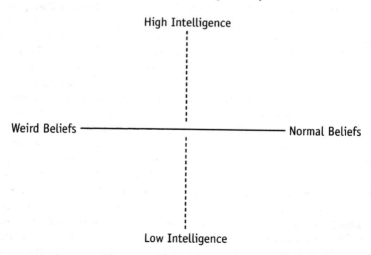

Magic is a useful analogue for this relationship. Folk wisdom has it that smart people are harder for magicians to fool because they are cleverer at figuring out how the tricks are done. But ask any magician (I have asked lots) and they will tell you that there is no better audience than a room full of scientists, academics, or, best of all, members of the high IQ club Mensa. Members of such cohorts, by virtue of their intelligence and education, think they will be better at discerning the secrets of the magician, but since they aren't they are easier to fool because in watching the tricks so intensely they more easily fall for the misdirection cues. The magician James "the Amazing" Randi, one of the smartest people I know, gleefully deceives Nobel laureates with the simplest of magic, knowing that intelligence is unrelated (or perhaps in this case slightly inversely correlated) to the ability to discern the real magic behind the tricks. Tellingly, over the years I have given a number of lectures to Mensa groups around the country and have been struck by the number of weird beliefs such exceptionally smart people hold, including and especially ESP. At one conference there was much discussion about whether Mensa members also had higher Psi-Qs (Psychic Quotient) than regular people!

Another problem is that smart people might be smart in only one field. We say that their intelligence is domain specific. In the field of intelligence studies there is a long-standing debate about whether the brain is "domain general" or "domain specific." Evolutionary psychologists John Tooby, Leda Cosmides, and Steve Pinker, for example, reject the idea of a domain-general processor, focusing on brain modules that evolved to solve specific problems in our evolutionary history. On the other hand, many psychologists accept the notion of a global intelligence that could be considered domain general (Barkow et al. 1992). Archaeologist Steven Mithen (1996) goes so far as to say that it was a domain-general processor that made us human: "The critical step in the evolution of the modern mind was the switch from a mind designed like a Swiss army knife to one with cognitive fluidity, from a specialized to a generalized type of mentality. This enabled people to design complex tools, to create art and believe in religious ideologies. Moreover, the potential for other types of thought which are critical to the modern world can be laid at the door of cognitive fluidity." (See also, Jensen 1998; Pinker 1997; Sternberg 1996; and Gardner 1983.) It seems reasonable to argue that the brain consists of both domain specific and domain-general modules. David Noelle, of the Center for the Neural Basis of Cognition at Carnegie Mellon University, informs me that "modern neuroscience has made it clear that the adult brain does contain functionally distinct circuits. As our understanding of the brain advances, however, we find that these circuits rarely map directly onto complex domains of human experience, such as 'religion' or 'belief.' Instead, we find circuits for more basic things, such as recognizing our location in space, predicting when something good is going to happen (e.g., when we will be rewarded), remembering events from our own lives, and keeping focused on our current goal. Complex aspects of behavior, like religious practices, arise from the interaction of these systems—not from any one module" (personal correspondence; see also Karmiloff-Smith 1995).

What happens when smart people may be smart in one field (domain specificity) but are not smart in an entirely different field, out of which may arise weird beliefs. When Harvard marine biologist Barry Fell jumped fields into archaeology and wrote a best-selling book, *America* B.C.: *Ancient Settlers in the New World* (1976), about all the people who discovered America before Columbus, he was woefully unprepared and obviously unaware that archaeologists had already considered his different hypotheses of who first discovered America (Egyptians, Greeks, Roman, Phoenicians, etc.) but rejected them for lack of credible evidence. This is a splendid example of the social aspects of science, and why being smart in one field does not make one smart in another. Science is a social process, where one is trained in a certain paradigm and works with others in the field. A community of scientists reads the same journals, goes to the same conferences, reviews one anothers' papers and books, and generally exchanges ideas about the facts, hypotheses, and theories in that field. Through vast experience they know, fairly quickly, which new ideas stand a chance of succeeding and which are obviously wrong. Newcomers from other fields, who typically dive in with both feet without the requisite training and experience, proceed to generate new ideas that they think—because of their success in their own field—will be revolutionary. Instead, they are usually greeted with disdain (or, more typically, simply ignored) by the professionals in the field. This is not because (as they usually think is the reason) insiders don't like outsiders (or that all great revolutionaries are persecuted or ignored), but because in most cases those ideas were considered years or decades before and rejected for perfectly legitimate reasons.

2. Gender and Belief

In many ways the orthogonal relationship of intelligence and belief is not unlike that of gender and belief. With the surge of popularity of psychic mediums like John Edward, James Van Praagh, and Sylvia Browne, it has become obvious to observers, particularly among journalists assigned to cover them, that at any given group gathering (usually at large hotel conference rooms holding several hundred people, each of whom paid several hundred dollars to be there), the vast majority (at least 75 percent) are women. Understandably, journalists inquire whether women, therefore, are more superstitious or less rational than men, who typically disdain such mediums and scoff at the notion of talking to the dead. Indeed, a number of studies have found that women hold more superstitious beliefs and accept more paranormal phenomena as real than men. In one study of 132 men and women in New York City, for example, scientists found that more women than men believed that knocking on wood or walking under a ladder brought bad luck (Blum and Blum 1974). Another study showed that more college women than men professed belief in precognition (Tobacyk and Milford 1983).

Although the general conclusion from such studies seems compelling, it is wrong. The problem here is with limited sampling. If you attend any meeting of creationists, Holocaust "revisionists," or UFOlogists, for instance, you will find almost no women at all (the few that I see at such conferences are the spouses of attending members and, for the most part, they look bored out of their skulls). For a variety of reasons related to the subject matter and style of reasoning, creationism, revisionism, and UFOlogy are guy beliefs. So, while gender is related to the target of one's beliefs, it appears to be unrelated to the process of believing. In fact, in the same study that found more women than men believe in precognition, it turned out that more men than women believe in Big Foot and the Loch Ness monster. Seeing into the future is a woman's thing, tracking down chimerical monsters is a man's thing. There are no differences between men and women in the power of belief, only in what they choose to believe.

3. Age and Belief

The relationship between age and belief is also mixed. Some studies, such as a 1990 Gallup poll indicating that people under thirty were more superstitious than older age groups, show that older people are more skeptical than younger people (http://www.gallup.com/poll/releases/pr010608.asp). Another study showed that younger police officers were more likely to believe in the full-moon effect (where allegedly crime rates are higher during full moons) than older police officers. Other studies are less clear about the relationship. British folklorist Gillian Bennett (1987) discovered that older retired English women were more likely to believe in premonition than younger women. Psychologist Seymour Epstein (1993) surveyed three different age groups (9–12, 18–22, 27–65) and discovered that the percentage of belief in each age division depended on the specific phenomena under question. For telepathy and precognition there were no age group differences. For good luck charms more older adults said they had one than did college students or children. The belief that wishing something to happen will make it so decreased steadily with age (Vyse 1997). Finally, Frank Sulloway and I found that religiosity and belief in God steadily decreased with age, until about age seventy-five, when it went back up (Shermer and Sulloway, in press).

These mixed results are due to what is known as person-by-situation effects, where a simple linear causal relationship between two variables rarely exists. Instead, to the question "does X cause Y?" the answer is often "it depends." Bennett, for example, concluded that the older women in her study had lost power, status, and especially loved ones, for which belief in the supernatural helped them recover. Sulloway and I concluded in our study that age and religiosity vary according to one's situation in relation to both early powerful influences and the later perceived impending end of life.

4. Education and Belief

Studies on the relationship between education and belief are, like intelligence, gender, and age, mixed. Psychologist Chris Brand (1981), for example, discovered a powerful inverse correlation of −.50 between IQ and authoritarianism (as IQ increases authoritarianism decreases). Brand concluded that authoritarians are characterized not by an affection for authority, but by "some simple-minded way in which the world has been divided up for them." In this case, authoritarianism was being expressed through prejudice by dividing the world up by race, gender, and age. Brand attributes the correlation to "crystallized intelligence," a relatively flexible form of intelligence shaped by education and life experience. But Brand is quick to point out that only when this type of intelligence is modified by a liberal education does one see a sharp decrease in authoritarianism. In other words, it is not so much that smart people are less prejudiced and authoritarian, but that educated people are less so.

Psychologists S. H. and L. H. Blum (1974) found a negative correlation between education and superstition (as education increased superstitious beliefs decreased). Laura Otis and James Alcock (1982) showed that college professors are more skeptical than either college students or the general public (with the latter two groups showing no difference in belief), but that within college professors there was variation in the types of beliefs held, with English professors more likely to believe in ghosts, ESP, and fortune-telling. Another study (Pasachoff et al. 1971) found, not surprisingly, that natural and social scientists were more skeptical than their colleagues in the arts and humanities; most appropriately, in this context, psychologists were the most skeptical of all (perhaps because they best understand the psychology of belief and how easy it is to be fooled).

Finally, Richard Walker, Steven Hoekstra, and Rodney Vogl (2001) discovered that there was no relationship between science education and belief in the paranormal among three groups of science students at three different colleges. That is, "having a strong scientific knowledge base is not enough to insulate a person against irrational beliefs. Students that scored well on these tests were no more or less skeptical of pseudoscientific claims than students that scored very poorly. Apparently, the students were not able to apply their scientific knowledge to evaluate these pseudoscientific claims. We suggest that this inability stems in part from the way that science is traditionally presented to students: Students are taught what to think but not how to think."

Whether teaching students how to think will attenuate belief in the paranormal remains to be seen. Supposedly this is what the critical thinking movement has been emphasizing for three decades now, yet polls show that paranormal beliefs continue to rise. A June 8, 2001, Gallup Poll, for example, reported a significant increase in belief in a number of paranormal phenomena since 1990, including haunted houses, ghosts, witches, communicating with the dead, psychic or spiritual healing, that extraterrestrial beings have visited earth, and clairvoyance. In support of my claim that the effects of gender, age, and education show content dependent effects, the Gallup poll found:

> *Gender:* Women are slightly more likely than men to believe in ghosts and that people can communicate with the dead. Men, on the other hand, are more likely than women to believe in only one of the dimensions tested: that extraterrestrials have visited earth at some point in the past.

> *Age:* Younger Americans—those 18 to 29—are much more likely than those who are older to believe in haunted houses, in witches, in ghosts, that extraterrestrials have visited earth, and in clairvoyance. There is little significant difference in belief in the other items by age group. Those 30 and older are somewhat more likely to believe in possession by the devil than are the younger group.

> *Education:* Americans with the highest levels of education are more likely than others to believe in the power of the mind to heal the body. On the other hand, belief in three of the phenomena tested goes up as the educational level of the respondent goes down: possession by the devil, astrology and haunted houses.

Additional results from the survey included:

	Believe	Not Sure	Don't Believe
ESP:	50%	20%	27%
Haunted Houses:	42%	16%	41%
Possession by the devil:	41%	6%	41%
Ghosts and spirits:	38%	17%	44%
Telepathy:	36%	26%	35%
Extraterrestrial contact:	33%	27%	38%
Clairvoyance:	32%	23%	45%
Talking to the dead:	28%	26%	46%
Astrology:	28%	18%	52%
Witches:	26%	15%	59%
Reincarnation:	25%	20%	54%
Channeling:	15%	21%	62%

An even more striking poll result was reported by Gallup on March 5, 2001, about the surprising lack of belief in and understanding of the theory of evolution. Specifically, of those Americans polled:

45% agreed with the statement: "God created human beings pretty much in their present form at one time within the last 10,000 years or so."

37% agreed with the statement: "Human beings have developed over millions of years from less advanced forms of life, but God guided this process."

12% agreed with the statement: "Human beings have developed over millions of years from less advanced forms of life, but God had no part in this process."

Despite enormous funds and efforts allocated toward the teaching of evolution in public schools, and the proliferation of documentaries, books, and magazines presenting the theory on all levels, Americans have not noticeably changed their opinion on this question since Gallup started asking it in 1982. Gallup did find that individuals with more education and people with higher incomes are more likely to think that evidence supports the theory of evolution, and that younger people are also more likely than older people to think that evidence supports Darwin's theory (again confounding the age variable). Nevertheless, only 34 percent of Americans consider themselves to be "very informed" about the theory of evolution, while a slightly greater percentage—40 percent—consider themselves to be "very informed" about the theory of creation. Younger people, people with more education, and people with higher incomes are more likely to say they are very informed about both theories.

5. Personality and Belief

Clearly, human thought and behavior are complex and thus studies such as those reported above rarely show simple and consistent findings. Studies on the causes and effects of mystical experiences, for example, show mixed findings. The religious scholar Andrew Greeley (1975), and others (Hay and Morisy, 1978), have found a slight but significant tendency for mystical experiences to increase with age, education, and income, but there were no gender differences. J. S. Levin (1993), by contrast, in analyzing the 1988 General Social Survey data, found no significant age trends in mystical experiences.

But within any group, as defined by intelligence, gender, age, or education, are there any personality characteristics related to belief or disbelief in weird things? First, we note that personality is best characterized by traits, or relatively stable dispositions. The assumption is that these traits, in being "relatively stable," are not provisional states, or conditions of the environment, the altering of which changes the personality. Today's most popular trait theory is what is known as the Five Factor model, or the "Big Five": (1) Conscientiousness (competence, order, dutifulness), (2) Agreeableness (trust, altruism, modesty), (3) Openness to Experience (fantasy, feelings, values), (4) Extroversion (gregariousness, assertiveness, excitement seeking), and (5) Neuroticism (anxiety, anger, depression). In the study on religiosity and belief in God Frank Sulloway and I conducted, we found openness to experience to be the most significant predictor, with higher levels of openness related to lower levels of religiosity and belief in God. In studies of individual scientists' personalities and their receptivity to fringe ideas like the paranormal, I found that a healthy balance between high conscientiousness and high openness to experience led to a moderate amount of skepticism. This was most clearly expressed in the careers of paleontologist Stephen Jay Gould and astronomer Carl Sagan (Shermer, in press). They were nearly off the scale in both conscientiousness and openness to experience, giving them that balance between being open-minded enough to accept the occasional extraordinary claim that turns out to be right, but not so open that one blindly accepts every crazy claim that anyone makes. Sagan, for example, was open to the search for extraterrestrial intelligence which, at the time, was considered a moderately heretical idea; but he was too conscientious to accept the even more controversial claim that UFOs and aliens have actually landed on earth (Shermer 2001).

The psychologist David Wulff (2000), in a general survey of the literature on the psychology of mystical experiences (a subset of weird things), concluded that there were some consistent personality differences:

Persons who tend to score high on mysticism scales tend also to score high on such variables as complexity, openness to new experience, breadth of interests, innovation, tolerance of ambiguity, and creative personality. Furthermore, they are likely to score high on measures of hypnotizability, absorption, and fantasy proneness, suggesting a capacity to suspend the judging process that distinguishes imaginings and real events and to commit their mental resources to representing the imaginal object as vividly as possible. Individuals high on hypnotic susceptibility are also more likely to report having undergone religious conversion, which for them is primarily an experiential rather than a cognitive phenomenon—that is, one marked by notable alterations in perceptual, affective, and ideomotor response patterns.

6. Locus of Control and Belief

One of the most interesting areas of research on the psychology of belief is in the area of what psychologists call locus of control. People who measure high on external locus of control tend to believe that circumstances are beyond their control and that things just happen to them. People who measure high on internal locus of control tend to believe they are in control of their circumstances and that they make things happen (Rotter 1966). External locus of control leads to greater anxiety about the world, whereas internal locus of control leads one to be more confident in one's judgment, skeptical of authority, and less compliant and conforming to external influences. In relation to beliefs, studies show that skeptics are high in internal locus of control whereas believers are high in external locus of control (Marshall et al. 1994). A 1983 study by Jerome Tobacyk and Gary Milford of introductory psychology students at Louisiana Tech University, for example, found that those who scored high in external locus of control tended to believe in ESP, witchcraft, spiritualism, reincarnation, precognition, and were more superstitious than those students who scored high in internal locus of control.

An interesting twist to this effect, however, was found by James McGarry and Benjamin Newberry in a 1977 study of strong believers in and practitioners of ESP and psychic power. Surprisingly, this group scored high in internal locus of control. The authors offered this explanation: "These beliefs [in ESP] may render such a person's problems less difficult and more solvable, lessen the probability of unpredictable occurrences, and offer hope that political and governmental decisions can be influenced." In other words, a deep commitment to belief in ESP, which usually entails believing that one has it, changes the focus from external to internal locus of control.

The effect of locus of control on belief is also mitigated by the environment, where there is a relationship between the uncertainty of an environment and the level of superstitious belief (as uncertainty goes up so too do superstitions). The anthropologist Bronislaw Malinowski (1954), for example, discovered that among the Trobriand Islanders (off the coast of New Guinea), the farther out to sea they went to fish the more they developed superstitious rituals. In the calm waters of the inner lagoon, there were very few rituals. By the time they reached the dangerous waters of deep sea fishing, the Trobrianders were also deep into magic. Malinowski concluded that magical thinking derived from environmental conditions, not inherent stupidities: "We find magic wherever the elements of chance and accident, and the emotional play between hope and fear have a wide and extensive range. We do not find magic wherever the pursuit is certain, reliable, and well under the control of rational methods and technological processes. Further, we find magic where the element of danger is conspicuous." Think of the superstitions of baseball players. Hitting a baseball is exceedingly difficult, with the best succeeding barely more than three out of every ten times at bat. And hitters are known for their extensive reliance on rituals and superstitions that they believe will bring them good luck. These same superstitious players, however, drop the superstitions when they take the field, since most of them succeed in fielding the ball more than 90 percent of the time. Thus, as with the other variables that go into shaping belief that are themselves orthogonal to intelligence, the context of the person and the belief system are important.

7. Influence and Belief

Scholars who study cults (or, as many prefer to call them by the less pejorative term, "New Religious Movements") explain that there is no simple answer to the question "Who joins cults?" The only consistent variable seems to be age—young people are more likely to join cults than older people—but beyond that, variables such as family background, intelligence, and gender are orthogonal to belief in and commitment to cults. Research shows that two-thirds of cult members come from normal functioning families and showed no psychological abnormalities whatsoever when they joined the cult (Singer, 1995). Smart people and non-smart people both readily join cults, and while women are more likely to join such groups as J. Z. Knight's "Ramtha"-based cult (she allegedly channels a 35,000-year old guru named "Ramtha" who doles out life wisdom and advice, in English with an Indian accent no less!), men are more likely to join militias and other anti-government groups.

Again, although intelligence may be related to how well one is able to justify one's membership in a group, and while gender may be related to which group is chosen for membership, intelligence and gender are unrelated to the general process of joining, the desire for membership in a cult, and belief in the cult's tenets. Psychiatrist Marc Galanter (1999), in fact, suggests that joining such groups is an integral part of the human condition to which we are all subject due to our common evolutionary heritage. Banding together in closely knit groups was a common practice in our evolutionary history because it reduced risk and increased survival by being with others of our perceived kind. But if the process of joining is common among most humans, why do some people join while others do not?

The answer is in the persuasive power of the principles of influence and the choice of what type of group to join. Cult experts and activists Steve Hassan (1990) and Margaret Singer outline a number of psychological influences that shape people's thoughts and behaviors that lead them to join more dangerous groups (and that are quite independent of intelligence): cognitive dissonance; obedience to authority; group compliance and conformity; and especially the manipulation of rewards, punishments, and experiences with the purpose of controlling behavior, information, thought, and emotion (what Hassan 2000 calls the "BITE model"). Social psychologist Robert Cialdini (1984) demonstrates in his enormously persuasive book on influence, that all of us are influenced by a host of social and psychological variables, including physical attractiveness, similarity, repeated contact or exposure, familiarity, diffusion of responsibility, reciprocity, and many others.

Smart Biases in Defending Weird Beliefs

In 1620 English philosopher and scientist Francis Bacon offered his own Easy Answer to the Hard Question:

> The human understanding when it has once adopted an opinion (either as being the received opinion or as being agreeable to itself) draws all things else to support and agree with it. And though there be a greater number and weight of instances to be found on the other side, yet these it either neglects and despises, or else by some distinction sets aside and rejects; in order that by this great and pernicious predetermination the authority of its former conclusions may remain inviolate. . . . And such is the way of all superstitions, whether in astrology, dreams, omens, divine judgments, or the like; wherein men, having a delight in such vanities, mark the events where they are fulfilled,

but where they fail, although this happened much oftener, neglect and pass
them by.

Why do smart people believe weird things? Because, to restate my thesis in light of Bacon's insight, *smart people believe weird things because they are
skilled at defending beliefs they arrived at for non-smart reasons.*

As we have already seen, there is a wealth of scientific evidence in support of this thesis, but none more so than two extremely powerful cognitive
biases that make it difficult for any of us to objectively evaluate a claim.
These biases, in fact, are especially well manipulated by smart people: the
Intellectual Attribution Bias and the Confirmation Bias.

Intellectual Attribution Bias. When Sulloway and I asked our subjects
why they believe in God, and why they think other people believe in God
(and allowed them to provide written answers), we were inundated with
thoughtful and lengthy treatises (many stapled multipage, typewritten
answers to their survey) and we discovered that they could be a valuable
source of data. Classifying the answers into categories, here were the top
reasons given:

Why People Believe in God

1. Arguments based on good design/natural beauty/perfection/
   complexity of the world or universe. (28.6%)
2. The experience of God in everyday life/a feeling that God is in
   us. (20.6%)
3. Belief in God is comforting, relieving, consoling, and gives
   meaning and purpose to life. (10.3%)
4. The Bible says so. (9.8%)
5. Just because/faith/or the need to believe in something. (8.2%)

Why People Think Other People Believe in God

1. Belief in God is comforting, relieving, consoling, and gives
   meaning and purpose to life. (26.3%)
2. Religious people have been raised to believe in God. (22.4%)
3. The experience of God in everyday life/a feeling that God is in
   us. (16.2%)
4. Just because/faith/or the need to believe in something. (13.0%)
5. People believe because they fear death and the unknown. (9.1%)

6. Arguments based on good design/natural beauty/perfection/
   complexity of the world or universe. (6.0%)

Note that the intellectually based reasons for belief in God of "good
design" and "experience of God," which were in 1st and 2nd place in the
first question of why do you believe in God?, dropped to 6th and 3rd place
for the second question of why do you think other people believe in God?
Taking their place as the two most common reasons given for why other
people believe in God were the emotionally based categories of religion
being judged as "comforting" and people having been "raised to believe"
in God. Grouping the answers into two general categories of rational reasons and emotional reasons for belief in God, we performed a Chi-Square
test and found the difference to be significant (Chi-Square[1] = 328.63
[r = .49], N = 1,356, p < .0001). With an odds ratio of 8.8 to 1, we may
conclude that people are nearly nine times more likely to attribute their
own belief in God to rational reasons than they are other people's belief in
God, which they will attribute to emotional reasons.

One explanation for this finding is the attribution bias, or the attribution of causes of our own and others' behaviors to either a situation or a
disposition. When we make a situational attribution, we identify the cause
in the environment ("my depression is caused by a death in the family");
when we make a dispositional attribution, we identify the cause in the person as an enduring trait ("her depression is caused by a melancholy personality"). Problems in attribution may arise in our haste to accept the first
cause that comes to mind (Gilbert et al. 1988). Plus, social psychologists
Carol Tavris and Carole Wade (1997) explain that there is a tendency for
people "to take credit for their good actions (a dispositional attribution)
and let the situation account for their bad ones." In dealing with others,
for example, we might attribute our own success to hard work and intelligence, whereas the other person's success is attributed to luck and circumstance (Nisbett and Ross 1980).

We believe that we found evidence for an intellectual attribution bias,
where we consider our own actions as being rationally motivated, whereas
we see those of others as more emotionally driven. Our commitment to a
belief is attributed to a rational decision and intellectual choice ("I'm
against gun control because statistics show that crime decreases when gun
ownership increases"); whereas the other person's belief is attributed to
need and emotion ("he's for gun control because he's a bleeding-heart liberal who needs to identify with the victim"). This intellectual attribution
bias applies to religion as a belief system and to God as the subject of
belief. As pattern-seeking animals, the matter of the apparent good design

of the universe, and the perceived action of a higher intelligence in the day-to-day contingencies of our lives, is a powerful one as an intellectual justification for belief. But we attribute other people's religious beliefs to their emotional needs and upbringing.

Smart people, because they are more intelligent and better educated, are better able to give intellectual reasons justifying their beliefs that they arrived at for nonintellectual reasons. Yet smart people, like everyone else, recognize that emotional needs and being raised to believe something are how most of us most of the time come to our beliefs. The intellectual attribution bias then kicks in, especially in smart people, to justify those beliefs, no matter how weird they may be.

Confirmation Bias. At the core of the Easy Answer to the Hard Question is the confirmation bias, or the tendency to seek or interpret evidence favorable to already existing beliefs, and to ignore or reinterpret evidence unfavorable to already existing beliefs. Psychologist Raymond Nickerson (1998), in a comprehensive review of the literature on this bias, concluded: "If one were to attempt to identify a single problematic aspect of human reasoning that deserves attention above all others, the confirmation bias would have to be among the candidates for consideration. . . . it appears to be sufficiently strong and pervasive that one is led to wonder whether the bias, by itself, might account for a significant fraction of the disputes, altercations, and misunderstandings that occur among individuals, groups, and nations."

Although lawyers purposefully employ a type of confirmation bias in the confrontational style of reasoning used in the courtroom by purposefully selecting evidence that best suits their client and ignoring contradictory evidence (where winning the case trumps the truth or falsity of the claim), psychologists believe that, in fact, we all do this, usually unconsciously. In a 1989 study, psychologists Bonnie Sherman and Ziva Kunda presented students with evidence that contradicted a belief they held deeply, and with evidence that supported those same beliefs; the students tended to attenuate the validity of the first set of evidence and accentuate the value of the second. In a 1989 study with both children and young adults who were exposed to evidence inconsistent with a theory they preferred, Deanna Kuhn found that they "either failed to acknowledge discrepant evidence or attended to it in a selective, distorting manner. Identical evidence was interpreted one way in relation to a favored theory and another way in relation to a theory that was not favored." Even in recall after the experiment, subjects could not remember what the contradictory evidence was that was presented. In a subsequent study in 1994, Kuhn exposed subjects to an audio recording of an actual murder trial and discovered that instead of evaluating the evidence objectively, most subjects first composed a story of what happened, and then sorted through the evidence to see what best fit that story. Interestingly, those subjects most focused on finding evidence for a single view of what happened (as opposed to those subjects willing to at least consider an alternative scenario) were the most confident in their decision.

Even in judging something as subjective as personality, psychologists have found that we see what we are looking for in a person. In a series of studies subjects were asked to assess the personality of someone they were about to meet, some given a profile of an introvert (shy, timid, quiet), others given a profile of an extrovert (sociable, talkative, outgoing). When asked to make a personality assessment, those told that the person would be an extrovert asked questions that would lead to that conclusion; the group given the introvert profile did the same. They both found in the person the personality they were seeking to find (Snyder 1981). Of course, the confirmation bias works both ways in this experiment. It turns out that the subjects whose personalities were being evaluated tended to give answers that would confirm whatever hypothesis the interrogator was holding.

The confirmation bias is not only pervasive, but its effects can be powerfully influential on people's lives. In a 1983 study, John Darley and Paul Gross showed subjects a video of a child taking a test. One group was told that the child was from a high socioeconomic class while the other group was told that the child was from a low socioeconomic class. The subjects were then asked to evaluate the academic abilities of the child based on the results of the test. Not surprisingly, the group told of the high socioeconomic class rated the child's abilities as above grade level, while the group that was told the child was from a low socioeconomic class rated the child's abilities as below grade level. In other words, the same data were seen by one group of evaluators differently than the other group, depending on what their expectations were. The data then confirmed those expectations.

The confirmation bias can also overwhelm one's emotional states and prejudices. Hypochondriacs interpret every little ache and pain as indications of the next great health calamity, whereas normal people simply ignore such random bodily signals (Pennebaker and Skelton 1978). Paranoia is another form of confirmation bias, where if you strongly believe that "they" are out to get you, then you will interpret the wide diversity of anomalies and coincidences in life to be evidence in support of that paranoid hypothesis. Likewise, prejudice depends on a type of confirmation bias, where the prejudged expectations of a group's characteristics leads one to evaluate an individual who is a member of that group in terms of those expectations (Hamilton et al. 1985). Even in depression, people tend to focus on those events and information that further reinforce the depression, and suppress evidence that things are, in fact, getting better (Beck 1976). As Nickerson noted in summary: "the presumption of a relationship predisposes one to find evidence of that relationship, even when there is none to be found or, if there is evidence to be found, to overweight it and arrive at a conclusion that goes beyond what the evidence justifies."

Even scientists are subject to the confirmation bias. Often in search of a particular phenomenon, scientists interpreting data may see (or select) those data most in support of the hypothesis under question and ignore (or toss out) those data not in support of the hypothesis. Historians of science have determined, for example, that in one of the most famous experiments in the history of science, the confirmation bias was hard at work. In 1919, the British astronomer Arthur Stanley Eddington tested Einstein's prediction for how much the sun would deflect light coming from a background star during an eclipse (the only time you can see stars behind the sun). It turns out that Eddington's measurement error was as great as the effect he was measuring. As Stephen Hawking (1988) described it, "The British team's measurement had been sheer luck, or a case of knowing the result they wanted to get, not an uncommon occurrence in science." In going through Eddington's original data, historians S. Collins and J. Pinch (1993) found that "Eddington could only claim to have confirmed Einstein because he used Einstein's derivations in deciding what his observations really were, while Einstein's derivations only became accepted because Eddington's observation seemed to confirm them. Observation and prediction were linked in a circle of mutual confirmation rather than being independent of each other as we would expect according to the conventional idea of an experiment test." In other words, Eddington found what he was looking for. Of course, science contains a special self-correcting mechanism to get around the confirmation bias: other people will check your results or rerun the experiment. If your results were entirely the product of the confirmation bias, someone will sooner or later catch you on it. That is what sets science apart from all other ways of knowing.

Finally, and most importantly for our purposes here, the confirmation bias operates to confirm and justify weird beliefs. Psychics, fortune tellers, palm readers, and astrologers, for example, all depend on the power of the confirmation bias by telling their clients (some would call them "marks") what to expect in their future. By offering them one-sided events (instead of two-sided events in which more than one outcome is possible), the occurrence of the event is noticed while the nonoccurrence of the event is not. Consider numerology. The search for meaningful relationships in various measurements and numbers available in almost any structure in the world (including the world itself, as well as the cosmos) has led numerous observers to find deep meaning in the relationship between these numbers. The process is simple. You can start off with the number you seek and try to find some relationship that ends in that number, or one close to it. Or, more commonly, you crunch through the numbers and see what pops out of the data that looks familiar. In the Great Pyramid, for example (as discussed in chapter 16), the ratio of the pyramid's base to the width of a casing stone is 365, the number of days in the year. Such number crunching with the confirmation bias in place has led people to "discover" in the pyramid the earth's mean density, the period of precession of the earth's axis, and the mean temperature of the earth's surface. As Martin Gardner (1957) wryly noted, this is a classic example of "the ease with which an intelligent man, passionately convinced of a theory, can manipulate his subject matter in such a way as to make it conform to precisely held opinions." And the more intelligent the better.

So, in sum, being either high or low in intelligence is orthogonal to and independent of the normalness or weirdness of beliefs one holds. But these variables are not without some interaction effects. High intelligence, as noted in my Easy Answer, makes one skilled at defending beliefs arrived at for non-smart reasons. In chapter 3 I discuss a study conducted by psychologist David Perkins (1981), in which he found a positive relationship between intelligence and the ability to justify beliefs, and a negative relationship between intelligence and the ability to consider other beliefs as viable. That is to say, smart people are better at rationalizing their beliefs with reasoned arguments, but as a consequence they are less open to considering other positions. So, although intelligence does not affect what you believe, it does influence how beliefs are justified, rationalized, and defended after the beliefs are acquired for non-smart reasons.

Enough theory. As the architect Mies van der Rohe noted, God dwells in the details. The following examples of the difference between intelligence and belief are carefully chosen not from the lunatic fringe or culturally marginalized, but from the socially mainstream and especially from the academy. That is what makes the Hard Question so hard. It is one thing to evaluate the claims of a government coverup from a raving conspiratorialist publishing a newsletter out of his garage in Fringeville, Idaho; it is quite another when it comes from a Columbia University political science professor, or from a Temple University history professor, or from an Emory University social scientist, or from a multimillionaire business genius from Silicon Valley, or from a Pulitzer Prize–winning professor of psychiatry at Harvard University.

UFOs and Alien Abductions
A Weird Belief with Smart Supporters

UFOs and alien abductions meet my criteria for a weird thing because the claim that such sightings and experiences represent actual encounters with extraterrestrial intelligences is (1) unaccepted by most people in astronomy, exobiology, and the Search for Extra-Terrestrial Intelligence (despite the near universal desire by practitioners to find life of any grade somewhere other than earth), (2) extremely unlikely (although not logically impossible), and (3) is largely based on anecdotal and uncorroborated evidence. Are UFO and alien abduction claims supported by smart people? While the community of believers used to be populated largely by those in the nooks and crannies of society's fringes, they have successfully migrated into the cultural mainstream. In the 1950s and 1960s, those who told stories of alien encounters were, at best, snickered at behind closed doors (and sometimes when the doors were wide open) or, at worst, sent to psychiatrists for mental health evaluations. And they were always the butt of jokes among scientists. But in the 1970s and 1980s a gradual shift occurred in the credentials of the believers, and in the 1990s they received a boost from the academy that has helped metastasize their beliefs into society's main body.

Consider Jodi Dean's widely reviewed 1998 book *Aliens in America*. Dean is a Columbia University Ph.D., a professor of political science at Hobart and William Smith Colleges, and a noted feminist scholar. Her book is published by Cornell University Press and begins as if it is going to be a thoughtful sociology of UFOlogy with a thesis that abductees feel "alienated" from modern American society because of economic insecurities, threats of environmental destruction, worldwide militarism, colonialism, racism, misogyny, and other cultural bogeymen: "My argument is that the aliens infiltrating American popular cultures provide icons through which to access the new conditions of democratic politics at the millennium." Since Dean rejects science and rationality as methods of discriminating between sense and nonsense, we "have no criteria for choosing among policies and verdicts, treatments and claims. Even further, we have no recourse to procedures, be they scientific or juridical, that might provide some 'supposition of reasonableness.' " For Dean, not only is science not a solution, it is part of the problem: " 'Scientists' are the ones who have problems with the 'rationality' of those in the UFO community. 'Scientists' are the ones who feel a need to explain why some people believe in flying saucers, or who dismiss those who do so as 'distorted' or 'prejudiced' or 'ignorant.' " Indeed, Dean concludes, since postmodernism has shown all truth to be relative and consensual, then the UFOlogists' claims are as true as anyone's claims: "The early ufologists fought against essentialist understandings of truth that would inscribe truth in objects (and relations between objects) in the world. Rejecting this idea, they relied on an understanding of truth as consensual. If our living in the world is an outcome of a consensus on reality, then stop and notice that not everyone is consenting to the view of reality espoused by science and government."

With this relativist view of truth Dean never tells us whether she believes the UFO/abduction narratives told by her subjects. So I asked her just that in a radio interview, to which she replied: "I believe that they believe their stories." I acknowledged the clarification but pressed the point: "But what do you believe?" Dean refused to answer the question. Fair enough, I suppose, since she is trying to take a nonjudgmental perspective (although I could not get her to offer an opinion even off the air and off the record). But my point here is that by so doing this smart person is lending credence to a weird belief, adding to its credibility as an acceptable tenet of truth that should be part of acceptable social dialogue when, in fact, there is no more evidence for the existence of aliens on earth than there is for fairies (which, in the 1920s, enjoyed their own cultural heyday and the backing of smart people like the creator of Sherlock Holmes, Arthur Conan Doyle; see Randi 1982).

Where Dean equivocates on the veracity question, Temple University history professor David Jacobs does not. Jacobs, who earned his doctorate from the University of Wisconsin and subsequently published his dissertation in 1975 as *The UFO Controversy in America* through Indiana University Press, in 1992 wrote *Secret Life: Firsthand Accounts of UFO Abductions* (even landing a mainstream trade publisher in Simon & Schuster, one of the largest and most prestigious publishing houses in the world). In 1998 he ratcheted up the stakes with *The Threat: The Secret Agenda—What the Aliens Really Want . . . and How They Plan to Get It*. He admits in this latest book that "when I talk about the subject to my colleagues in the academic community, I know they think that my intellectual abilities are seriously impaired." Shortly after *The Threat* was released, I interviewed Jacobs on my weekly NPR radio show in Los Angeles. His intellectual abilities are not impaired in the least. I found him to be bright, articulate, and completely committed to his belief. He spoke like an academic, explained his theory and evidence with the cool dispatch of a seasoned scholar, and acted as if this claim were no different than discussing some other aspect of twentieth-century American history, which he teaches.

Yet Jacobs' books resound with the anthem "I know this sounds weird, but I'm a smart guy." His first book includes a foreword by Harvard's John Mack (more on Mack below), who praises Jacobs as "scholarly and dispassionate," the product of "rigorous scholarship," "careful observation," and "meticulous documentation." In his second book his Ph.D. graces not only the cover, but appears as a header on every page, again punching home the message to the reader that no matter how weird it all seems, a Doctor of Philosophy is endorsing it. Jacobs' narrative style is designed to sound scholarly and scientific. He speaks of his "research," the "methodologies" used, his fellow "investigators," their "huge database," the "documentation" in support of the database, the numerous "theories," "hypotheses," and "evidence" that confirm not only the fact that the aliens are here, but enlighten us about their agenda. Even though this field of study has not one iota of physical evidence—all claims depend entirely on blurry photographs, grainy videos, recovered memories through hypnosis, and endless anecdotes about things that go bump in the night—Jacobs admits these limitations of his "data," but argues that if you combine them you can make the leap from skepticism to belief: "Our encounters with the abduction phenomenon have often come through the haze of confabulation, channeling, and unreliable memories reported by inexperienced or incompetent researchers. It smacks so much of cultural fantasy and psychogenesis that the barriers to acceptance of its reality seem unsurmountable." Indeed, but never underestimate the power of belief. "Yet, I am persuaded that the abduction phenomenon is real. And as a result, the intellectual safety net with which I operated for so many years is now gone. I am as vulnerable as the abductees themselves. I should 'know better,' but I embrace as real a scenario that is both embarrassing and difficult to defend." If the evidence is so weak for this phenomenon, then how can a smart guy like Jacobs believe in it? His answer, coming in the final pages of the book, closes the belief off to counter evidence: "The aliens have fooled us. They lulled us into an attitude of disbelief, and hence complacency, at the very beginning of our awareness of their presence." It is the perfect circular (and impenetrable) argument. The aliens have either caused your belief or your skepticism. Either way, aliens exist.

Whereas Jacobs admits that his evidence is anecdotal and thus nonfalsifiable, Emory University's Courtney Brown, a professor of political science with a couple of bestselling books on aliens and UFOs by mainstream publishers, grounds his beliefs on a method of "data collection" he calls "Scientific Remote Viewing." SRV (both the name and the abbreviation are "registered service marks of Farsight, Inc.," so noted on his copyright page). SRV, more commonly known as Remote Viewing, is the process employed by a group of researchers hired by the CIA to try to close the "psi gap" (similar to the missile gap) between the United States and the Soviet Union in the 1980s (one of them, Ed Dames, was Brown's mentor). During the cold war there was fear on the part of some American government officials that the Russians might have made greater advances in psychic power. So the CIA established a small department that spent $20 million in ten years to determine if they could "remote view" the location of missile silos, MIAs, and gather other intelligence information. The name is almost self-explanatory. To remote view you sit in a room and attempt to "see" (in your mind's eye, of sorts) the target object whose location could be anywhere in the world. After learning the RV ropes, from his home in the suburbs of Atlanta and then from his own institute dedicated to promoting SRV—The Farsight Institute—Brown began to remote view aliens and extraterrestrials.

Like Jacobs' degree, Brown's Ph.D. is prominently displayed on his books. Interestingly, however, his Emory University connection is nowhere to be found in his second book, *Cosmic Explorers: Scientific Remote Viewing, Extraterrestrials, and a Message for Mankind*. I asked him about this in a 1999 radio interview. Emory, it would seem, wants nothing to do with UFOlogy and alien encounters—Brown had to sign a document specifying that when he is discussing his encounters with aliens to the media and the public, no mention of the university is to be made. And, like Jacobs, Brown came off on the air as a thoughtful and intelligent scientist "just following the data" (as they are all wont to say) wherever that might lead.

The claims in Brown's two books are nothing short of spectacularly weird. Through his numerous SRV sessions he says he has spoken with Jesus and Buddha (both, apparently, are advanced aliens), visited other inhabited planets, time traveled to Mars back when it was fully inhabited by intelligent ETs, and has even determined that aliens are living among us—one group in particular resides underground in New Mexico. When I asked him about these unusual claims on the air he balked, redirecting the conversation to the "scientific" aspects of remote viewing, how valid and reliable a method it is for collecting data, how as a social scientist he has applied the rigorous methodologies of the statistical sciences to his newfound research methodology, and that this should all be taken very seriously by scientists. (His first book, published in 1996, was entitled *Cosmic Voyage: A Scientific Discovery of Extraterrestrials Visiting Earth*.) The rhetoric of his written narrative also wafts with scientism meant to convey the message that this weird thing is being presented by a very smart person. Consider just one randomly chosen passage:

A P4 1/2S is the same as a P4 1/2, but it is a sketch rather than a verbal description. When the viewer perceives some visual data in Phase 4 that can be sketched, the viewer writes "P4 1/2S" in either the physicals or the sub-space column, depending on whether the sketch is to be of something in physical reality or subspace reality. The viewer then takes another piece of paper, positions it lengthwise, labels it P4 1/2S centered at the top, and gives it a page number that is the same as the matrix page containing the column entry "P4 1/2S," with an A appended to it. Thus, if the entry for the P4 1/2S is located on page 9, then the P4 1/2S sketch is located on page 9A.

What this passage describes is different methods a remote viewer can use to record different aspects of the fantasy trip: either it is a voyage through the physical world or through "subspace" existence. My point is not to ridicule through obfuscation but to reveal the lengths smart people will go to in order to rationalize a weird belief. When Brown appears on Art Bell's late night radio show he can wax poetic about alien invasions and Jesus' advice. But when he's on my show—by definition a science show broadcast in Southern California and listened to by many from the Caltech, JPL, and aerospace communities, he wants only to discuss the rigors of his scientific methodologies.

In like manner did the multimillionaire Silicon Valley business genius Joe Firmage (1999) respond when I interviewed him on the radio. The 28-year-old founder of the $3 billion Internet company USWeb (who had already sold his first Internet company for $24 million when he was only 19) requested that he be introduced as the founder and chairman of the International Space Sciences Organization (ISSO) and was interested only in discussing his love of science and his new work as a "scientist" for ISSO (to my knowledge he has no formal training as a scientist). What about all those press reports that erupted immediately following his announcement that he was quitting USWeb to pursue his belief that UFOs have landed and that the United States government had captured some of the alien technology and "back-engineered" it and fed it to the American science and technology industries? They exaggerated and distorted what he really believes, Firmage explained. He never actually said that he believed the U.S. government stole alien technologies. Nor did he really want to elaborate upon a 1997 experience he had (he seemed genuinely uncomfortable when I brought it up) with an alien intelligence. The media, he explained, exaggerated that one as well. This I found odd, even disingenuous, since it was his own public relations company that generated all the media attention, including the stories of stolen alien technology and his life-changing alien encounter.

In the fall of 1997, Firmage says that he was awakened in the early morning to see "a remarkable being, clothed in brilliant white light hovering over my bed." The being asked Firmage: "Why have you called me here?" Firmage says he replied: "I want to travel in space." The alien questioned his desire and inquired why such a wish should be granted. "Because I'm willing to die for it," Firmage answered. At this point, says Firmage, out of the alien being "emerged an electric blue sphere, just smaller than a basketball. . . . It left his body, floated down and entered me. Instantly I was overcome by the most unimaginable ecstasy I have ever experienced, a pleasure vastly beyond orgasm. . . . Something had been given to me." The result was Firmage's ISSO and his 1999 Internet electronic book immodestly entitled *The Truth*, a rambling 244-page manuscript filled with warnings to humanity that could have been taken out of a 1950s B science fiction film. The book is heavily sprinkled with the jargon of physics and aeronautics, including Firmage's goal to convince the "scientific establishment" of the reality of UFOs and such advanced technologies as Zero Point Energy from the vacuum of space, "propellantless propulsion" and "gravitational propulsion" for "greater-than-light" travel, "vacuum fluctuations" to alter "gravitational and inertial masses," and the like.

Again, my point is not to belittle, but to understand. Why would a smart man like Joe Firmage give up such a remarkably lucrative and successful career as a Silicon Valley wizard to chase the chimera of aliens? Well, he was raised as a Mormon but in his teen years he "began to have questions about the more dogmatic aspects of the religion." Mormons believe in direct human-angel contact based on the claim that the Church's founder, Joseph Smith, was contacted by the angel Moroni and guided to the sacred golden tablets from which the Book of Mormon was written. In *The Truth*, Firmage explains that the revelation "was received by a man named Joseph Smith, whose descriptions of encounters with brilliant, white-clothed beings are almost indistinguishable from many modern-day accounts of first-hand encounters with 'visitors.'" So, Joseph Smith had a close encounter of the third kind. And apparently he was by no means the first. Eighteen hundred years earlier St. John the Divine received his "revelation" from which the last book in the Bible was written, and shortly before that a carpenter from the tiny hamlet of Nazareth experienced his own visions and epiphanies from on high. Although he does not say it directly, the inference is clear: Jesus the Christ, St. John the Divine, Joseph Smith, and Joseph Firmage each made contact with one of these higher beings, and as a consequence changed the world. Firmage found his calling, and the meaning of his close encounters:

One of the purposes of this Internet book is to share with each of you funda-
mentally new ideas—ideas that one day could transform the world. In this
work, I wish to propose a way to completely restructure over time our economic institutions to operate in a manner compatible with a living Earth,
while preserving the proven entrepreneurial creativity that has built a remarkable modern civilization. . . . . Is this a radical proposal? Absolutely. Is it
insane? Yes. Is it a utopian fantasy? Totally. Radical and insane proposals are
necessary to save a short-sighted and dangerously hubris nation from self-
destruction. . . . My business partner and I built USWeb Corporation, the
largest Internet services company on the planet, so I know what I am talking
about creating here.

Indeed he does. He is a smart man with a weird belief and a lot of
money to legitimize it. But neither the smarts nor the money alter one iota
the fact that there exists not one piece of tangible evidence of alien visita-
tion. And where evidence is lacking, the mind fills in the gaps, and smart
minds are better at gap filling.

Cornell University, Emory University, Temple University, and Silicon
Valley are impressive venues from which to launch weird salvos, but
UFOlogists and the alien experiencer (the preferred term to "abduction")
community received its biggest boost in 1994 with the publication of
Abduction: Human Encounters with Aliens by Harvard Medical School psychiatrist John Mack. Mack's M.D. is boldly emblazoned on the cover,
along with "Winner of the Pulitzer Prize" (awarded for a biography of
T. E. Lawrence, not a book on psychiatry), thereby establishing credibility.
The publisher might as well have printed at the bottom of the dust jacket:
"smart man endorses weird belief." Mack admits in his introduction that
when he first heard about abductee proponent and pioneer Budd Hopkins,
and of people claiming to have been abducted by aliens, "I then said something to the effect that he must be crazy and so must they." But when
Mack met some of them "they seemed in other respects quite sane."
Further, as far as he could tell, these folks had nothing to gain and everything to lose in coming forth with such stories, therefore "they were troubled as a consequence of something that had apparently happened to
them." Mack's skepticism morphed into belief after interviewing over a
hundred alien experiencers, concluding that "there was nothing to suggest
that their stories were delusional, a misinterpretation of dreams, or the
product of fantasy. None of them seemed like people who would concoct a
strange story for some personal purpose."

Agreed, but is "concoct" the right word? I think not. "Experiencer" is an
apt description because there is no doubt that the experiences these people
have had are very real. The core question is, does the experience represent

something exclusively inside the mind or outside in the real world? Since there is no physical evidence to confirm the validity of the latter hypothesis, the logical conclusion to draw, knowing what we do about the fantastic imagery the brain is capable of producing, is that experiencer's experiences are nothing more than mental representations of strictly internal brain phenomena. Their motivation for telling Mack and others about these experiences, assuming (naively perhaps) that they do not do it for the public attention, fame, or money, is external validation of an internal process. And the more prestigious the source of that validation—the "smarter" the validator is, so to speak—the more valid becomes the experience: "Hey, I'm not losing my mind—that smart guy at Harvard says it's real."

The Harvard affiliation with such fringe elements was not lost on the university's administration, who made motions to reign in Mack and squelch his alien agenda, but he retained a lawyer, held his ground on the issue of academic freedom (Mack is tenured), and won the right to continue his academic center called PEER, Program for Extraordinary Experience Research. Many questioned his motives. "He enjoys being the center of attention," said Arnold S. Relman, professor emeritus at Harvard Medical School, who led the formal academic investigation of Mack's research. "He's not taken seriously by his colleagues anymore," Relman continued, "but in the interests of academic freedom, Harvard can afford to have a couple of oddballs" (quoted in Lucas 2001).

The consequences of this shift in belief for Mack—his own form of validation in a way—were profound: "What the abduction phenomenon has led me . . . to see is that we participate in a universe or universes that are filled with intelligences from which we have cut ourselves off, having lost the senses by which we might know them." However, allow me to fill in the ellipses: "I would now say inevitably." (Read it again with the ellipses filled.) Why inevitably? Mack's answer is enlightening: "It has become clear to me also that our restricted worldview or paradigm lies behind most of the major destructive patterns that threaten the human future—mindless corporate acquisitiveness that perpetuates vast differences between rich and poor and contributes to hunger and disease; ethnonational violence resulting in mass killing which could grow into a nuclear holocaust; and ecological destruction on a scale that threatens the survival of the earth's living systems."

The story is as old as the science fiction genre from which it sprang, and reveals the deeper mythic motif behind encounter narratives as a type of secular theology, with UFOs and aliens as gods and messiahs coming down to rescue us from our self-imposed destruction—think of Robert Wise's 1951 *The Day the Earth Stood Still*, where the superior alien intelligence in this Christ allegory (the alien's Earth name was "Mr. Carpenter") comes to save the planet from nuclear armageddon. Here we glimpse a possible motive for Mack. Is he a secular saint, Moses come down from the Harvard mountain to mingle with the masses and enlighten us to the true meaning of the cosmos? This is, perhaps, an exaggeration, but there is something deeper in Mack's story that he reveals toward the end of the introduction to his book, and that is his fascination with Thomas Kuhn's concept of the paradigm, and the revolutionary paradigm shift:

> I knew Tom Kuhn since childhood, for his parents and mine were friends in New York and I had often attended eggnog parties at Christmastime in the Kuhns' home. What I found most hopeful was Kuhn's observation that the Western scientific paradigm had come to assume the rigidity of a theology, and that this belief system was held in place by the structures, categories, and polarities of language, such as real/unreal, exists/does not exist, objective/subjective, intrapsychic/external world, and happened/did not happen. He suggested that in pursuing my investigations I suspend to the degree that I was able all of these language forms and simply collect raw information, putting aside whether or not what I was learning fit any particular worldview. Later I would see what I had found and whether any coherent theoretical formulation would be possible.

There is remarkable irony in this statement—one I find difficult to believe Kuhn would endorse—because one of the main points of Kuhn's revolutionary 1962 book, *The Structure of Scientific Revolutions*, is that it is virtually impossible for any of us to "suspend . . . language forms and simply collect raw information." We are all embedded in a worldview, locked in a paradigm, and ensconced in a culture. And, as we saw, the attribution and confirmation biases are all powerful and pervasive that none of us can escape. The language forms of alien abduction narratives are very much a part of a larger culture in twentieth-century America that includes science fiction literature about aliens, the actual exploration of space, films and television programs about spacecraft and aliens, and especially the Search for Extra-Terrestrial Intelligence (SETI) being conducted by mainstream scientists. This is, in large part, the explanation skeptics offer for the consistency of the abduction stories—the memory motifs come from these commonly experienced cultural inputs. But the point is that Mack's alleged unsullied collection of "raw information" seems disingenuous from what we know about how beliefs are formed. (I would also point out—though there is no way that Mack would know this from his one foray into the paranormal—that the identification of the Kuhnian paradigm and the call for a revolutionary shift to the believer's radical idea is made by nearly every claimant who is out of the mainstream, from UFOlogists and psychic investigators to proponents of cold fusion and perpetual motion machines.) Joe Friday's "Just the facts, ma'am" sounds good in principle, but is never conducted in practice. All observations are filtered through a model or theory, so at some point Mack's observations within a skeptical paradigm became data in support of a believing paradigm. How did this happen?

John Mack is smart enough to realize that the data and data collection techniques he and others use in drawing out these abduction narratives are questionable to say the least. Hypnotic regression, fantasy role playing, and suggestive talk therapy all leading to so-called recovered memories, is now well known to actually generate false memories. Of the alleged disappearance of abductees, Mack admits that "there is no firm proof that abduction was the cause of their absence." The scars from alien surgeries, Mack admits, are "usually too trivial by themselves to be medically significant." Of the missing babies from alien-human sexual encounters, Mack notes that there is "not yet a case where a physician has documented that a fetus has disappeared in relation to an abduction." And of the evidence in total, Mack confesses that it is "maddeningly subtle and difficult to corroborate with as much supporting data as firm proof would require."

To accept these shortcomings and continue his work, Mack must make a reality leap of Kuhnian proportions. The limitation is not in our methodologies of research, it is in the subjects themselves: "If the abduction phenomenon, as I suspect, manifests itself in our physical space/time world but is not of it in a literal sense, our notions of accuracy of recall regarding what did or did not 'happen' (Kuhn's advice about suspending categories seems relevant here) may not apply, at least not in the literal physical sense." These aliens may not be from "space," as in outer space, but may be from another dimension, accessible only through these ephemeral mental states and thus immune to skeptics' demand for a body or artifact from the spacecraft. This may be a Kuhnian model of science, but it is not Popperian since there is no way to falsify the claims. Mack's retreat to allowing "aliens" to be inner dimensional beings capable of detection only in the minds of experiencers is indistinguishable from my own hypothesis that they are entirely the product of neural activity. With no way to distinguish between these two hypotheses, we are out of the realm of science and into the field of creative literature. Science fiction, I think, would more adequately describe this entire field.

The epistemological problems from the beginning, then, are enormous, as Mack himself confesses in giving up the game of science entirely: "In this work, as in any clinically sound investigation, the psyche of the investigator, or, more accurately, the interaction of the psyches of the client and the clinician, is the means of gaining knowledge. . . . Thus experience, the reporting of that experience, and the receiving of that experience through the psyche of the investigator are, in the absence of physical verification or 'proof' . . . the only ways that we can know about abductions." Four hundred pages later, in a final section entitled "Paradigm Shift," Mack once again calls for a change comparable to the Copernican revolution (a favorite analogy among paranormalists and fringers of all stripes): "It would appear that what is required is a kind of cultural ego death, more profoundly shattering (a word that many abductees use when they acknowledge the actuality of their experiences) than the Copernican revolution. . . ." How else are we to understand these alien intelligences? "It is an intelligence that provides enough evidence that something profoundly important is at work, but it does not offer the kinds of proof that would satisfy an exclusively empirical, rationalistic way of knowing."

As Mack told Robert Boynton (1994) in *Esquire* magazine, "People always think that aliens are either real or psychological, and I ask them to consider the possibility that they are somehow both. But that means our entire definition of reality has to change." Boynton notes that Mack has long been searching for that alternate reality through such trendy New Age beliefs as EST and holotropic breathing techniques: "He uses the latter to attain a trancelike state. During one session, he had a past-life experience in which he was a sixteenth-century Russian who had to watch while a band of Mongols decapitated his four-year-old son." In fact, Mack admitted to Carl Sagan (1996) that "I wasn't looking for this. There's nothing in my background that prepared me. It's completely persuasive because of the emotional power of these experiences." In a revealing interview in *Time* magazine Mack said, "I don't know why there's such a zeal to find a conventional physical explanation. We've lost all that ability to know a world beyond the physical. I am a bridge between those two worlds."

Mack's bridge has expanded into another book (1999), *Passport to the Cosmos*, in which he once again pleads that "I am not in this book seeking to establish the material reality of the alien abduction phenomenon . . . rather, I am more concerned with the meaning of these experiences for the so-called abductees and for humankind more generally." In this sense, Mack's abduction belief system operates much like religion and other faith-based beliefs, in that for those who believe proof is not necessary, for those who do not believe, proof is not possible. In other words, the belief in UFOs and alien abductions, like that of other weird beliefs, is orthogonal to and independent of the evidence for or against it, or the intelligence of its proponents, which makes my point. Q.E.D.

BIBLIOGRAPHY

Adams, R. L., and B. N. Phillips. 1972. Motivation and Achievement Differences Among Children of Various Ordinal Birth Positions. *Child Development* 43:155–164.

Alcock, J. E., and Otis, L. P. 1980. Critical Thinking and Belief in the Paranormal. *Psychological Reports*. 46:479–482.

Allen, S. 1993. The Jesus Cults: A Personal Analysis by the Parent of a Cult Member. *Skeptic* 2, no. 2:36–49.

Altea, R. [pseud.]. 1995. *The Eagle and the Rose: A Remarkable True Story.* New York: Warner.

Amicus Curiae Brief of Seventy-two Nobel Laureates, Seventeen State Academies of Science, and Seven Other Scientific Organizations, in Support of Appellees, Submitted to the Supreme Court of the United States, October Term, 1986, as Edwin W. Edwards, in His Official Capacity as Governor of Louisiana, et al., Appellants v. Don Aguillard et al., Appellees. 1986.

Anti-Defamation League. 1993. *Hitler's Apologists: The Anti-Semitic Propaganda of Holocaust "Revisionism."* New York: Anti-Defamation League.

App, A. 1973. *The Six Million Swindle: Blackmailing the German People for Hard Marks with Fabricated Corpses.* Tacoma Park, Md.

Applebaum, E. 1994. Rebel Without a Cause. *The Jewish Week*, April 8–14.

Aretz, E. 1970. *Hexeneinmaleins einer Lüge.*

Ayala, F. 1986. Press Statement by Dr. Francisco Ayala. *Los Angeles Skeptics Evaluative Report* 2, no. 4:7.

Bacon, F. 1620 (1939). Novum Organum. In *The English Philosophers from Bacon to Mill*, ed. E. A. Burtt. New York: Random House.

Bacon, F. 1965. *Francis Bacon: A Selection of His Works.* Ed. S. Warhaft. New York: Macmillan.

Baker, R. A. 1987/1988. The Aliens Among Us: Hypnotic Regression Revisited. *Skeptical Inquirer* 12, no. 2:147–162.

———. 1990. *They Call It Hypnosis.* Buffalo, N.Y.: Prometheus.

———. 1996. Hypnosis. In *The Encyclopedia of the Paranormal*, ed. G. Stein. Buffalo, N.Y.: Prometheus.

Baker, R. A., and J. Nickell. 1992. *Missing Pieces.* Buffalo, N.Y.: Prometheus.

315

Baldwin, L. A., N. Koyama, and G. Teleki. 1980. Field Research on Japanese Monkeys: An Historical, Geographical, and Bibliographical Listing. *Primates* 21, no. 2:268–301.

Ball, J. C. 1992. *Air Photo Evidence: Auschwitz, Treblinka, Majdanek, Sobibor, Bergen Belsen, Belzec, Babi Yar, Katyn Forest.* Delta, Canada: Ball Resource Services.

Bank, S. P., and M. D. Kahn. 1982. *The Sibling Bond.* New York: Basic.

Barkow, J. H., L. Cosmides, and J. Tooby. 1992. *The Adapted Mind.* Oxford: Oxford University Press.

Barrow, J., and F. Tipler. 1986. *The Anthropic Cosmological Principle.* Oxford: Oxford University Press.

Barston, A. 1994. *Witch Craze: A New History of European Witch Hunts.* New York: Pandora/HarperCollins.

Bass, E., and L. Davis. 1988. *The Courage to Heal: A Guide for Women Survivors of Child Sexual Abuse.* New York: Reed Consumer Books.

Bauer, Y. 1994. *Jews for Sale? Nazi-Jewish Negotiations, 1933–1945.* New Haven, Conn.: Yale University Press.

Beck, A. T. 1976. *Cognitive Therapy and the Emotional Disorders.* New York: International Universities Press.

Behe, M. 1996. *Darwin's Black Box.* New York: Free Press.

Bennett, G. 1987. *Traditions of Belief: Women, Folklore, and the Supernatural Today.* London: Penguin Books.

Bennetta, W. 1986. Looking Backwards. In his *Crusade of the Credulous: A Collection of Articles About Contemporary Creationism and the Effects of That Movement on Public Education.* San Francisco: California Academy of Science Press.

Berenbaum, M. 1994. Transcript of Interview by M. Shermer, April 13.

Berkeley, G. 1713. In *The Guardian*, June 23. Quoted in H. L. Mencken, ed. 1987. *A New Dictionary of Quotations on Historical Principles from Ancient and Modern Sources.* New York: Knopf.

Berra, T. M. 1990. *Evolution and the Myth of Creationism: A Basic Guide to the Facts in the Evolution Debate.* Stanford, Calif.: Stanford University Press.

Beyerstein, B. L. 1996. Altered States of Consciousness. In *The Encyclopedia of the Paranormal*, ed. G. Stein. Buffalo, N.Y.: Prometheus.

Blackmore, S. 1991. Near-Death Experiences: In or Out of the Body? *Skeptical Inquirer* 16, no. 1:34–45.

———. 1993. *Dying to Live: Near-Death Experiences.* Buffalo, N.Y.: Prometheus.

———. 1996. Near-Death Experiences. In *The Encyclopedia of the Paranormal*, ed. G. Stein. Buffalo, N.Y.: Prometheus.

Blum, S. H., and L. H. Blum. 1974. Do's and Don'ts: An Informal Study of Some Prevailing Superstitions. *Psychological Reports* 35:567–571.

Bowers, K. S. 1976. *Hypnosis*. New York: Norton.

Bowler, P. J. 1989. *Evolution: The History of an Idea*, rev. ed. Berkeley: University of California Press.

Boynton, R. S. 1994. Professor Mack, Phone Home. *Esquire*, March, 48.

Brand, C. 1981. Personality and Political Attitudes. In *Dimensions of Personality; Papers in Honour of H. J. Eysenck*, ed. R. Lynn. Oxford: Pergamon Press., 7–38, 28.

Branden, B. 1986. *The Passion of Ayn Rand*. New York: Doubleday.

Branden, N. 1989. *Judgment Day: My Years with Ayn Rand*. Boston: Houghton Mifflin.

Braudel, F. 1981. *Civilization and Capitalism: Fifteenth to Eighteenth Century*, vol. 1, *The Structures of Everyday Life*. Trans. S. Reynolds. New York: Harper & Row.

Briggs, R. 1996. *Witches and Witchcraft: The Social and Cultural Context of European Witchcraft*. New York: Viking.

Broszat, M. 1989. Hitler and the Genesis of the "Final Solution": An Assessment of David Irving's Theses. In *The Nazi Holocaust*, vol. 3, *The Final Solution*, ed. M. Marrus. Westport, Conn.: Meckler.

Brown, C. 1996. *Cosmic Voyage: A Scientific Discovery of Extraterrestrials Visiting Earth*. New York: Dutton.

———. 1999. *Cosmic Explorers: Scientific Remote Viewing, Extraterrestrials, and a Message for Mankind*. New York: Dutton.

Brugioni, D. A., and R. G. Poirer. 1979. *The Holocaust Revised: A Retrospective Analysis of the Auschwitz-Birkenau Extermination Complex*. Washington, D.C.: Central Intelligence Agency (available from National Technical Information Service).

Butz, A. 1976. *The Hoax of the Twentieth Century*. Newport Beach, Calif.: Institute for Historical Review.

Bynum, W. F., E. J. Browne, and R. Porter. 1981. *Dictionary of the History of Science*. Princeton, N.J.: Princeton University Press.

Campbell, J. 1949. *The Hero with a Thousand Faces*. Princeton, N.J.: Princeton University Press.

———. 1988. *The Power of Myth*. New York: Doubleday.

Capra, F. 1975. *The Tao of Physics: An Exploration of the Parallels Between Modern Physics and Eastern Mysticism*. New York: Bantam.

———. 1982. *The Turning Point: Science, Society, and the Rising Culture*. New York: Bantam.

Carlson, M. 1995. The Sex-Crime Capital. *Time*, November 13.

Carporael, L. 1976. Ergotism: Satan Loosed in Salem. *Science*, no. 192:21–26.

Carter, B. 1974. Large Number Coincidences and the Anthropic Principle in Cosmology. In *Confrontation of Cosmological Theories with Observational Data*, ed. M. S. Longair. Dordrecht, Netherlands: Reidel.

Cavalli-Sforza, L. L., and F. Cavalli-Sforza. 1995. *The Great Human Diaspora: The History of Diversity and Evolution.* Trans. S. Thorne. Reading, Mass.: Addison-Wesley.

Cavalli-Sforza, L. L., P. Menozzi, and A. Piazza. 1994. *The History and Geography of Human Genes.* Princeton, N.J.: Princeton University Press.

Cerminara, G. 1967. *Many Mansions: The Edgar Cayce Story on Reincarnation.* New York: Signet.

Christenson, C. 1971. *Kinsey: A Biography.* Indianapolis: Indiana University Press.

Christophersen, T. 1973. *Die Auschwitz Lüge.* Koelberhagen.

Cialdini, R. 1984. *Influence: The New Psychology of Modern Persuasion.* New York: William Morrow.

Cobden, J. 1991. An Expert on "Eyewitness" Testimony Faces a Dilemma in the Demjanjuk Case. *Journal of Historical Review* 11, no. 2:238–249.

Cohen, I. B. 1985. *Revolution in Science.* Cambridge, Mass.: Harvard University Press.

Cole, D. 1994. Transcript of Interview by M. Shermer, April 26.

———. 1995. Letter to the Editor. *Adelaide Institute Newsletter* 2, no. 4:3.

Collins, S., and J. Pinch. 1993. *The Golem: What Everyone Should Know About Science.* New York: Cambridge University Press.

Cowen, R. 1986. Creationism and the Science Classroom. *California Science Teacher's Journal* 16, no. 5:8–15.

Crews, F., et al. 1995. *The Memory Wars: Freud's Legacy in Dispute.* New York: New York Review of Books.

Curtius, M. 1996. Man Won't Be Retried in Repressed Memory Case. *Los Angeles Times,* July 3.

Darley, J. M., and P. H. Gross. 1983. A Hypothesis-Confirming Bias in Labelling Effects. *Journal of Personality and Social Psychology,* 44:20–33.

Darwin, C. 1859. *On the Origin of Species by Means of Natural Selection: Or the Preservation of Favoured Races in the Struggle for Life. A Facsimile of the First Edition.* Cambridge, Mass.: Harvard University Press, 1964.

———. 1871. *The Descent of Man and Selection in Relation to Sex.* 2 vols. London: J. Murray.

———. [1883]. In Box 106, Darwin archives, Cambridge University Library.

Darwin, M., and B. Wowk. 1989. *Cryonics: Beyond Tomorrow.* Riverside, Calif.: Alcor Life Extension Foundation.

Davies, P. 1991. *The Mind of God.* New York: Simon & Schuster.

Dawkins, R. 1976. *The Selfish Gene.* Oxford: Oxford University Press.

———. 1986. *The Blind Watchmaker.* New York: Norton.

———. 1995. Darwin's Dangerous Disciple: An Interview with Richard Dawkins. *Skeptic* 3, no. 4:80–85.

———. 1996. *Climbing Mount Improbable.* New York: Norton.

Dean, J. 1998. *Aliens in America: Conspiracy Cultures from Outerspace to Cyberspace.* New York: Cornell University Press.

Dembski, W. 1998. *The Design Inference: Eliminating Chance Through Small Probabilities.* Cambridge: Cambridge University Press.

Demos, J. P. 1982. *Entertaining Satan: Witchcraft and the Culture of Early New England.* New York: Oxford University Press.

Dennett, D. C. 1995. *Darwin's Dangerous Idea: Evolution and the Meanings of Life.* New York: Simon & Schuster.

Desmond, A., and J. Moore. 1991. *Darwin: The Life of a Tormented Evolutionist.* New York: Warner.

De Solla Price, D. J. 1963. *Little Science, Big Science.* New York: Columbia University Press.

Dethier, V. G. 1962. *To Know a Fly.* San Francisco: Holden-Day.

Drexler, K. E. 1986. *Engines of Creation.* New York: Doubleday.

Dyson, F. 1979. *Disturbing the Universe.* New York: Harper & Row.

Eddington, A. S. 1928. *The Nature of the Physical World.* New York: Macmillan.

———. 1958. *The Philosophy of Physical Science.* Ann Arbor: University of Michigan Press.

Ehrenreich, B., and D. English. 1973. *Witches, Midwives and Nurses: A History of Women Healers.* New York: Feminist Press.

Eldredge, N. 1971. The Allopatric Model and Phylogeny in Paleozoic Invertebrates. *Evolution* 25:156–167.

———. 1985. *Time Frames: The Rethinking of Darwinian Evolution and the Theory of Punctuated Equilibria.* New York: Simon & Schuster.

Eldredge, N., and S. J. Gould. 1972. Punctuated Equilibria: An Alternative to Phyletic Gradualism. In *Models in Paleobiology,* ed. T. J. M. Schopf. San Francisco: Freeman, Cooper.

Epstein, S. 1993. Implications of Cognitive-Experiential Self-Theory for Personality and Developmental Psychology. In *Studying Lives Through Time: Personality and Developmental Psychology,* eds. D. C. Funder et al. Washington, D.C.: American Psychological Association. 399–438.

Erikson, K. T. 1966. *Wayward Puritans: A Study in the Sociology of Deviance.* New York: Wiley.

Eve, R. A., and F. B. Harrold. 1991. *The Creationist Movement in Modern America.* Boston: Twayne.

Faurisson, R. 1980. *Mémoire en defense: contre ceux qui m'accusent de falsifier l'histoire: la question des chambers à gaz* (Treatise in Defense Against Those Who Accuse Me of Falsifying History: The Question of the Gas Chambers). Paris: Vieille Taupe.

Feynman, R. P. 1959. There's Plenty of Room at the Bottom. Lecture given at the annual meeting of the American Physical Society, California Institute of Technology.

———. 1988. *What Do You Care What Other People Think?* New York: Norton.

Firmage, J. 1999. *The Truth.* Internet electronic book produced by the International Space Sciences Organization. When printed out in the web page format it came out at 244 pages.

Futuyma, D. J. 1983. *Science on Trial: The Case for Evolution.* New York: Pantheon.

Galanter, M. 1999. *Cults: Faith, Healing, and Coercion.* 2nd Edition. New York: Oxford University Press.

Gallup, G. 1982. *Adventures in Immortality.* New York: McGraw-Hill.

Gallup, G. H., Jr., and F. Newport. 1991. Belief in Paranormal Phenomena Among Adult Americans. *Skeptical Inquirer* 15, no. 2:137–147.

Gardner, H. 1983. *Frames of Mind: The Theory of Multiple Intelligences.* New York: Basic Books.

Gardner, M. 1952. *Fads and Fallacies in the Name of Science.* New York: Dover.

———. 1957. *Fads and Fallacies in the Name of Science.* New York: Dover.

———. 1981. *Science: Good, Bad, and Bogus.* Buffalo, N.Y.: Prometheus.

———. 1983. *The Whys of a Philosophical Scrivener.* New York: Quill.

———. 1991a. *The New Age: Notes of a Fringe Watcher.* Buffalo, N.Y.: Prometheus.

———. 1991b. Tipler's Omega Point Theory. *Skeptical Inquirer* 15, no. 2:128–134.

———. 1992. *On the Wild Side.* Buffalo, N.Y.: Prometheus.

———. 1996. Transcript of Interview by M. Shermer, August 11.

Gell-Mann, M. 1986. Press Statement by Dr. Murray Gell-Mann. *Los Angeles Skeptics Evaluative Report* 2, no. 4:5.

———. 1990. Transcript of Interview by M. Shermer.

———. 1994a. What Is Complexity? *Complexity* 1, no. 1:16–19.

———. 1994b. *The Quark and the Jaguar.* New York: Freeman.

George, J., and L. Wilcox. 1992. *Nazis, Communists, Klansmen, and Others on the Fringe: Political Extremism in America.* Buffalo, N.Y.: Prometheus.

Getzels, J. W., and P. W. Jackson. 1962. *Creativity and Intelligence: Explorations with Gifted Students.* New York: John Wiley.

Gilbert, D. T., B. W. Pelham, and D. S. Krull. 1988. On Cognitive Busyness: When Person Perceivers Meet Persons Perceived. *Journal of Personality and Social Psychology* 54:733–739.

Gilkey, L., ed. 1985. *Creationism on Trial: Evolution and God at Little Rock.* New York: Harper & Row.

Gish, D. T. 1978. *Evolution: The Fossils Say No!* San Diego: Creation-Life.

Godfrey, L. R., ed. 1983. *Scientists Confront Creationism.* New York: Norton.

Goldhagen, D. J. 1996. *Hitler's Willing Executioners: Ordinary Germans and the Holocaust.* New York: Knopf.

Goodman, L. S., and A. Gilman, eds. 1970. *The Pharmacological Basis of Therapeutics.* New York: Macmillan.

Gould, S. J. 1983a. *Hen's Teeth and Horse's Toes.* New York: Norton.

———. 1983b. A Visit to Dayton. In *Hen's Teeth and Horse's Toes.* New York: Norton.

———. 1985. *The Flamingo's Smile.* New York: Norton.

———. 1986a. Knight Takes Bishop? *Natural History* 5:33–37.

———. 1986b. Press Statement by Dr. Stephen Jay Gould. *Los Angeles Skeptics Evaluative Report* 2, no. 4:5.

———. 1987a. Darwinism Defined: The Difference Between Fact and Theory. *Discover,* January, 64–70.

———. 1987b. *An Urchin in the Storm.* New York: Norton.

———. 1989. *Wonderful Life.* New York: Norton.

———. 1991. *Bully for Brontosaurus.* New York: Norton.

Grabiner, J. V., and P. D. Miller. 1974. Effects of the Scopes Trial. *Science,* no. 185:832–836.

Greeley, A. M. 1975. *The Sociology of the Paranormal: A Reconnaissance.* Beverly Hills, Calif: Sage.

Gribbin, J. 1993. *In the Beginning: The Birth of the Living Universe.* Boston: Little, Brown.

Grinfeld, M. J. 1995. Psychiatrist Stung by Huge Damage Award in Repressed Memory Case. *Psychiatric Times* 12, no. 10.

Grinspoon, L., and J. Bakalar. 1979. *Psychedelic Drugs Reconsidered.* New York: Basic Books.

Grobman, A. 1983. *Genocide: Critical Issues of the Holocaust.* Los Angeles: Simon Wiesenthal Center.

Grof, S. 1976. *Realms of the Human Unconscious.* New York: Dutton.

Grof, S., and J. Halifax. 1977. *The Human Encounter with Death.* New York: Dutton.

Gutman, Y., ed. 1990. *Encyclopedia of the Holocaust.* 4 vols. New York: Macmillan.

Gutman, Y., and M. Berenbaum, eds. 1994. *Anatomy of the Auschwitz Death Camp.* Bloomington: Indiana University Press.

Gutman, Y. 1996. Transcript of Interview by M. Shermer and A. Grobman, May 10.

Hamilton, D. L., P. M. Dugan, and T. K. Trolier. 1985. The Formation of Stereotypic Beliefs: Further Evidence for Distinctiveness-Based Illusory Correlations. *Journal of Personality and Social Psychology* 48: 5–17.

Hardison, R. C. 1988. *Upon the Shoulders of Giants.* New York: University Press of America.

Harré, R. 1970. *The Principles of Scientific Thinking.* Chicago: University of Chicago Press.

———. 1985. *The Philosophies of Science.* Oxford: Oxford University Press.

Harris, M. 1974. *Cows, Pigs, Wars, and Witches: The Riddles of Culture.* New York: Vintage.

Harwood, R. 1973. *Did Six Million Really Die?* London.

Hassan, S. 1990. *Combatting Cult Mind Control.* Rochester, Vt.: Park Street Press.

———. *Releasing the Bonds: Empowering People to Think for Themselves.* Somerville, Mass: Freedom of Mind Press.

Hawking, S. W. 1988. *A Brief History of Time: From the Big Bang to Black Holes.* New York: Bantam.

Hay, D., and A. Morisy. 1978. Reports of Ecstatic, Paranormal, or Religious Experience in Great Britain and the United States—A Comparison of Trends. *Journal for the Scientific Study of Religion* 17: 255–268.

Headland, R. 1992. *Messages of Murder: A Study of the Reports of the Einsatzgruppen of the Security Police and the Security Service, 1941–1943.* Rutherford, N.J.: Fairleigh Dickinson University Press.

Herman, J. 1981. *Father-Daughter Incest.* Cambridge, Mass.: Harvard University Press.

Herrnstein, R. J., and C. Murray. 1994. *The Bell Curve: Intelligence and Class Structure in American Life.* New York: Free Press.

Hilberg, R. 1961. *The Destruction of the European Jews.* Chicago: Quadrangle.

———. 1994. Transcript of Interview by M. Shermer, April 10.

Hilgard, E. R. 1977. *Divided Consciousness: Multiple Controls in Human Action and Thought.* New York: Wiley.

Hilton, I. 1967. Differences in the Behavior of Mothers Toward First and Later Born Children. *Journal of Personality and Social Psychology* 7:282–290.

Hobbes, T. [1651] 1968. *Leviathan.* Ed. C. B. Macpherson. New York: Penguin.

———. 1839–1845. *The English Works of Thomas Hobbes of Malmesbury.* Ed. W. Molesworth. 11 vols. London: J. Bohn.

Hochman, J. 1993. Recovered Memory Therapy and False Memory Syndrome. *Skeptic* 2, no. 3:58–61.

Hook, S. 1943. *The Hero in History: A Study in Limitation and Possibility.* New York: John Day.

Horner, J. R., and J. Gorman. 1988. *Digging Dinosaurs*. New York: Workman.

House, W. R. 1989. *Tales of the Holohoax*. Champaign, Ill.: John McLaughlin/Wiswell Ruffin House.

Hudson, L. 1966. *Contrary Imaginations: A Psychological Study of the English Schoolboy*. London: Methuen.

Hume, D. [1758] 1952. *An Enquiry Concerning Human Understanding*. Great Books of the Western World. Chicago: University of Chicago Press.

Huxley, A. 1954. *The Doors of Perception*. New York: Harper.

Imanishi, K. 1983. Social Behavior in Japanese Monkeys. In *Primate Social Behavior*, ed. C. A. Southwick. Toronto: Van Nostrand.

Ingersoll, R. G. 1879. Interview in the *Chicago Times*, November 14. Quoted in H. L. Mencken, ed. 1987. *A New Dictionary of Quotations on Historical Principles from Ancient and Modern Sources*. New York: Knopf.

Irving, D. 1963. *The Destruction of Dresden*. London: W. Kimber.

———. 1967. *The German Atomic Bomb: The History of Nuclear Research in Nazi Germany*. New York: Simon & Schuster.

———. 1977. *Hitler's War*. New York: Viking.

———. 1977. *The Trail of the Fox*. New York: Dutton.

———. 1987. *Churchill's War*. Bullsbrook, Australia: Veritas.

———. 1989. *Goering: A Biography*. New York: Morrow.

———. 1994. Transcript of Interview by M. Shermer, April 25.

———. 1996. *Goebbels: Mastermind of the Third Reich*. London: Focal Point.

Jäckel, E. 1989. Hitler Orders the Holocaust. In *The Nazi Holocaust*, vol. 3, *The Final Solution*, ed. M. Marrus. Westport, Conn.: Meckler.

———. 1993. *David Irving's Hitler: A Faulty History Dissected: Two Essays*. Trans. H. D. Kirk. Brentwood Bay, Canada: Ben-Simon.

Jacobs, D. 1975. *The UFO Controversy in America*. Indianapolis: Indiana University Press.

———. 1992. *Secret Life: Firsthand Accounts of UFO Abductions*. New York: Simon & Schuster.

———. 1998. *The Threat: The Secret Agenda: What the Aliens Really Want . . . and How They Plan to Get it*. New York: Simon & Schuster.

Jensen, A. R. 1998. *The g Factor: The Science of Mental Ability*. Westport, Conn: Praeger.

Johnson, D. M. 1945. The "Phantom Anesthetist" of Mattoon. *Journal of Abnormal and Social Psychology* 40:175–186.

Johnson, P. 1991. *Darwin on Trial*. Downers Grove, Ill.: InterVarsity Press.

Karmiloff-Smith, A. 1995. *Beyond Modularity: A Developmental Perspective on Cognitive Science*. London: Bradford.

Kauffman, S. A. 1993. *The Origins of Order: Self-Organization and Selection in Evolution*. New York: Oxford University Press.

Kaufman, B. 1986. SCS Organizes Important *Amicus Curiae* Brief for United States Supreme Court. *Los Angeles Skeptics Evaluative Report* 2, no. 3:4–6.

Kawai, M. 1962. On the Newly Acquired Behavior of a Natural Troop of Japanese Monkeys on Koshima Island. *Primates* 5:3–4.

Keyes, K. 1982. *The Hundredth Monkey*. Coos Bay, Oreg.: Vision.

Kidwell, J. S. 1981. Number of Siblings, Sibling Spacing, Sex, and Birth Order: Their Effects on Perceived Parent-Adolescent Relationships. *Journal of Marriage and Family*, May, 330–335.

Kihlstrom, J. F. 1987. The Cognitive Unconscious. *Science*, no. 237:1445–1452.

Killeen, P., R. W. Wildman, and R. W. Wildman II. 1974. Superstitiousness and Intelligence. *Psychological Reports* 34:1158.

Kinsey, A. C., W. B. Pomeroy, and C. E. Martin. 1948. *Sexual Behavior in the Human Male*. Philadelphia: Saunders.

Klaits, J. 1985. *Servants of Satan: The Age of the Witch Hunts*. Bloomington: Indiana University Press.

Klee, E., W. Dressen, and V. Riess, eds. 1991. *"The Good Old Days": The Holocaust as Seen by Its Perpetrators and Bystanders*. Trans. D. Burnstone. New York: Free Press.

Knox, V. J., A. H. Morgan, and E. R. Hilgard. 1974. Pain and Suffering in Ischemia. *Archives of General Psychiatry* 80:840–847.

Kofahl, R. 1977. *Handy Dandy Evolution Refuter*. San Diego: Beta.

Kremer, J. P. 1994. *KL Auschwitz Seen by the SS*. Oswiecim, Poland: Auschwitz-Birkenau State Museum.

Kübler-Ross, E. 1969. *On Death and Dying*. New York: Macmillan.

———. 1981. Playboy Interview: Elisabeth Kübler-Ross. *Playboy*.

Kuhn, D. 1989. Children and Adults as Intuitive Scientists. *Psychological Review* 96:674–689.

Kuhn, D., M. Weinstock, and R. Flaton. 1994. How Well Do Jurors Reason? Competence Dimensions of Individual Variation in a Juror Reasoning Task. *Psychological Science* 5:289–296.

Kuhn, T. 1962. *The Structure of Scientific Revolutions*. Chicago: University of Chicago Press.

———. 1977. *The Essential Tension: Selected Studies in Scientific Tradition and Change*. Chicago: University of Chicago Press.

Kulaszka, B. 1992. *Did Six Million Really Die? Report of the Evidence in the Canadian "False News" Trial of Ernst Zündel*. Toronto: Samisdat.

Kusche, L. 1975. *The Bermuda Triangle Mystery—Solved.* New York: Warner.

Lea, H. 1888. *A History of the Inquisition of the Middle Ages.* 3 vols. New York: Harper & Brothers.

Lederer, W. 1969. *The Fear of Women.* New York: Harcourt.

Leeper, R. 1935. A Study of a Neglected Portion of the Field of Learning—The Development of Sensory Organization. *Journal of Genetics and Psychology* 46:41–75.

Lefkowitz, M. 1996. *Not Out of Africa: How Afrocentrism Became an Excuse to Teach Myth as History.* New York: Basic Books.

Lehman, J. 1989. Transcript of Interview by M. Shermer, April 12.

Leuchter, F. 1989. *The Leuchter Report.* London: Focal Point.

Levin, J. S. 1993. Age Differences in Mystical Experience. *The Gerontologist* 33:507–13.

Lindberg, D. C., and R. L. Numbers. 1986. *God and Nature.* Berkeley: University of California Press.

Linde, A. 1991. *Particle Physics and Inflationary Cosmology.* New York: Gordon & Breach.

Loftus, E., and K. Ketcham. 1991. *Witness for the Defense: The Accused, the Eyewitnesses, and the Expert Who Puts Memory on Trial.* New York: St. Martin's.

———. 1994. *The Myth of Repressed Memory: False Memories and the Allegations of Sexual Abuse.* New York: St. Martin's.

Lucas, Michael. 2001. Venturing from Shadows into Light: They claim to have been abducted by aliens. A Harvard research psychiatrist backs them. *Los Angeles Times,* September 4.

Macfarlane, A. J. D. 1970. *Witchcraft in Tudor and Stuart England.* New York: Harper.

Mack, J. 1994. *Abduction: Human Encounters with Aliens.* New York: Scribner's.

———. 2001. *Passport to the Cosmos: Human Transformation and Alien Encounters.* New York: Crown.

Malinowski, B. 1954. *Magic, Science, and Religion.* New York: Doubleday, 139–140.

Mander, A. E. 1947. *Logic for the Millions.* New York: Philosophical Library.

Marcellus, T. 1994. An Urgent Appeal from IHR. Institute for Historical Review mailing.

Markus, H. 1981. Sibling Personalities: The Luck of the Draw. *Psychology Today* 15, no. 6:36–37.

Marrus, M. R., ed. 1989. *The Nazi Holocaust.* 9 vols. Westport, Conn.: Meckler.

Marshall, G. N., C. B. Wortman, R. R. Vickers, Jr., J. W. Kusulas, and L. K. Hervig. 1994. The Five-Factor Model of Personality as a Framework for Personality-Health Research. *Journal of Personality and Social Psychology* 67:278–286.

Masson, J. 1984. *The Assault on Truth: Freud's Suppression of the Seduction Theory.* New York: Farrar, Straus & Giroux.

Mayer, A. J. 1990. *Why Did the Heavens Not Darken? The "Final Solution" in History.* New York: Pantheon.

Mayr, E. 1970. *Populations, Species, and Evolution.* Cambridge, Mass.: Harvard University Press.

———. 1982. *Growth of Biological Thought.* Cambridge, Mass.: Harvard University Press.

———. 1988. *Toward a New Philosophy of Biology.* Cambridge, Mass.: Harvard University Press.

McDonough, T., and D. Brin. 1992. The Bubbling Universe. *Omni,* October.

McGarry, J., JU. and B. H. Newberry. 1981. Beliefs in Paranormal Phenomena and Locus of Control: A Field Study. *Journal of Personality and Social Psychology* 41:725–736.

McIver, T. 1994. The Protocols of Creationists: Racism, Antisemitism, and White Supremacy in Christian Fundamentalists. *Skeptic* 2, no. 4:76–87.

Medawar, P. B. 1969. *Induction and Intuition in Scientific Thought.* Philadelphia: American Philosophical Society.

Messer, W. S., and R. A. Griggs. 1989. Student Belief and Involvement in the Paranormal and Performance in Introductory Psychology. *Teaching of Psychology* 16:187–191.

Midelfort, H. C. E. 1972. *Witch Hunting in Southwest Germany, 1562–1684.* Palo Alto, Calif.: Stanford University Press.

Mithen, S. 1996. *The Prehistory of the Mind: The Cognitive Origins of Art, Religion, and Science.* London: Thames and Hudson, 163.

Moody, R. 1975. *Life After Life.* Covinda, Ga.: Mockingbird.

Müller, F. 1979. *Eyewitness Auschwitz: Three Years in the Gas Chambers.* With H. Freitag; ed. and trans. S. Flatauer. New York: Stein and Day.

Neher, A. 1990. *The Psychology of Transcendence.* New York: Dover.

Nelkin, D. 1982. *The Creation Controversy: Science or Scripture in the Schools.* New York: Norton.

Newton, I. [1729] 1962. Sir Isaac Newton's Mathematical Principles of Natural Philosophy and His System of the World. Trans. A. Motte; trans. rev. F. Cajoni. 2 vols. Berkeley: University of California Press.

Nickerson, R. S. 1998. Confirmation Bias: A Ubiquitous Phenomenon in Many Guises. *Review of General Psychology* 2, no. 2:175–220, 175.

Nisbett, R. E. 1968. Birth Order and Participation in Dangerous Sports. *Journal of Personality and Social Psychology* 8:351–353.

Nisbett, R. E., and L. Ross. 1980. *Human Inference: Strategies and Shortcomings of Social Judgment.* Englewood Cliffs, N.J.: Prentice-Hall.

Numbers, R. 1992. *The Creationists.* New York: Knopf.

Obert, J. C. 1981. Yockney: Profits of an American Hitler. *The Investigator* (October).

Official Transcript Proceedings Before the Supreme Court of the United States, Case N. 85-1513, Title: Edwin W. Edwards, Etc., et al., Appellants v. Don Aguillard et al., Appellees. December 10, 1986.

Olson, R. 1982. *Science Deified and Science Defied: The Historical Significance of Science in Western Culture from the Bronze Age to the Beginnings of the Modern Era, ca. 3500 B.C. to A.D. 1640.* Berkeley: University of California Press.

————. 1991. *Science Deified and Science Defied: The Historical Significance of Science in Western Culture from the Early Modern Age Through the Early Romantic Era, ca. 1640 to 1820.* Berkeley: University of California Press.

————. 1993. Spirits, Witches, and Science: Why the Rise of Science Encouraged Belief in the Supernatural in Seventeenth-Century England. *Skeptic* 1, no. 4:34–43.

Otis, L. P., and J. E. Alcock. 1982. Factors Affecting Extraordinary Belief. *The Journal of Social Psychology* 118:77–85.

Overton, W. R. 1985. Memorandum Opinion of United States District Judge William R. Overton in *McLean v. Arkansas*, 5 January 1982. In *Creationism on Trial*, ed. L. Gilkey. New York: Harper & Row.

Padfield, P. 1990. *Himmler.* New York: Henry Holt.

Paley, W. 1802. *Natural Theology, or, Evidences of the Existence and Attributes of the Deity: Collected from the Appearances of Nature.* Philadelphia: Printed for John Morgan by H. Maxwell.

Pasachoff, J. M., R. J. Cohen, and N. W. Pasachoff. 1971. Belief in the Supernatural Among Harvard and West African University Students. *Nature* 232:278–279.

Pasley, L. 1993. Misplaced Trust: A First Person Account of How My Therapist Created False Memories. *Skeptic* 2, no. 3:62–67.

Pearson, R. 1991. *Race, Intelligence, and Bias in Academe.* New York: Scott Townsend.

————. 1995. Transcript of Interview by M. Shermer, December 5.

————. 1996. *Heredity and Humanity: Race, Eugenics, and Modern Science.* Washington, D.C.: Scott Townsend.

Pendergrast, M. 1995. *Victims of Memory: Incest Accusations and Shattered Lives.* Hinesberg, Va.: Upper Access.

————. 1996. First of All, Do No Harm: A Recovered Memory Therapist Recants— An Interview with Robin Newsome. *Skeptic* 3, no. 4:36–41.

Pennebaker, J. W., and J. A. Skelton. 1978. Psychological Parameters of Physical Symptoms. *Personality and Social Psychology Bulletin* 4:524–530.

Perkins, D. N. 1981. *The Mind's Best Work.* Cambridge: Harvard University Press.

Pinker, S. 1997. *How the Mind Works.* New York: W. W. Norton.

Pirsig, R. M. 1974. *Zen and the Art of Motorcycle Maintenance.* New York: Morrow.

Planck, M. 1936. *The Philosophy of Physics.* New York: Norton.

Plato. 1952. *The Dialogues of Plato.* Trans. B. Jowett. Great Books of the Western World. Chicago: University of Chicago.

Polkinghorne, J. 1994. *The Faith of a Physicist.* Princeton, N.J.: Princeton University Press.

Rand, A. 1943. *The Fountainhead.* New York: Bobbs-Merrill.

———. 1957. *Atlas Shrugged.* New York: Random House.

———. 1962. Introducing Objectivism. *Objectivist Newsletter,* August, 35.

Randi, J. 1982. *Flim-Flam!* Buffalo, N.Y.: Prometheus.

Rassinier, P. 1978. *Debunking the Genocide Myth: A Study of the Nazi Concentration Camps and the Alleged Extermination of European Jewry.* Trans. A. Robbins. Los Angeles: Noontide.

Ray, O. S. 1972. *Drugs, Society, and Human Behavior.* St. Louis, Mo.: Mosby.

Richardson, J., J. Best, and D. Bromley, eds. 1991. *The Satanism Scare.* Hawthorne, N.Y.: Aldine de Gruyter.

Rohr, J, ed. 1986. *Science and Religion.* St. Paul, Minn.: Greenhaven.

Roques, H. 1995. Letter to the Editor. *Adelaide Institute Newsletter* 2, no. 4:3.

Ross, H. 1993. *The Creator and the Cosmos: How the Greatest Scientific Discoveries of the Century Reveal God.* Colorado Springs, Colo.: Navpress.

———. 1994. *Creation and Time: A Biblical and Scientific Perspective on the Creation-Date Controversy.* Colorado Springs, Colo.: Navpress.

———. 1996. *Beyond the Cosmos: What Recent Discoveries in Astronomy and Physics Reveal About the Nature of God.* Colorado Springs, Colo.: Navpress.

Rotter, J. B. 1966. Generalized Expectancies for Internal versus External Control of Reinforcement. *Psychological Monographs* 80, no. 609:1–28.

Ruse, M. 1982. *Darwinism Defended.* Reading, Mass.: Addison-Wesley.

———. 1989. *The Darwinian Paradigm.* London: Hutchinson.

Rushton, J. P. 1994. Sex and Race Differences in Cranial Capacity from International Labour Office Data. *Intelligence* 19:281–294.

Russell of Liverpool, Lord. 1963. *The Record: The Trial of Adolf Eichmann for His Crimes Against the Jewish People and Against Humanity.* New York: Knopf.

Saavedra-Aguilar, J. C., and J. S. Gomez-Jeria. 1989. A Neurobiological Model for Near-Death Experiences. *Journal of Near-Death Studies* 7:205–222.

Sabom, M. 1982. *Recollections of Death.* New York: Harper & Row.

Sagan, C. 1973. *The Cosmic Connection: An Extraterrestrial Perspective.* New York: Doubleday.

———. 1979. *Broca's Brain.* New York: Random House.

———. 1980. *Cosmos.* New York: Random House.

———. 1996. *The Demon Haunted World: Science as a Candle in the Dark.* New York: Random House.

Sagan, C., and T. Page, eds. 1974. *UFO's: A Scientific Debate.* New York: Norton.

Sagi, N. 1980. *German Reparations: A History of the Negotiations.* Trans. D. Alon. Jerusalem: Hebrew University/Magnes Press.

Sarich, V. 1995. In Defense of *The Bell Curve:* The Reality of Race and the Importance of Human Differences. *Skeptic* 3, no. 4:84–93.

Sarton, G. 1936. *The Study of the History of Science.* Cambridge, Mass.: Harvard University Press.

Scheidl, F. 1967. *Geschicte der Verfemung Deutschlands.* 7 vols. Vienna: Dr. Scheidl-Verlag.

Schmidt, M. 1984. *Albert Speer: The End of a Myth.* Trans. J. Neugroschel. New York: St. Martin's.

Schoonmaker, F. 1979. Denver Cardiologist Discloses Findings After 18 Years of Near-Death Research. *Anabiosis* 1:1–2.

Sebald, H. 1996. Witchcraft/Witches. In *The Encyclopedia of the Paranormal,* ed. G. Stein. Buffalo, N.Y.: Prometheus.

Segraves, K. 1975. *The Creation Explanation: A Scientific Alternative to Evolution.* San Diego: Creation-Science Research Center.

Segraves, N. 1977. *The Creation Report.* San Diego: Creation-Science Research Center.

Sereny, G. 1995. *Albert Speer: His Battle with Truth.* New York: Knopf.

Sheils, D. 1978. A Cross-Cultural Study of Beliefs in Out of the Body Experiences. *Journal of the Society for Psychical Research* 49:697–741.

Sherman, B., and Z. Kunda. 1989. Motivated Evaluation of Scientific Evidence. Paper presented at the annual meeting of the American Psychological Society, Arlington, Va.

Shermer, M. 1991. Heretic-Scientist: Alfred Russel Wallace and the Evolution of Man. Ann Arbor, Mich.: UMI Dissertation Information Service.

———. 1993. The Chaos of History: On a Chaotic Model That Represents the Role of Contingency and Necessity in Historical Sequences. *Nonlinear Science Today* 2, no. 4:1–13.

———. 1994. Satanic Panic over in UK. *Skeptic* 4, no. 2:21.

———. 1995. Exorcising Laplace's Demon: Chaos and Antichaos, History and Metahistory. *History and Theory* 34, no. 1:59–83.

———. 1999. *How We Believe: The Search for God in an Age of Science.* New York: W. H. Freeman.

———. 2001. *The Borderlands of Science: Where Sense Meets Nonsense.* New York: Oxford University Press.

———. 2002. This View of Science: Stephen Jay Gould as Historian of Science and Scientific Historian. In press.

Shermer, M., and A. Grobman. 1997. *Denying History: Who Says the Holocaust Never Happened and Why Do They Say It?* Jerusalem: Yad Vashem; Los Angeles: Martyrs' Memorial and Museum of the Holocaust.

Shermer, M., and F. Sulloway. 2001. Belief in God: An Empirical Study. In press.

Siegel, R. K. 1977. Hallucinations. *Scientific American*, no. 237:132–140.

Simon Wiesenthal Center. 1993. *The Neo-Nazi Movement in Germany.* Los Angeles: Simon Wiesenthal Center.

Simonton, D. K. 1999. *Origins of Genius: Darwinian Perspectives on Creativity.* Oxford: Oxford University Press.

Singer, B., and G. Abell, eds. 1981. *Science and the Paranormal.* New York: Scribner's.

Singer, M. 1995. *Cults in Our Midst: The Hidden Menace in Our Everyday Lives.* San Francisco: Jossey-Bass Publishers.

Smith, B. 1994. *Smith's Report,* no. 19 (Winter).

Smith, W. 1994. The Mattoon Phantom Gasser: Was the Famous Mass Hysteria Really a Mass Hoax? *Skeptic* 3, no. 1:33–39.

Smolin, L. 1992. Did the Universe Evolve? *Classical and Quantum Gravity* 9:173.

Snelson, J. S. 1993. The Ideological Immune System. *Skeptic* 1, no. 4:44–55.

Snyder, L., ed. 1981. *Hitler's Third Reich.* Chicago: Nelson-Hall.

Snyder, M. 1981. Seek and Ye Shall Find: Testing Hypotheses About Other People. In *Social Cognition: The Ontario Symposium on Personality and Social Psychology,* eds. E. T. Higgins, C. P. Heiman, and M. P. Zanna. Hillsdale, N.J.: Erlbaum, 277–303.

Somit, A., and S. A. Peterson. 1992. *The Dynamics of Evolution.* Ithaca, N.Y.: Cornell University Press.

Speer, A. 1976. *Spandau: The Secret Diaries.* New York: Macmillan.

Starkey, M. L. 1963. *The Devil in Salem.* New York: Time Books.

Stearn, J. 1967. *Edgar Cayce—The Sleeping Prophet.* New York: Bantam.

Sternberg, R. J. 1996. *Successful Intelligence: How Practical and Creative Intelligence Determine Succcess in Life.* New York: Simon & Schuster.

Strahler, A. N. 1987. *Science and Earth History: The Evolution/Creation Controversy.* Buffalo, N.Y.: Prometheus.

Strieber, W. 1987. *Communion: A True Story.* New York: Avon.

Sulloway, F. J. 1990. Orthodoxy and Innovation in Science: The Influence of Birth Order in a Multivariate Context. Preprint.

———. 1991. "Darwinian Psychobiography." Review of *Charles Darwin: A New Life*, by John Bowlby. *New York Review of Books*, October 10.

———. 1996. *Born to Rebel: Birth Order, Family Dynamics, and Creative Lives*. New York: Pantheon.

Swiebocka, T., ed. 1993. *Auschwitz: A History in Photographs*. English ed. J. Webber and C. Wilsack. Bloomington: Indiana University Press.

Syllabus from the Supreme Court of the United States in Edwards v. Aguillard. 1987.

Taubes, G. 1993. *Bad Science*. New York. Random House.

Tavris, C., and C. Wade. 1997. *Psychology in Perspective*. Second Edition. New York: Longman/Addison-Wesley.

Taylor, J. 1859. *The Great Pyramid: Why Was It Built? And Who Built It?* London: Longman.

Thomas, K. 1971. *Religion and the Decline of Magic*. New York: Scribner's.

Thomas, W. A. 1986. Commentary: Science v. Creation-Science. *Science, Technology, and Human Values* 3:47–51.

Tipler, F. 1981. Extraterrestrial Intelligent Beings Do Not Exist. *Quarterly Journal of the Royal Astronomical Society* 21:267–282.

———. 1994. *The Physics of Immortality: Modern Cosmology, God and the Resurrection of the Dead*. New York: Doubleday.

———. 1995. Transcript of Interview by M. Shermer, September 11.

Tobacyk, J., and G. Milford. 1983. Belief in Paranormal Phenomena: Assessment Instrument Development and Implications for Personality Functioning. *Journal of Personality and Social Psychology* 44:1029–1037.

Toumey, C. P. 1994. *God's Own Scientists: Creationists in a Secular World*. New Brunswick, N.J.: Rutgers University Press.

Trevor-Roper, H. R. 1969. *The European Witch-Craze of the Sixteenth and Seventeenth Centuries and Other Essays*. New York: Harper Torchbooks.

Tucker, W. H. 1994. *The Science and Politics of Racial Research*. Urbana: University of Illinois Press.

Turner, J. S., and D. B. Helms. 1987. *Lifespan Development*, 3rd ed. New York: Holt, Rinehart & Winston.

Vankin, J., and J. Whalen. 1995. *The Fifty Greatest Conspiracies of All Time*. New York: Citadel.

Victor, J. 1993. *Satanic Panic: The Creation of a Contemporary Legend*. Chicago: Open Court.

Voltaire. 1985. *The Portable Voltaire*. Ed. B. R. Redman. New York: Penguin.

Vyse, S. A. 1997. *Believing in Magic: The Psychology of Superstition*. New York: Oxford University Press.

Walker, D. P. 1981. *Unclean Spirits: Possession and Exorcism in France and England in the Late Sixteenth and Early Seventeenth Centuries.* Philadelphia: University of Pennsylvania Press.

Walker, W. R., S. J. Hoekstra, and R. J. Vogl. 2001. Science Education is No Guarantee for Skepticism. *Skeptic* 9, no. 3.

Wallace, A. R. 1869. Sir Charles Lyell on Geological Climates and Origin of Species. *Quarterly Review* 126:359–394.

Watson, L. 1979. *Lifetide.* New York: Simon & Schuster.

Weaver, J. H., ed. 1987. *The World of Physics: A Small Library of the Literature of Physics from Antiquity to the Present,* vol. 2, *The Einstein Universe and the Bohr Atom.* New York: Simon & Schuster.

Weber, M. 1992. The Nuremberg Trials and the Holocaust. *Journal of Historical Review* 12, no. 3:167–213.

———. 1993a. *Auschwitz: Myths and Facts,* brochure. Newport Beach, Calif.: Institute for Historical Review.

———. 1993b. *The Zionist Terror Network.* Newport Beach, Calif.: Institute for Historical Review.

———. 1994a. *The Holocaust: Let's Hear Both Sides,* brochure. Newport Beach, Calif.: Institute for Historical Review.

———. 1994b. Transcript of Interview by M. Shermer, February 11.

———. 1994c. The Jewish Role in the Bolshevik Revolution and Russia's Early Soviet Regime. *Journal of Historical Review* 14, no. 1:4–14.

Webster, R. 1995. *Why Freud Was Wrong: Sin, Science, and Psychoanalysis.* New York: Basic Books.

Whitcomb, J., Jr., and H. M. Morris. 1961. *The Genesis Flood: The Biblical Record and Its Scientific Implications.* Philadelphia: Presbyterian and Reformed Publishing.

Wikoff, J., ed. 1990. *Remarks: Commentary on Current Events and History.* Aurora, N.Y.

Wulff, D. M. 2000. Mystical Experience. In *Varieties of Anomalous Experience: Examining the Scientific Evidence,* eds. E. Cardena, S. J. Lynn, and S. Krippner. Washington, D.C.: American Psychological Association, 408.

Yockey, F. P. [U. Varange, pseud.]. [1948] 1969. *Imperium: The Philosophy of History and Politics.* Sausalito, Calif.: Noontide.

Zukav, G. 1979. *The Dancing Wu Li Masters: An Overview of the New Physics.* New York: Bantam.

Zündel, E. 1994. Transcript of Interview by M. Shermer, April 26.

Index